Advances in Intelligent Systems and Computing

Volume 1121

The series "Advances in Intelligent Systems and Computing" contains publications on theory, applications, and design methods of Intelligent Systems and Intelligent Computing. Virtually all disciplines such as engineering, natural sciences, computer and information science, ICT, economics, business, e-commerce, environment, healthcare, life science are covered. The list of topics spans all the areas of modern intelligent systems and computing such as: computational intelligence, soft computing including neural networks, fuzzy systems, evolutionary computing and the fusion of these paradigms, social intelligence, ambient intelligence, computational neuroscience, artificial life, virtual worlds and society, cognitive science and systems, Perception and Vision, DNA and immune based systems, self-organizing and adaptive systems, e-Learning and teaching, human-centered and human-centric computing, recommender systems, intelligent control, robotics and mechatronics including human-machine teaming, knowledge-based paradigms, learning paradigms, machine ethics, intelligent data analysis, knowledge management, intelligent agents, intelligent decision making and support, intelligent network security, trust management, interactive entertainment, Web intelligence and multimedia.

The publications within "Advances in Intelligent Systems and Computing" are primarily proceedings of important conferences, symposia and congresses. They cover significant recent developments in the field, both of a foundational and applicable character. An important characteristic feature of the series is the short publication time and world-wide distribution. This permits a rapid and broad dissemination of research results.

** Indexing: The books of this series are submitted to ISI Proceedings, EI-Compendex, DBLP, SCOPUS, Google Scholar and Springerlink **

More information about this series at http://www.springer.com/series/11156

Hoai An Le Thi · Hoai Minh Le ·
Tao Pham Dinh · Ngoc Thanh Nguyen
Editors

Advanced Computational Methods for Knowledge Engineering

Proceedings of the 6th International Conference
on Computer Science, Applied Mathematics
and Applications, ICCSAMA 2019

 Springer

Editors
Hoai An Le Thi
Computer Science and Applications
Department LGIPM
University of Lorraine
Metz Cedex 03, France

Tao Pham Dinh
Laboratory of Mathematics
National Institute for Applied Sciences
Saint-Étienne-du-Rouvray Cedex, France

Hoai Minh Le
Computer Science and Applications
Department LGIPM
University of Lorraine
Metz Cedex 03, France

Ngoc Thanh Nguyen
Department of Information Systems
Wrocław University of Science
and Technology
Wrocław, Poland

ISSN 2194-5357   ISSN 2194-5365   (electronic)
Advances in Intelligent Systems and Computing
ISBN 978-3-030-38363-3   ISBN 978-3-030-38364-0   (eBook)
https://doi.org/10.1007/978-3-030-38364-0

This Springer imprint is published by the registered company Springer Nature Switzerland AG
The registered company address is: Gewerbestrasse 11, 6330 Cham, Switzerland

Preface

ICCSAMA 2019, held on December 19–20, 2019, in Hanoi, Vietnam, was the sixth event of the series of international scientific conferences on computer science, applied mathematics and applications. The conference is co-organized by the Computer Science and Applications Department, LGIPM, University of Lorraine, France; the Institute for research and applications of optimization (VinOptima), VinTech, Vingroup; the International School, Vietnam National University, Hanoi, Vietnam; the Laboratory of Mathematics, National Institute for Applied Sciences-Rouen, France, and the Department of Information Systems Wroclaw University of Science and Technology, Poland.

The aim of ICCSAMA 2019 is to bring together leading academic scientists, researchers and scholars to discuss and share their newest results in the fields of computer science, applied mathematics and their applications. These two fields are very close and related to each other. It is also clear that the potentials of computational methods for knowledge engineering and optimization algorithms are to be exploited, and this is an opportunity and a challenge for researchers.

For the ICCSAMA 2019 edition, the Program Committee received more than 75 submissions. Each paper was peer-reviewed by at least two members of the International Program Committee and the International Reviewer board. After the review process, 37 high-quality papers were selected for oral presentation and publication in this book. The selected papers cover several topics in applied mathematics and computer science, and they are divided into four parts: nonconvex optimization, DC programming and DCA and applications; data mining and data processing; machine learning methods and applications, and knowledge information and engineering systems. Extended versions of selected papers will be considered for publication in post-conference special issues including *Journal of Global Optimization*.

ICCSAMA 2019 was attended by about 100 scientists and practitioners. The conference program is composed of four plenary lectures and one semi-plenary lecture of world-class speakers and the oral presentation of 37 selected papers as well as several selected abstracts.

ICCSAMA 2019 has created numerous interesting interactions between two communities computer science and applied mathematics, and we hope that researchers and practitioners can find here many inspiring ideas and useful tools and techniques for their works. Many such challenges are suggested by particular approaches and models presented in individual chapters of this book.

We would like to thank the chairs and the members of International Program Committee as well as the reviewers for their hard work in the review process, which helped us to guarantee the highest quality of the selected papers for the conference. We also would like to express our thanks to the keynote speakers for their interesting and informative talks. Our sincere thanks go to all the authors for their valuable contributions and to the other participants who enriched the conference success.

We wish to thank all members of the Organizing Committee for their excellent work to make the conference a success.

We cordially thank Prof. Janusz Kacprzyk and Dr. Thomas Ditzinger from Springer for their help in publishing this book.

Finally, we would like to express our special thanks to the main sponsor VinTech City, VinTech, VinGroup for their considerable support.

December 2019   Hoai An Le Thi
                                                                                           Hoai Minh Le
                                                                                          Tao Pham Dinh
                                                                                     Ngoc Thanh Nguyen

Organization

ICCSAMA 2019 is co-organized by the Computer Science and Applications Department, LGIPM, University of Lorraine, France, the Institute for research and applications of optimization (VinOptima), VinTech, Vingroup, and in collaboration with the International School, Vietnam National University, Hanoi, Vietnam.

Conference Chair

Hoai An Le Thi   University of Lorraine, France

Program Chairs

Hoai An Le Thi   University of Lorraine, France
Tao Pham Dinh   National Institute of Applied Sciences of Rouen,
                                  France, and Institute for Research and
                                  Applications of Optimization (VinOptima),
                                  VinTech, Vingroup, Hanoi, Vietnam
Ngoc Thanh Nguyen   Wroclaw University of Technology, Poland

Honorary Chair

Hoang Hai Nguyen   Vietnam National University, Hanoi, Vietnam

Organizing Chairs

Hoai Minh Le   University of Lorraine, France, and Institute for
                                  Research and Applications of Optimization
                                  (VinOptima), VinTech, Vingroup, Hanoi,
                                  Vietnam
Trung Thanh Le   International School, Vietnam National
                                  University, Hanoi, Vietnam

Publicity Chair

Ho Vinh Thanh   Institute for Research and Applications
                                 of Optimization (VinOptima), VinTech,
                                 Vingroup, Hanoi, Vietnam

International Program Committee Members

Alain Bui   University of Versailles
                                 Saint-Quentin-en-Yvelines, France
Minh Phong Bui   Eotvos Lorand University, Hungary
Nam Hoai Do   Budapest University of Technology
                                 and Economics, Hungary
Van Tien Do   Budapest University of Technology
                                 and Economics, Hungary
Thanh Nghi Do   Can Tho University, Vietnam
Quang Thuy Ha   Vietnam National University, Vietnam
Ferenc Hain   Budapest College of Communications
                                 and Business, Hungary
Vinh Thanh Ho   Institute for Research and Applications
                                 of Optimization (VinOptima), VinTech,
                                 Vingroup, Hanoi, Vietnam
Chi Hieu Le   University of Greenwich, UK
Hoai Minh Le   University of Lorraine, France, and Institute for
                                 Research and Applications of Optimization
                                 (VinOptima), VinTech, Vingroup, Hanoi,
                                 Vietnam
Nguyen Thinh Le   Humboldt Universitat zu Berlin, Germany
Anh Linh Nguyen   Warsaw University, Poland
Canh Nam Nguyen   Hanoi University of Science and Technology,
                                 Vietnam
Benjamin Nguyen   University of Versailles
                                 Saint-Quentin-en-Yvelines, France
Duc Cuong Nguyen   School of Computer Science and Engineering
                                 of International University, Vietnam
Duc Khuong Nguyen   IPAG Business School, Paris, France
Duc Manh Nguyen   Hanoi National University of Education, Vietnam
Giang Nguyen   Slovak Academy of Sciences, Slovakia
Hung Son Nguyen   Warsaw University, Poland
Luu Lan Anh Nguyen   Eotvos Lorand University, Hungary
Manh Cuong Nguyen   Hanoi University of Industry, Vietnam
Quang Thuan Nguyen   International School, Vietnam National
                                 University, Hanoi, Vietnam
Thanh Binh Nguyen   International Institute for Applied Systems
                                 Analysis (IIASA), Austria

Thanh Thuy Nguyen	National University of Hanoi, Vietnam
Van Sinh Nguyen	International University, Vietnam National University HCM, Vietnam
Viet Hung Nguyen	LIP6, Sorbonne University, France
Cong Duc Pham	University of Pay and Pays de l'Adour, France
Duc Truong Pham	University of Birmingham, UK
Hoang Pham	Rutgers, The State University of New Jersey, USA
Ngoc Anh Pham	Posts and Telecommunications Institute of Technology, Vietnam
Viet Nga Pham	Vietnam National University of Agriculture, Vietnam
Duong Hieu Phan	Université Paris 8, France
Duy Nhat Phan	University of Lorraine, France
Thong Vinh Ta	Budapest University of Technology and Economics, Hungary
Anh Son Ta	Hanoi University of Science and Technology, Vietnam, and Institute for Research and Applications of Optimization (VinOptima), VinTech, Vingroup, Hanoi, Vietnam
Duc Quynh Tran	Vietnam National University of Agriculture, Vietnam
Dinh Viet Tran	Slovak Academy of Sciences, Slovakia
Gia Phuoc Tran	University of Wuerzburg Am Hubland, Germany
Hoai Linh Tran	Hanoi University of Science and Technology, Vietnam
Thi Thuy Tran	FPT University, Vietnam
Anh Tuan Tran	Budapest University of Technology and Economics, Hungary
Trong Tuong Truong	Cergy-Pontoise University, France
Niu Yi-Shuai	Shanghai Jiao Tong University, China

External Reviewers

Viet Anh Nguyen	University of Lorraine, France
Xuan Thanh Vo	Phuoc Quang, Tuy Phuoc, Binh Dinh, Vietnam

Organizing Committee Members

Vinh Thanh Ho	Institute for Research and Applications of Optimization (VinOptima), VinTech, Vingroup, Hanoi, Vietnam
Duc Thinh Le	International School, Vietnam National University, Hanoi, Vietnam

Tuyet Hoa Mai Nguyen   International School, Vietnam National
                              University, Hanoi, Vietnam
Quang Thuan Nguyen   International School, Vietnam National
                              University, Hanoi, Vietnam
Bach Tran   University of Lorraine, France
Anh Son Ta   Institute for Research and Applications
                              of Optimization (VinOptima), VinTech,
                              Vingroup, Hanoi, Vietnam
Thi Thuy Linh Vu   Institute for Research and Applications
                              of Optimization (VinOptima), VinTech,
                              Vingroup, Hanoi, Vietnam

Contents

Machine Learning Methods and Applications

Knowledge Information and Engineering Systems

Nonconvex Optimization, DC Programming and DCA, and Applications

A New Efficient Algorithm
for Maximizing the Profit
and the Compactness in Land Use
Planing Problem

Tran Duc Quynh[(✉)]

Vietnam National University, Hanoi-International School, Hanoi, Vietnam
quynhtd@isvnu.vn, ducquynh@vnu.edu.vn

Abstract. This paper deals with a land-use planing problem in which the objective is to maximize the profit (or to minimize the cost) while ensuring the compactness. The original mathematical model is a multi-objective optimization problem with binary integer variables. It is then transformed to a single objective optimization problem. One may use a commercial software to solve such problem but the computation time is expensive especially in large scale problem. Hence, finding new efficient algorithms for the problem is necessary. Recently, two alternatives method based on genetic algorithm (GA) and non dominated sorting genetic algorithm (NSGA-II) are proposed. In this work, we propose a new local method based on difference of convex functions algorithm (DCA). The numerical results are compared with the one provided by GA. It shows that the proposed algorithm is much better and the obtained solutions are close to the global solutions.

Keywords: DCA · Mixed integer linear optimization · Land use planing problem · Profit · Compactness

1   Introduction

Land use planing problem is an important problem because the land area is limited while the population is continuously increasing. The area of agricultural land is about 46% of the earth's land. It may decrease and the food demand is increasing [10] because of the population's augmentation. It is estimated that the food demand in 2050 will increase by 70% compared to the present. Therefore, finding a solution to optimize the use of agricultural land attired the interest of scientists in mathematics, computer science and agronomy. In literature, the researchers often formulate the problem in the form of optimization problem and then develop solution methods for it. In recent 20 years, many mathematical models have been proposed. Each model considers a specific case, objective and constraint. We can classify the proposed models by 3 groups [10]: maximizing the profit [3] optimizing the management of water resources [1], optimizing

© Springer Nature Switzerland AG 2020
H. A. Le Thi et al. (Eds.): ICCSAMA 2019, AISC 1121, pp. 3–13, 2020.
https://doi.org/10.1007/978-3-030-38364-0_1

the protect of the environment and ecosystem [2]. Some research simultaneously consider 2 or 3 objectives, we then have multiple objective optimization problems.

In this research, we tackle a model used in [13] that is based on the one introduced by Jeroen et al. [4]. The aim to maximize the total profit while ensuring that the cells with the same land use are close as possible (compactness). The original mathematical model is a bi-objective optimization problem. One can transform the original model to a single objective optimization problem by using scalar technique. The objective of the resulting problem is the combination of the profit and the compactness. The difficulty of the problem comes from the mixed binary variables. The solution method often need a large executing time. Thus, developing efficient local methods for it is necessary. In [13], the author proposed two local methods called GA and NSGA-II to solve the problem. The experimentation showed that NSGA-II is better than GA by 9% but the computation time of NSGA-II is much longer. In this work, we develop a deterministic method based on DC programming and DCA to solve the mixed integer linear optimization model in [13]. The idea is to reformulate the problem as a DC program by using penalty technique and then develop DC algorithm (DCA) for solving it.

To evaluate the efficiency of the proposed algorithm, we consider 15 instances and compare the results provided by DCA and local method GA. The gap between the objective value obtained by DCA and the optimal value is also estimated. The results on simulation data show that the gap of DCA is smaller than 5%. It is quite good result with a local method.

The paper is organized as follows. In Sect. 2, we state the problem and present the mathematical model. Section 3 presents the solution method via DC programming and DCA. The computational results are reported in Sects. 4 and 5 concludes the paper.

2   Problem Statement

We consider the mathematical model of land use planing problem that has been addressed in [13]. It is a variant of the one in [4]. The difference is the replacement of minimizing the cost by maximizing the profit and we do not use the buffer for the cells in borders. The problem is stated as follows: consider a rectangular area which has to be allocated with different land uses. First, we divide the area into $N.M$ cells by N rows and M columns, the cell in row i and column j will be called (i,j). Suppose there are K different land uses, symbol k indicates a specific land use, $k \in 1, ..., K$. The following parameters are known:

- B_{ijk}: the profit generated by cell (i,j) if it is allocated to land use k.
- T_k: the total number of cell will be allocated to land use k.

The problem is to find the allocation such that the total profit generated by the considered area is the largest and the cells with the same allocated land use are placed close together to form a block (*compactness*).

In [4], the author proposed a mathematical model in the form of bi-objective linear optimization problem with binary 0–1 variables. Let x_{ijk} be the decision variables which equal to 1 if cell (i,j) is used for land use k, 0 otherwise. It is easy to see that the total profit is expressed as:

$$Profit = \sum_{i=1}^{N}\sum_{j=1}^{M}\sum_{k=1}^{K} B_{ijk}x_{ijk}$$

There are some following constraints

$$\sum_{k=1}^{K} x_{ijk} = 1 \quad \forall(i,j) \tag{1}$$

Constraint (1) ensures that each cell is allocated to only one land use.

$$\sum_{i=1}^{N}\sum_{j=1}^{M} x_{ijk} = T_k \quad \forall k \tag{2}$$

Constraint (2) ensures that the number of cells allocated to land use k is T_k.

To measure the compactness, variables y_{ijk} are introduced. The value of y_{ijk} equals to 0 if cell (i,j) is not allocated to land use k ($x_{ijk} = 0$). In the case where cell (i,j) is allocated to land use k ($x_{ijk} = 1$) then y_{ijk} is the number of cells close to cell (i,j) by row or collum, which are allocated to land use k. Variable y_{ijk} can be expressed as:

In the case where cell (i,j) is not on the borders.

$$y_{ijk} \leq 4.x_{ijk} \quad \forall i,j,k \tag{3}$$

$$y_{ijk} \leq x_{i-1jk} + x_{i+1jk} + x_{ij-1k} + x_{ij+1k} \quad \forall k, 2 \leq i \leq N-1, 2 \leq j \leq M-1 \tag{4}$$

$$y_{ijk} \geq x_{i-1jk} + x_{i+1jk} + x_{ij-1k} + x_{ij+1k} - 4.(1-x_{ijk}) \forall k, 2 \leq i \leq N-1, 2 < j \leq M-1 \tag{5}$$

In the case where cell (i,j) is on the borders but it is not a corner.

$$y_{ijk} \leq x_{i+1jk} + x_{ij-1k} + x_{ij+1k} \quad \forall k, i = 1, 2 \leq j \leq M-1 \tag{6}$$

$$y_{ijk} \geq x_{i+1jk} + x_{ij-1k} + x_{ij+1k} - 3.(1-x_{ijk}) \quad \forall k, i = 1, 2 \leq j \leq M-1 \tag{7}$$

$$y_{ijk} \leq x_{i-1jk} + x_{ij-1k} + x_{ij+1k} \quad \forall k, i = N, 2 \leq j \leq M-1 \tag{8}$$

$$y_{ijk} \geq x_{i-1jk} + x_{ij-1k} + x_{ij+1k} - 3.(1-x_{ijk}) \quad \forall k, i = N, 2 \leq j \leq M-1 \tag{9}$$

$$y_{ijk} \leq x_{i-1jk} + x_{i+1jk} + x_{ij+1k} \quad \forall k, 2 \leq i \leq N-1, j = 1 \tag{10}$$

$$y_{ijk} \geq x_{i-1jk} + x_{i+1jk} + x_{ij+1k} - 3.(1-x_{ijk}) \quad \forall k, 2 \leq i \leq N-1, j = 1 \tag{11}$$

$$y_{ijk} \leq x_{i-1jk} + x_{i+1jk} + x_{ij-1k} \quad \forall k, 2 \leq i \leq N-1, j = M \tag{12}$$

$$y_{ijk} \geq x_{i-1jk} + x_{i+1jk} + x_{ij-1k} - 3.(1-x_{ijk}) \quad \forall k, 2 \leq i \leq N-1, j = M \tag{13}$$

In the case where cell (i, j) is a corner.

$$y_{ijk} \leq x_{i+1jk} + x_{ij+1k} \quad \forall k, i = 1, j = 1 \tag{14}$$

$$y_{ijk} \geq x_{i+1jk} + x_{ij+1k} - 2.(1 - x_{ijk}) \quad \forall k, i = 1, j = 1 \tag{15}$$

$$y_{ij1k} \leq x_{i+1jk} + x_{ij-1k} \quad \forall k, i = 1, j = M \tag{16}$$

$$y_{ijk} \geq x_{i+1jk} + x_{ij-1k} - 2.(1 - x_{ijk}) \quad \forall k, i = 1, j = M \tag{17}$$

$$y_{ij1k} \leq x_{i-1jk} + x_{ij+1k} \quad \forall k, i = N, j = 1 \tag{18}$$

$$y_{ijk} \geq x_{i-1jk} + x_{ij+1k} - 2.(1 - x_{ijk}) \quad \forall k, i = N, j = 1 \tag{19}$$

$$y_{ijk} \leq x_{i-1jk} + x_{ij-1k} \quad \forall k, i = N, j = M \tag{20}$$

$$y_{ijk} \geq x_{i-1jk} + x_{ij-1k} - 2.(1 - x_{ijk}) \quad \forall k, i = N, j = M. \tag{21}$$

The function that measures the compactness is given by

$$f_2(x, y) = \sum_{i=1}^{N} \sum_{j=1}^{M} \sum_{k=1}^{K} y_{ijk}.$$

We can see that the measurement of compactness $f_2(x, y)$ is calculated based on the number of pair of two consecutive cells (by row or column) which are allocated the same land use. The aim is to maximize the compactness.

We also need the non-negativity and binary constraints

$$x_{ijk}, y_{ijk} \geq 0 \quad \forall i, j, k. \tag{22}$$

$$x_{ijk} \in \{0, 1\} \quad \forall i, j, k. \tag{23}$$

Hence, we obtain a multi-objective optimization problem

$$
\begin{aligned}
\max \; f_1(x, y) &= \sum_{i=1}^{N} \sum_{j=1}^{M} \sum_{k=1}^{K} B_{ijk} x_{ijk} \\
\max \; f_2(x, y) &= \sum_{i=1}^{N} \sum_{j=1}^{M} \sum_{k=1}^{K} y_{ijk} \\
\text{s.t.} \quad & (3) - (23)
\end{aligned}
\tag{P}
$$

A technique to solve multi-objective optimization problem is to transform it to a single optimization one. By using a coefficient $w > 0$, the single objective optimization problem is written as follows:

$$
\begin{aligned}
\max \; f(x, y) &= f_1(x, y) + w.f_2(x, y) \\
\text{s.t.} \quad & (3) - (23)
\end{aligned}
\tag{P'}
$$

Problem (P') is a mixed integer linear program. It can be solved by using a commercial software but the computation time is very long in the case of large number of integer variables. In [13], the author proposed two methods based on genetic scheme to solve the two objectives optimization problem and the single one. In this work, we propose a local approach based on DC programming and DCA. The work is motivated by the rapidity and the efficiency of DCA.

3   DC Programming and Solution Method

3.1   A Brief Presentation of DC Programming and DCA

DC programming and DCA is backbone of non convex programming. DCA was first introduced by Pham Dinh Tao in 1985 and has been extensively developed since 1994 by Le Thi Hoai An and Pham Dinh Tao in their common works. It has been successfully applied to many large-scale (smooth or nonsmooth) nonconvex programs in various domains of applied science, and has now become classic and popular. In this section, we briefly present DC programming and DCA (see [5–7] and references therein for more detail).

Let $\Gamma_0(\mathbb{R}^n)$ denotes the convex cone of all lower semi-continuous proper convex functions on $\mathbb{R}^n$. Consider the following primal DC program:

$$(P_{dc}) \quad \alpha = \inf\{f(z) := g(z) - h(z) \ : \ z \in \mathbb{R}^n\}, \tag{24}$$

where $g, h \in \Gamma_0(\mathbb{R}^n)$ and function $f(z)$ is called a DC function (difference of convex functions).

Let C be a nonempty closed convex set. The indicator function on C, denoted χ_C, is defined by $\chi_C(z) = 0$ if $z \in C$, ∞ otherwise. Then, the problem

$$\inf\{f(z) := g(z) - h(z) \ : \ x \in C\}, \tag{25}$$

can be transformed into an unconstrained DC program by using the indicator function of C, i.e.,

$$\inf\{f(z) := \phi(z) - h(z) \ : \ z \in \mathbb{R}^n\}, \tag{26}$$

where $\phi := g + \chi_C$ is in $\Gamma_0(\mathbb{R}^n)$.

Recall that, for $h \in \Gamma_0(\mathbb{R}^n)$ and $z_0 \in \mathrm{dom}\ h := \{z \in \mathbb{R}^n | h(z_0) < +\infty\}$, the subdifferential of h at z_0, denoted $\partial h(z_0)$, is defined as

$$\partial h(z_0) := \{\xi \in \mathbb{R}^n : h(z) \geq h(z_0) + \langle z - z_0, \xi \rangle, \forall z \in \mathbb{R}^n\}, \tag{27}$$

which is a closed convex set in $\mathbb{R}^n$. It generalizes the derivative in the sense that h is differentiable at z_0 if and only if $\partial h(z_0)$ is reduced to a singleton which is exactly $\{\nabla h(z_0)\}$.

The idea of DCA is simple: each iteration of DCA approximates the concave part $-h$ by its affine majorization (that corresponds to taking $\xi^k \in \partial h(z^k)$) and minimizes the resulting convex problem (P_k).

Generic DCA scheme
Initialization: Let $z^0 \in \mathbb{R}^n$ be a best guess, $0 \leftarrow k$.
Repeat
    Calculate $\xi^k \in \partial h(z^k)$
    Calculate $z^{k+1} \in \arg\min\{g(z) - h(z^k) - \langle z - z^k, \xi^k \rangle : x \in \mathbb{R}^n\}$ (P_k)
    $k + 1 \leftarrow k$
Until convergence of z^k.

Convergence properties of the DCA and its theoretical bases are described in [5,9,11,12].

3.2   Reformulation and DC Algorithm

To use DCA for solving (P'), we transform it into a DC program by using a penalty technique given in [8]. The work is based on the following theorem.

Theorem 1. *[8] Let Ω be a nonempty bounded polyhedral convex set, f be a finite DC function on Ω and p be a finite nonnegative concave function on Ω. Then there exists $\eta_0 \geq 0$ such that for $\eta > \eta_0$ the following problems have the same optimal value and the same solution set*

$$(P_\eta) \qquad \alpha(\eta) = \min\{f(z) + \eta.p(z) : z \in \Omega\},$$

$$(P) \qquad \alpha = \min\{f(z) : z \in \Omega, p(z) \leq 0\}.$$

Proof. see [8].

Denote by L the number of variables of problem (P'), $L = 2.N.M.K$ and $S = \{z = (x,y) \in \mathbb{R}^L \quad s.t. \quad (3) - (23)\}$. Set D is the relaxed domain of S, say $D = \{z = (x,y) \in \mathbb{R}^L \quad s.t. \quad (3) - (22); \quad 0 \leq x \leq 1\}$.

We consider function $p(z) = \sum_{i=1}^{N}\sum_{j=1}^{M}\sum_{k=1}^{K}(1 - x_{ijk})x_{ijk}$. It is clear that $p(z) \geq 0 \quad \forall z \in D$. Problem (P') can be written as:

$$(P') \quad
\begin{aligned}
&\min -f(z) = -\sum_{i=1}^{N}\sum_{j=1}^{M}\sum_{k=1}^{K} B_{ijk}x_{ijk} - w.\sum_{i=1}^{N}\sum_{j=1}^{M}\sum_{k=1}^{K} y_{ijk} \\
&s.t. \ z \in D \\
&\quad\ p(z) \leq 0.
\end{aligned}$$

By using Theorem 1, Problem (P') is transformed to the equivalent one

$$(P_{eq}) \quad
\begin{aligned}
&\min F(z) = -f(z) + \eta p(z) \\
&s.t. \ z \in D
\end{aligned}$$

where η is a sufficiently large number. It can be seen that (P_{eq}) is a DC program. The DC decomposition $F(z) = G(z) - H(z)$ is described as

$$G(z) = -\sum_{i=1}^{N}\sum_{j=1}^{M}\sum_{k=1}^{K} B_{ijk}x_{ijk} - w.\sum_{i=1}^{N}\sum_{j=1}^{M}\sum_{k=1}^{K} y_{ijk}$$

$$H(z) = \eta \sum_{i=1}^{N}\sum_{j=1}^{M}\sum_{k=1}^{K}(x_{ijk}^2 - x_{ijk}).$$

From the definition of H, it is easy to see that H is differentiable and

$$\begin{cases} \dfrac{\partial H}{\partial x_{ijk}} = 2.\eta.x_{ijk} - \eta \quad \forall i,j,k. \\ \dfrac{\partial H}{\partial y_{ijk}} = 0 \quad \forall i,j,k. \end{cases} \tag{28}$$

DCA applied to land use problem (P_{eq}) can be described as follows:

DCA-LU

Initialization

Let ϵ be a sufficiently small positive number. Set $\ell = 0$ and the initial point $z^0 \in \mathbb{R}^L$.

Repeat

Calculate $\beta_{ijk}^\ell = \frac{\partial H}{\partial x_{ijk}} = 2.\eta.x_{ijk} - \eta \quad \forall i, j, k.$

Solve the linear program

$$\min - \sum_{i=1}^N \sum_{j=1}^M \sum_{k=1}^K (-B_{ijk} - \beta_{ijk}^\ell) x_{ijk} - w. \sum_{i=1}^N \sum_{j=1}^M \sum_{k=1}^K y_{ijk}$$
$$s.t. \ z \in D$$

to obtain $z^{\ell+1}$.

$\ell \longleftarrow \ell + 1$

Until $\|z^{\ell+1} - z^\ell\| \le \epsilon$ or $\|F(z^{\ell+1}) - F(z^\ell)\| \le \epsilon$.

In the case where the solution provided by DCA does not satisfy the integer constraints, we change the value of penalty coefficient η and the initial point and then rerun **DCA-LU**. We obtain a multi-restart DC algorithm as follows:

ResDCA-LU

Initialization

Let η^0 be the initial value of the penalty coefficient. Set $\ell = 0$ and the initial point $z^0 = (x^0, y^0) \in \mathbb{R}^L$.

Repeat

Launch DCA-LU with the initial point z^ℓ to obtain $z^{\ell+1} = (x^{\ell+1}, y^{\ell+1})$. Set $IntVar = x^{\ell+1}$.

If $x_{ijk}^{\ell+1}$ is not integer then reset $x_{ijk}^{\ell+1}$ by the rule

$$x_{ijk}^{\ell+1} = \begin{cases} 0 & if \ x_{ijk}^{\ell+1} < 0.5 \\ 1 & otherwise \end{cases} \quad \forall i, j, k.$$

$\eta^{\ell+1} = 10 * \eta^\ell$

$\ell \longleftarrow \ell + 1$

Until $IntVar$ is integer.

4 Numerical Results

To evaluate the efficiency of the proposed algorithm, we compare the result provided by ResDCA-LU and GA. Because of the lack of the real data, we use 15 simulation instances by changing the size of the area and profits generated by each land use. There are 3 sizes (N = 10, M = 10), (N = 20, M = 20) and (N = 50, M = 50). For all instances, we suppose that there are 4 land uses (K = 4). If cell (i, j) is suitable for land use k then the corresponding profit $B_{ijk} = cof > 1$ and $B_{ijk} = 1$ otherwise. Five cases corresponding to $(cof = 1.5; 2; 3; 4; 5)$ are investigated. Assume that the top left corner, the top right corner, the bottom left corner, the bottom right corner are suitable for the first land use, the second land use, the third land use, the fourth land use respectively. Both algorithms ResDCA-LU and GA are implemented in Matlab 2017, run on CPU Intel core i5 2.8 GHz, RAM 8 GB. The free software CVX is used to solve the linear programs. The setting for GA is similar to the one in [13].

We run ResDCA-LU with the initial penalty coefficient of 200. The initial point for the first run of DCA is $z^0 = 0$, parameter w is fixed 0.5 for all runs. For each instance, we run 10 times of GA and pick up the highest quality solution to compared with ResDCA. Table 1 presents the results given by ResDCA-LU and GA. In the table, some notations are used:

⋄ *Size*: the size of the area. It is given by the number of rows and columns.

⋄ T_k; the number of cells being allocated to land use k.

⋄ *cof*: the coefficient reflects the suitability of cells for land uses. It is described in the first paragraph.

⋄ val_{DCA}: the objective value given by ResDCA-LU.

⋄ RN: the number of rerunning DCA in ResDCA-LU.

⋄ LB: the objective value of the relaxed problem that is obtained from problem (P') by removing integer constraints. It is a lower bound of the optimal objective value.

⋄ T_{DCA}: the executing time in seconds of ResDCA-LU.

⋄ G_{DCA}: the gap of DCA. It is calculated by $G_{DCA} = 100|\frac{val_{DCA}-LB}{LB}|$.

⋄ val_{GA}: the best objective value given by GA.

⋄ T_{GA}: the executing time in seconds of GA.

⋄ G_{GA}: the gap of GA. It is calculated by $G_{GA} = 100|\frac{val_{GA}-LB}{LB}|$.

Table 1. Results provided by ResDCA and GA

Size	$T_1; T_2; T_3; T_4$	cof	val_{DCA}	LB	RN	T_{DCA}	G_{DCA}	val_{GA}	T_{GA}	G_{GA}
10 × 10	20; 30; 30; 20	1.5	−298.0	−316.6	1	43.7	5.9	−212.0	161.8	33.0
10 × 10	20; 30; 30; 20	2	−345.0	−360.3	0	25.2	4.2	−245.0	162.8	32.0
10 × 10	20; 30; 30; 20	3	−435.0	−450.0	3	82.9	3.3	−302.0	163.3	32.9
10 × 10	20; 30; 30; 20	4	−525.0	−540.3	3	85.3	2.8	−376.0	161.9	30.4
10 × 10	20; 30; 30; 20	5	−615.0	−630.3	3	81.2	2.4	−438.0	163.3	30.5
20 × 20	80; 120; 120; 80	1.5	−1291.0	−1322.8	0	158.0	2.4	−748.0	660.3	43.5
20 × 20	80; 120; 120; 80	2	−1471.0	−1502.5	0	119.5	2.1	−827.0	663.4	45.0
20 × 20	80; 120; 120; 80	3	−1829.0	−1862.5	3	473.9	1.8	−993.0	657.6	46.7
20 × 20	80; 120; 120; 80	4	−2193.0	−2222.5	0	154.6	1.3	−1166.0	667.4	47.5
20 × 20	80; 120; 120; 80	5	−2553.0	−2580.5	0	158.0	1.1	−1344.0	667.0	47.9
50 × 50	500; 750; 750; 500	1.5	−8388.0	−8489.8	1	2390.0	1.2	−4314.0	9304.7	49.2
50 × 50	500; 750; 750; 500	2	−9517.0	−9614.5	1	2387.9	1.0	−4687.0	9277.7	51.3
50 × 50	500; 750; 750; 500	3	−11767.0	−11864.5	1	2245.0	0.8	−5479.0	9323.1	53.8
50 × 50	500; 750; 750; 500	4	−14010.0	−14114.5	3	3229.2	0.7	−6244.0	9154.1	55.8
50 × 50	500; 750; 750; 500	5	−16260.0	−16364.6	3	2901.0	0.6	−7033.0	9259.1	57.0

From the results, we observe that:

- ResDCA-LU provides an integer solution for all instances although DCA-LU is a local algorithm and works on continuous domain.
- The number of rerunning DCA-LU is less than or equal to 3. In some cases, It does not need to recall DCA-LU.
- The quality of solution given by ResDCA-LU is much higher than the one furnished by GA. The DCA's solutions are very close to the global optimal solutions. The gap is smaller than 3% for almost instances (12/15 instances). We can consider the obtained solutions as a global solution.
- ResDCA-LU is much faster than GA. The executing time of GA is about 4 times of the executing time of DCA.

Figure 1 presents the gap provided by ResDCA-LU and GA. The gap of ResDCA-LU decreases when the size of the problem is augmented and the gap of GA increases. It reflects that ResDCA-LU is more efficient for larger scale problems.

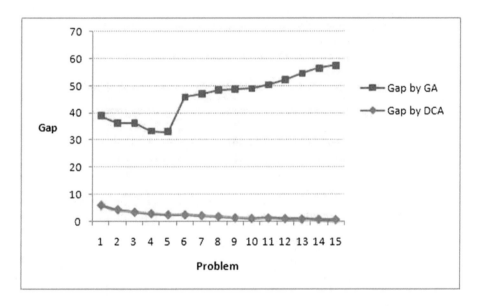

Fig. 1. The gaps by DCA and GA.

5   Conclusion

In this paper, we investigate a mixed integer linear model for land use planning problem in which the objective is to maximize the combination of the profit and the compactness. A local algorithm based on DC programming is proposed by using the reformulation and exact penalty techniques. The new algorithm is compared with a genetic algorithm (a recent stochastic local algorithm). The experimentation shows that the results are promising. For 15 simulation instances, DCA dominates GA for both objective value and executing time. The solutions provided by DCA are very close to the global optimal solutions. The limitation of this research is only the lack of results on real data. In future work, we plan to investigate more deeply DCA by considering some others data scenarios, combine DCA with a global scheme to globally solve the problem, or develop a variant of the existing model by integrating some others criterion.

References

1. Altinakar, M., Qi, H.: Numerical-simulation based multiobjective optimization of agricultural land-use with uncertainty. In: World Environmental and Water Resources Congress, Honolulu, Hawaii, United States, pp. 1–10 (2008). https://doi.org/10.1061/40976(316)481
2. Aerts, J.C.J.H., Herwijnen, M., Stewart, T.: Using simulated annealing and spatial goal programming for solving a multi site land use allocation problem. In: Fonseca, C.M., Fleming, P.J., Zitzler, E., Thiele, L., Deb, K. (eds.) Evolutionary Multi-criterion Optimization. EMO 2003, LNCS, vol. 2632, pp. 448–463. Springer, Heidelberg (2003)
3. Chetty, S., Adewumi, A.O.: Three new stochastic local search metaheuristics for the annual crop planning problem based on a new irrigation scheme. J. Appl. Math. 14 (2013). https://www.hindawi.com/journals/jam/2013/158538/
4. Aerts, J.C., Eisinger, E., Heuvelink, G.B., Stewart, T.J.: Using linear integer programming for multi-site land- use allocation. J. Geogr. Anal. **35**, 148–169 (2003)
5. Le Thi, H.A.: Contribution à l'optimisation non-convex and l'optimisation globale: Théorie, Algorithmes et Applications, Habilitation à Diriger des recherches, Université de Rouen (1997)
6. Le Thi, H.A., Pham Dinh, T.: A continuous approach for globally solving linearly constrained quadratic zero-one programming problem. Optimization **50**(1–2), 93–120 (2001)
7. Le Thi, H.A., Pham Dinh, T.: The DC(difference of convex functions) programming and DCA revisited with DC models of real world non-convex optimization problems. Ann. Oper. Res. **133**, 23–46 (2005)
8. Le Thi, H.A., Pham Dinh, T., Huynh, V.N.: Exact penalty and error bounds in DC programming. J. Glob. Optim. **52**(3), 509–535 (2012)
9. Le Thi, H.A., Pham Dinh, T.: The DC (difference of convex functions) programming and DCA revisited with DC models of real world non convex optimization problems. Ann. Oper. Res. **133**, 23–46 (2005)
10. Memmah, M.-M., Lescourret, F., Yao, X., Lavigne, C.: Metaheuristics for agricultural land use optimization. a review. Agron. Sustain. Dev. **35**, 975–998 (2015)
11. Pham Dinh, T., Le Thi, H.A.: Convex analysis approach to D.C programming: theory, algorithms and applications. Acta Math. Vietnamica **22**(1), 289–355 (1997). Dedicated to Professor Hoang Tuy on the occasion of his 70th birthday

12. Pham Dinh, T., Le Thi, H.A.: DC optimization algorithms for solving the trust region subproblem. SIAM J. Optim. **8**, 476–505 (1998)
13. Quynh, T.D.: A genetic algorithm and NSGA-II for a land use planing problem. In: Proceeding of REV-ECIT 2018, pp. 169–173 (2018). (in Vietnamese)

A New Solution Method for a Mean-Risk Mixed Integer Nonlinear Program in Transportation Network Protection

Luong Vuong Le[1,2], Quang Thuan Nguyen[3], and Duc Quynh Tran[3(✉)]

[1] Hanoi University of Science and Technology, Hanoi, Vietnam
[2] Industrial University of Ho Chi Minh City, Ho Chi Minh City, Vietnam
`leluongvuong@iuh.edu.vn`
[3] Vietnam National University Hanoi - International School, Hanoi, Vietnam
{`nguyenquangthuan,ducquynh`}`@vnu.edu.vn`

Abstract. The paper deals with a transportation network protection problem. The aim is to limit losses due to disasters by choosing an optimal retrofiting plan. The mathematical model given by Lu, Gupte, Huang [11] is a mixed integer non linear optimization problem. Existing solution methods are complicated and their computing time is long. Hence, it is necessary to develop efficient solution methods for the considered model. Our approach is based on DC (difference of two convex functions) programming and DC algorithm (DCA). The original model is first reformulated as a DC program by using exact penalty techniques. We then apply DCA to solve the resulting problem. Numerical results on a small network are reported to see the behavior of DCA. It shows that DCA is fast and the proposed approach is promising.

Keywords: DC programming · DC algorithm · Penalty function · Transportation · Retrofitting · CVaR

1 Introduction

In a transportation network, on roads the bridges are built to cross rivers or places with uneven terrain. Due to long-term use or outdated construction structures, these bridges are at risk of serious damage or collapse when natural disasters occur. Once the bridges are damaged as a result of extreme phenomena, they will lead to economic and social losses due to the cost of repairing and restoring. Moreover, the transportation network is affected by repair activities. These losses can be avoided or reduced if the risk bridges are identified and evaluated, and thus a proactive implementation strategy can be proposed. However, due to limited resources, it is not possible to retrofit all completed bridges in practice. So there should be a plan to improve bridges in the direction of priority to have economic efficiency. Choosing which risk bridge to retrofit should consider the impact on other risk bridges in the transportation network because of a

© Springer Nature Switzerland AG 2020
H. A. Le Thi et al. (Eds.): ICCSAMA 2019, AISC 1121, pp. 14–26, 2020.
https://doi.org/10.1007/978-3-030-38364-0_2

change in redistribution of traffic flows in the network. Therefore, it is necessary to consider strategies for retrofitting bridges at the network level.

Network-based bridge retrofitting problem is a general transportation network protection problem, and it can be divided into two broad categories, depending on whether bridges are considered as links or as paths. Therefore, in essence, the problem of transportation network protection is a network design problem. Typically, a network design is a bi-level mathematical optimal model. The upper level problem involves the retrofit decisions that are optimal for the best social wellfares while the lower-level one is concerned about the behavior of network users, which often present demand performance equilibrium.

Scenarios of natural phenomena are considered to be included in the transportation protection problems. Because we do not know for sure which scenario will occur, a method that can consider a lot of possible scenarios should be developed such as stochastic programming (SP) [10] or robust optimization (RO) method [1] to take into handle all scenarios. Stochastic programming methods take into account the expectation of a series of all scenarios. So it is suitable for problems with the goal of achieving long-term economic efficiency. However, it does not work well for extreme events. Therefore, when extreme events occur, the network will be affected. Meanwhile, RO methods consider the worst cases with low probability of occurrence and often offers costly solutions. Thus, it can be seen that SP and RO methods are not the best methods to consider the change of risk problem.

In [11], Lu, Gupte and Huang developed a mean-risk two-stage stochastic programming model that is more flexible in handling risks in a favorable way when resources are limited. The first stage minimizes the retrofitting cost by making strategic retrofit decisions whereas the second stage minimizes the travel cost. The conditional value-at-risk (CVaR) is included as the risk measure for the total system cost. The considered model is equivalent to a nonconvex mixed integer nonlinear program (MINLP), where the travel cost for bridge links is a nonlinear and non-convex function of retrofit decisions. According to [2], nonconvex MINLPs can be very difficult to solve. In [11], the model was solved by the Generalized Benders Decomposition method [3]. The authors derived a convex reformulation of the second-stage problem to overcome algorithmic challenges embedded in the non-convexity, nonlinearity, and non-separability of first- and second-stage variables. Thus, the model of the transportation protection problem is formulated as a convex mixed integer nonlinear program (CMINLP).

In [11], the authors proposed a method called generalized Benders decomposition to solve (CMINLP). We also use a commercial software for solving it but the executing time is quite long even for a small network. Therefore, developing efficient solution methods for CMINLP is still a challenge.

In this work, we introduce a new alternative solution method based on the mathematical technique in non-convex optimization, namely, DC programming and DC algorithm in conjunction with the use of the penalty function technique for solving Problem (CMINLP). This technique has been successfully applied to many non-convex optimization problems and showed the efficiency in particular for large-scale problems [5,8,9,12]. We tested on a nine-node network and found the algorithm running very fast. Moreover, we analyze the factors affecting the convergence time and optimal value of the DC algorithm such as choosing penalty functions, penalty parameters, starting point.

The structure of the paper is organized as follows. After the introduction section, we present the problem description in Sect. 2. Section 3 introduces the solution method. Experimental results are presented in Sect. 4. The conclusion is showed in the last section.

2   Problem Description

In this section we redescribe the model presented in [11]. This model focuses on transport network protection to prevent against extreme disasters such as earthquakes.

2.1   Parameters and Variables

To describe the problem, we use the following notations:

A transportation network with the set of nodes N and the set of directed arcs (or links) A, denoted by $G = (N, A)$;

R: the set of origins in the network;

S: the set of destinations in the network;

$\mathcal{OD}$: the set of network origin-destination (O-D) pairs;

$d^{rs} \in \mathbb{R}_+$: the given travel demand between O-D pair (r, s), $(r, s) \in \mathcal{OD}$;

$\overline{A}$ ($\overline{A} \subset A, \overline{A} \neq \emptyset$): the set of arcs that are directedly affected by hazards, primarily including risk bridges;

c_a: the practical capacity of arc a;

H: the finite set representing a list of retrofit strategies that can be applied to at-risk bridges to mitigate the adverse effects caused by future disaster events;

b_a^h: the retrofit cost for $a \in \overline{A}$ with strategy h;

b_0: the total budget is used for retrofitting bridges;

K: the set of hazard scenarios which can happen to the network;

$p_k \in (0, 1)$: the given probability of scenario k, $k \in K$;

$\theta_a^{h,k}$: the ratio of post-disaster arc capacity to the full arc capacity, with each $k \in K$ and for every $a \in \overline{A}$ $h \in H$, $\theta_a^{h,k} \in (0, 1]$. When a disaster occurs, the post-disaster capacity of arc $a \in \overline{A}$ that has been retrofitted with strategy $h \in H$ equals $c_a \theta_a^{h,k}$;

δ: the experimental data;

γ: the parameter converts the travel time into monetary value;

t_{0a}: the parameter indicates the travel time in case of the free-flow-rate of arc a.

We use some variables as follows:

u_a^h: the binary variable, takes a value of 1 if using strategy h for arc a and 0 otherwise, for every $a \in \overline{A}, h \in H$;

$x_a^{rs,k}$: the flow on arc a corresponds to the (r, s) pair for scenario k, for every $a \in A$, $(r, s) \in \mathcal{OD}$ and $k \in K$;

v_a^k: the total flow on arc $a \in A$, and $v_a^k = \sum_{(r,s) \in \mathcal{OD}} x_a^{rs,k}$ for all $a \in A$;

$q^{rs,k}$: the travel demand is not satisfied for the O-D pair (r, s).

The model allows for post-disaster travel demand that are not satisfied for a variety of reasons, such as turning off certain routes, increasing traffic congestion in the network, etc.

2.2   Mathematical Model

Let U be the set defined by

$$U := \left\{ u \in \{0,1\}^{|\overline{A}| \times |H|} \,\middle|\, \sum_{h \in H} u_a^h = 1\,, \forall a \in \overline{A}, b^T u \le b_0 \right\}. \quad (1)$$

For the kth scenario, let $f^k(u) = b^T u + Q^k(u)$ be the total cost function, where $Q^k(u)$ is the optimal value for the total travel cost, given the retrofitting vector u.

The two-stage SP is as

$$\text{(2-stage SP)}: \min_u \sum_{k \in K} p_k f^k(u) = \min_u b^T u + \sum_{k \in K} p_k Q^k(u) \text{ subject to } u \in U. \quad (2)$$

For the kth scenario, the recourse function is defined as

$$Q^k(u) = \min_{x^k, q^k} \gamma \sum_{a \in A} v_a^k t_a^k + M \sum_{(r,s) \in \mathcal{OD}} q^{rs,k} \quad (3)$$

$$= \min_{v^k, x^k, q^k} \gamma \sum_{a \in A} t_{0a} \left[v_a^k + \delta \frac{(v_a^k)^5}{\hat{c}_a^k(u)^4} \right] + M \sum_{(r,s) \in \mathcal{OD}} q^{rs,k} \quad (4)$$

$$\text{s.t. } v_a^k = \sum_{(r,s) \in \mathcal{OD}} x_a^{rs,k}, \forall a \in A, (x^k, q^k) \in X. \quad (5)$$

where

$$t_a^k = t_{0a} \left[1 + \delta \left(\frac{v_a^k}{\hat{c}_a^k(u)} \right)^4 \right] \text{ (Bureau of Public Records function [16])}$$

is the arc travel time per unit flow, and

$$\hat{c}_a^k(u) = \begin{cases} c_a \sum_{h \in H} \theta_a^{h,k} u_a^h & a \in \overline{A} \\ c_a & a \in A \setminus \overline{A} \end{cases}. \quad (6)$$

The objective function (3) consists of two terms. The first term is total travel cost. The second term is included to represent the penalty cost for unsatisfied demand. The set X is defined as:

$$X = \left\{ (x,q) \geq (0,0) \mid \sum_{j:(r,j)\in A} x_{rj}^{rs} - \sum_{j:(j,r)\in A} x_{jr}^{rs} + q^{rs} = d^{rs} \; \forall (r,s) \in \mathcal{OD}, \quad (7) \right.$$

$$\sum_{j:(s,j)\in A} x_{sj}^{rs} - \sum_{j:(j,s)\in A} x_{js}^{rs} - q^{rs} = -d^{rs} \; \forall (r,s) \in \mathcal{OD}, \quad (8)$$

$$\left. \sum_{j:(t,j)\in A} x_{tj}^{rs} - \sum_{j:(j,t)\in A} x_{jr}^{rs} = 0 \; \forall (r,s) \in \mathcal{OD}, t \in N\setminus\{r,s\} \right\}. \quad (9)$$

For each pair (r,s), Eqs. (7) and (8), respectively, allow a slack of q^{rs} in the flow balance at r and s to solve unsatisfied demand, whereas the preservation of flow at other nodes in network is shown by Eq. (9).

The recourse function $Q^k(u)$ is a nonlinear optimization problem in (3)–(5) for each scenario k. This problem is non-convex because of presence of the terms $t_a^k v_a^k$ in the objective function and the equality constraints defining t_a^k are non-linear. In [11], for every $u \in U$, the authors derived a reformulation to obtain a convex program and there is a separation of variables between the first and second stages.

To reformulate the problem, the following inequality is added by introducing an auxiliary second stage nonnegative continuous variable y_a^k for each $a \in \overline{A}$,

$$y_a^k \geq \frac{\left(v_a^k\right)^5}{\left(c_a \sum_{h\in H} u_a^h \theta_a^{h,k}\right)^4} \quad \forall a \in \overline{A}. \quad (10)$$

Hence, we have

$$Q^k(u) = \min_{v^k,x^k,q^k,y^k} \gamma \sum_{a\in A} t_{0a}\left[v_a^k + \delta y_a^k\right] + M \sum_{(r,s)\in\mathcal{OD}} q^{rs,k} \quad (11)$$

$$s.t. \quad (5),(10). \quad (12)$$

According to [11], the recourse function $Q^k(u)$ can be formulated as:

$$Q^k(u) = \min_{v^k,x^k,q^k,y^k,w^k} \gamma \sum_{a\in A} t_{0a}\left[v_a^k + \delta y_a^k\right] + M \sum_{(r,s)\in\mathcal{OD}} q^{rs,k} \quad (13)$$

$$s.t. \quad v_a^k = \sum_{(r,s)\in\mathcal{OD}} x_a^{rs,k}, \quad \forall a \in A, \; \left(x^k,q^k\right) \in X \quad (14)$$

$$\left(v_a^k\right)^5 \leq c_a^4 \sum_{h\in H} \omega_a^{h,k}, y_a^k = \sum_{h\in H} y_a^{h,k} \; \forall a \in \overline{A} \quad (15)$$

$$\omega_a^{h,k} \leq c_a^4 \left(\theta_a^{h,k}\right)^4 y_a^{h,k} \quad \forall h \in H, a \in \overline{A} \tag{16}$$

$$0 \leq y_a^{h,k} \leq \frac{c_a \varsigma_a^5}{\left(\theta_a^{h,k}\right)^4} u_a^h, 0 \leq \omega_a^{h,k} \leq c_a \varsigma_a^5 u_a^h \quad \forall h \in H, a \in \overline{A} \tag{17}$$

where ς_a is a positive constant large enough such that $\varsigma_a c_a$ is an upper bound on the travel flow of link a, for every $a \in \overline{A}$.

This proposition allows linear separation of the first stage variable $u \in U$ from the second stage variables.

According to [11], the mean risk problem with α-level is a convex MINLP.

$$\min_{\substack{u,g,z,v, \\ q,x,y,\omega}} (1+\lambda) b^T u + \sum_{k \in K} p_k \left[\gamma \sum_{a \in A} t_{0a} \left[v_a^k + \delta y_a^k\right] + M \sum_{(r,s) \in \mathcal{OD}} q^{rs,k} \right]$$

$$+ \lambda \left(g + \frac{1}{1-\alpha} \sum_{k \in K} p_k z^k \right) \tag{CMINLP}$$

$$\text{subject to} \quad u \in U, z^k \geq 0 \ \forall k \in K \tag{18}$$

$$z^k \geq \gamma \sum_{a \in A} t_{0a} \left[v_a^k + \delta y_a^k\right] + M \sum_{(r,s) \in \mathcal{OD}} q^{rs,k} - g \ \forall k \in K \tag{19}$$

$$(14) - (17) \ \forall k \in K, \tag{20}$$

where λ is a predefined weighting factor. The objective of the problem is to minimize the total cost of retrofitting bridges, expected travel cost, unsatisfied demand penalty and the risk term.

3 Solution Method

This section introduces a new alternative solution method based on the mathematical technique in non-convex optimization, namely, DC programming and DCA for solving Problem CMINLP. This technique has been successfully applied to many non-convex optimization problems and showed the efficiency in particular for large-scale problems [5,8,9,12].

3.1 DC Programming and DC Algorithm

DC Programming and DCA constitute the backbone of smooth/nonsmooth non-convex programming and global optimization. They were introduced by Pham Dinh Tao in 1985 in their preliminary form and have been extensively developed by Le Thi Hoai An and Pham Dinh Tao since 1994. DCA has been successfully applied to real world non-convex programs in different fields of applied sciences (see e.g. [5,13,14] and the references therein). DCA is one of rare efficient algorithms for non-smooth non-convex programming which allows solving large-scale

DC programs. Although DCA is a continuous approach, it has been efficiently investigated for solving nonconvex Linear/quadratic programming with binary variables via exact penalty techniques [4].

For a convex function f defined on $\mathbb{R}^n$ and $x_0 \in \mathrm{dom} f := \{x \in \mathbb{R}^n | f(x) < +\infty\}$, $\partial f(x_0)$ denotes the sub-differential of f at x_0 that is

$$\partial f(x_0) := \{y \in \mathbb{R}^n | f(x) \geq f(x_0) + \langle x - x_0, y \rangle, \forall x \in \mathbb{R}^n\}.$$

The sub-differential $\partial f(x_0)$ is a closed convex set in $\mathbb{R}^n$. It generalizes the derivative in the sense that f is differentiable at x_0 if and only if $\partial f(x_0)$ is reduced to a singleton that is exactly $\{f'(x_0)\}$.

A general DC program is of the form

$$\inf \{f(x) := g(x) - h(x) | x \in \mathbb{R}^n\}, \tag{P_{dc}}$$

with $g, h \in \Gamma_0(\mathbb{R}^n)$, the set of all lower semi-continuous proper convex functions on $\mathbb{R}^n$. Such a function f is called DC function, and g, h are its DC components. A generic DCA scheme is shown as follows:

Initialization: Let $x^0 \in \mathbb{R}^n$ be a good guess, $k = 0$;
Repeat

- Calculate $y^k \in \partial h(x^k)$;
- Calculate x^{k+1} by solving the convex problem

$$\min \{g(x) - h(x^k) - \langle x - x^k, y^k \rangle | x \in \mathbb{R}^n\}; \tag{P_k}$$

$k = k + 1$;
Until convergence of x^k.

Each DC function f has infinitely many DC decompositions which have crucial implications for the qualities (speed of convergence, robustness, efficiency, globality of computed solutions, ...) of DCA.

We now present the results of the penalty technique presented in [7] relating to exact penalty techniques in DC programming developed in [6].

Let K be a nonempty bounded polyhedral convex in $\mathbb{R}^n$ and f is a DC function. We consider the general $0-1$ problem (GZOP) in the form:

$$\min \{f(x) | x \in K; x \in \{0, 1\}\}. \tag{GZOP}$$

Thanks to the next theorem, we can reformulate a combinatorial optimization problem as a continuous one.

Theorem 1. *[7] Let K be a nonempty bounded polyhedral convex set in $\mathbb{R}^n$, f be a finite DC function on K and p be a finite nonnegative concave function on K. Then there exists $t_0 \geq 0$ such that for all $t > t_0$ the following problems have the same optimal value and the same solution set:*

$$(P_t) \quad \alpha(t) = \min\{f(x) + tp(x) | x \in K\} \tag{21}$$
$$(P) \quad \alpha = \min\{f(x) | x \in K, p(x) \leq 0\}. \tag{22}$$

Now, we are able to formulate (GZOP) as a continuous optimization problem. Let p be the finite function defined on K by

$$p(x) = \sum_{i=1}^{n} \min\{x_i, 1 - x_i\}.$$

It is obvious that on the set $K' = K \cap [0,1]^n$, p is nonnegative and concave function. Furthermore, we have

$$\{x \in K | x \in \{0,1\}^n\} = \{x \in K' | p(x) = 0\} = \{x \in K' | p(x) \leq 0\}.$$

Therefore, the problem (GZOP) can be rewritten as

$$\min\{f(x) | x \in K', p(x) \leq 0\}.$$

With a sufficiently large number t, from Theorem 1 it follows that the last problem is equivalent to

$$\min\{f(x) + tp(x) | x \in K'\}.$$

3.2   DCA for CMINLP

Now, let us get back to the original problem CMINLP.
Set $NV = |\overline{A}|.|H|$ and $T = |\overline{A}|.|H| + 2|A||K| + |\mathcal{OD}||K| + 1 + |K| + 2|\overline{A}||K||H|$.
Let $D \subset \mathbb{R}^T$ be the set defined by (18)–(20), $D' = D \cap \left([0,1]^{NV} \times \mathbb{R}^{T-NV}\right)$.
Set

$$
\begin{aligned}
(1+\lambda)\, b^T u &+ \sum_{k \subset K} p_k \left[\gamma \sum_{a \in A} t_{0a} \left[v_a^k + \delta y_a^k\right] + M \sum_{(r,s) \in \mathcal{OD}} q^{rs,k} \right] \\
&+ \lambda \left(g + \frac{1}{1-\alpha} \sum_{k \in K} p_k z^k \right) \\
&= (1+\lambda)\, b^T u + \sum_{i=1}^{T-NV} \alpha_i r_i = f(u,r) = f(\overline{x}).
\end{aligned}
\tag{23}
$$

Let $p_1(\overline{x}) = \sum_{i=1}^{NV} \min\{\overline{x}_i, 1 - \overline{x}_i\}$ and $p_2(\overline{x}) = \sum_{i=1}^{NV} \overline{x}_i(1 - \overline{x}_i)$ be two functions defined over D'.

Then according to Theorem 1, the problem is equivalent to

$$\min \left\{ F\left(\overline{x}\right) = f\left(\overline{x}\right) + tp\left(\overline{x}\right) : \overline{x} \in D' \right\}$$

with a sufficiently large number t and $p\left(\overline{x}\right) = p_1\left(\overline{x}\right)$ or $p\left(\overline{x}\right) = p_2\left(\overline{x}\right)$. We have a DC decomposition

$$F\left(\overline{x}\right) = g\left(\overline{x}\right) - h\left(\overline{x}\right)$$

where $g\left(\overline{x}\right) = \chi_D(\overline{x})$ and $h\left(\overline{x}\right) = -f\left(\overline{x}\right) - tp\left(\overline{x}\right)$. Here χ_D stands for the indicator function of D: $\chi_D(\overline{x}) = 0$ if $x \in D$, $\chi_D(\overline{x}) = +\infty$ otherwise.
The DC algorithm solves the problem as follows:
Initialization: Let $x_0 \in \mathbb{R}^T$ be a good guess, $k = 0$;
Repeat

- Compute $\overline{y}^k \in \partial h(\overline{x}^k) = \{-f'(\overline{x}^k) - tp'(\overline{x}^k)\}$
  With $p(\overline{x}) = p_1(\overline{x})$, we have

$$\overline{y}_\ell^k = \begin{cases} \begin{cases} t - (1+\lambda)\, b_l & if \ \ \overline{x}_\ell \geq 0.5 \\ -t - (1+\lambda)\, b_l & if \ \ \overline{x}_l < 0.5 \end{cases} & if \ \ \ell = 1, \ldots, NV \\ \overline{y}_\ell^k = -\alpha_\ell & if \ \ \ell = (NV+1), \ldots, T \end{cases},$$

and with $p(\overline{x}) = p_2(\overline{x})$, we have

$$\overline{y}_\ell^k = \begin{cases} t(2\overline{x}_\ell^k - 1) - (1+\lambda)b_\ell & if \ \ \ell = 1, \ldots, NV \\ \overline{y}_\ell^k = -\alpha_\ell & if \ \ \ell = (NV+1), \ldots, T \end{cases}.$$

- Take $\overline{x}^{k+1} \in \partial h(\overline{y}^k)$

$$\overline{x}^{k+1} \in argmin \left\{ g(\overline{x}) - h(\overline{x}^k) - \langle \overline{x} - \overline{x}^k, \overline{y}^k \rangle | \overline{x} \in D' \right\} \qquad \text{(CNLP)}$$
$$\equiv argmin \left\{ -\langle \overline{x}, \overline{y}^k \rangle | \overline{x} \in D' \right\}.$$

Until convergence of $\overline{x}^k$.

The problem (CNLP) is convex programming with the objective function as a linear function. It can be solved by CVX Solver. So instead of solving a discrete problem we will solve a series of continuous problems to obtain the solution.

4   Experimental Results

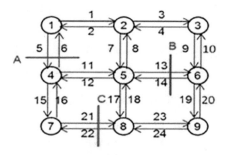

Fig. 1. Nine-node network [11]

We tested the proposed algorithm on the nine node network described in Fig. 1, which is used in [11]. It consists of nine nodes ($|N| = 9$), 24 directional links ($|A| = 24$), and 72 O-D pairs ($|\mathcal{OD}| = 72$). There are three bridges, labeled as A, B, and C, on both directions on the network. These bridges are susceptible to seismic disasters. There are 6 links that are directly affected by the passing bridges, i.e. $\overline{A} = \{a_1, \cdots a_6\} = \{5, 6, 11, 12, 21, 22\}$, $|\overline{A}| = 6$. Let the set $K = \{1, 2, 3, 4, 5, 6\}$ and each scenario $k \in K$, we randomly generated $p_k \in (0, 1)$. We consider five strategies, denoted as $h_1 - h_5$, we randomly generated $\theta_a^{h,k} \in (0, 1]$. Table 1 reports the ratios for two scenarios.

Table 1. Some sample values of $\theta_a^{h,k}$ for fixed scenarios $k = 1, 2$.

Link	Strategy					Link	Strategy				
	h_1	h_2	h_3	h_4	h_5		h_1	h_2	h_3	h_4	h_5
link 5	0.15	0.4	0.4	0.6	1	link 5	0.03	0.4	0.3	0.4	0.9
link 6	0.15	0.4	0.4	0.6	1	link 6	0.03	0.4	0.3	0.4	0.9
link 11	0.25	0.55	0.55	0.85	0.85	link 11	0.4	0.4	0.4	0.65	0.65
link 12	0.25	0.55	0.55	0.85	0.85	link 12	0.4	0.4	0.4	0.65	0.65
link 21	0.18	0.43	0.43	0.77	0.77	link 21	0.07	0.23	0.23	0.57	0.57
link 22	0.18	0.43	0.43	0.77	0.77	link 22	0.07	0.23	0.23	0.57	0.57

Other input parameters related to the algorithm are given as follows:

$$(c_a)_{1 \times 24} = 10^2 \times [10\ 12\ 14\ 16\ 14\ 12\ 16\ 15\ 12\ 14\ 18\ 12$$
$$14\ 13\ 14\ 16\ 10\ 14\ 18\ 12v14\ 18\ 16\ 14];$$

$$(b_a^h)_{1 \times 30} = [a_1^{h_1}, \cdots, a_1^{h_5}, a_2^{h_1}, \cdots, a_2^{h_5}, \cdots, a_6^{h_1}, \cdots, a_6^{h_5}]$$
$$= 10^5 \times [2.5\ 1\ 1.5\ 2\ 2.5\ 1.5\ 2\ 1.5\ 2\ 2.5\ 0.5\ 1\ 1.5\ 2\ 2.5$$
$$1.5\ 1\ 1.5\ 2\ 2.5\ 1.5\ 1\ 1.5\ 2\ 2.5\ 1.5\ 1\ 2.5\ 2\ 2.5];$$

$$(d^{rs})_{1\times 72} = [20\ 10\ 30\ 20\ 15\ 20\ 20\ 15\ 10\ 12\ 18\ 10\ 16\ 14\ 20\ 12\ 10\ 20\ 14\ 16\ 12\ 20$$
$$10\ 18\ 14\ 10\ 20\ 12\ 14\ 18\ 20\ 10\ 12\ 14\ 16\ 18\ 12\ 20\ 14\ 16\ 18\ 10\ 20\ 14\ 15\ 20\ 14$$
$$18\ 20\ 15\ 18\ 20\ 10\ 14\ 18\ 20\ 10\ 14\ 16\ 12\ 20\ 18\ 15\ 14\ 18\ 12\ 14\ 20\ 12\ 18\ 14\ 10];$$

$$(t_{0a})_{1\times 24} = [1\ 1.5\ 2\ 1.2\ 1.3\ 1.2\ 1.4\ 1.2\ 1.6\ 1.3\ 1.2\ 1.2$$
$$1\ 1.5\ 2\ 1.2\ 1.3\ 1.2\ 1.4\ 1.2\ 1.6\ 1.3\ 1.2\ 1.2];$$

$$\varsigma_a = [\varsigma_{a_1}, \cdots, \varsigma_{a_6}] = [10\ 12\ 10\ 8\ 14\ 12],\ M = 10^7,\ b_0 = 15 \times 10^5.$$

Other parameters taken from [11] are $\lambda = 1$, $\alpha = 0.7$, $\delta = 0.15$, $\gamma = 10^3$.

We take two starting points $x_0 = (u, r) = (\mathbf{1}, \mathbf{0})$ and $x_1 = (u, r)$ is a point as follows: for every $a \in \overline{A}, u_a^{h_1} = 1$ and $u_a^h = 0$ for $h \neq h_1$ and $r = 0$. The stop condition of the algorithm is $||x^{k+1} - x^k||_2 \leq \epsilon$ with $\epsilon = 10^{-3}$. DCA is implemented in Matlab 2017b with the number of variables and constraints respectively 3008 and 2221. We tested the nine-node instances on a computer with 8 GB RAM and Intel(R) Core(TM) i5-8400@2.80 GHz processor under Windows 10 pro environment. The results are shown in the Table 2.

A lower bound (LB) is calculated by solving a relaxation of the original problem in which the integer variables are ignored. This value calculated, say 64488×10^2, is used to compute $GAP = (Obj.Value - LB) \times 100\%/LB$.

From the results, we can see that:

– DCA always provides a feasible solution although it is a local algorithm and works on the relaxed domain.

Table 2. The results for different cases

P.f. $p(\overline{x})$	t	Ini. point	Obj. value	GAP%	TCPU (secs)	Iter.	P.f. $p(\overline{x})$	t	Ini. point	Obj. value	GAP%	TCPU (secs)	Iter.
$p_1(\overline{x})$	10^0	x_0	65088	0.9304	0.69	2	$p_2(\overline{x})$	10^0	x_0	65088	0.9304	0.67	2
	10^1		65088	0.9304	0.90	3		10^1		65088	0.9304	0.72	2
	10^2		65088	0.9304	1.15	4		10^2		65088	0.9304	1.67	7
	10^3		65088	0.9304	0.85	3		10^3		65088	0.9304	1.33	5
	10^4		65088	0.9304	0.87	3		10^4		65088	0.9304	1.64	7
	10^5		65088	0.9304	0.90	3		10^5		65088	0.9304	1.43	5
	10^6		65088	0.9304	0.84	3		10^6		65088	0.9304	1.55	5
	10^7		65088	0.9304	0.82	3		10^7		65088	0.9304	1.46	5
	10^8		65088	0.9304	0.81	3		10^8		65088	0.9304	1.19	4
$p_1(\overline{x})$	10^0	x_1	65088	0.9304	0.68	2	$p_2(\overline{x})$	10^0	x_1	65088	0.9304	0.66	2
	10^1		65088	0.9304	0.72	2		10^1		65088	0.9304	0.72	2
	10^2		65088	0.9304	0.67	2		10^2		65088	0.9304	1.05	4
	10^3		65088	0.9304	0.67	2		10^3		65088	0.9304	0.66	2
	10^4		65088	0.9304	0.70	2		10^4		65088	0.9304	0.71	2
	10^5		68088	5.5824	0.95	3		10^5		68088	5.5824	0.93	3
	10^6		71088	10.234	0.73	2		10^6		71088	10.234	0.73	2
	10^7		71088	10.234	0.70	2		10^7		71088	10.234	0.71	2
	10^8		71088	10.234	0.69	2		10^8		71088	10.234	0.67	2

Obj. value ($\times 10^2$)
TCPU: Total CPU time

– The computing time is good. DCA needs about 1 second to solve a problem of 3008 variables and 2221 constraints.
– The small GAPs show that the solutions obtained by DCA are good.
– The impact of the change of penalty functions is not seen but the influence of the starting points is clear. The obtained solutions with the starting point x_0 are stable. It does not depend on the penalty parameter.

5   Conclusions

In this paper, we proposed a new alternative method based on DC programing and DCA for a transportation network protection problem. The exact penalty technique is used to reformulate the original model and overcome the difficulties due to integer variables. The proposed algorithm was tested on a small network with the structure being similar to the one used in [11]. The impact of penalty parameter, penalty functions and starting point was reported. The results show that the first starting point is better and the algorithm is rapid. In future works, we may combine DCA with another method, for instance, branch and bound to globally solve the problem. The experimentation for larger scale setting should be investigated.

References

1. Atamturk, A., Zhang, M.: Two-stage robust network flow and design under demand uncertainty. Oper. Res. **55**, 662–673 (2007)
2. Burer, S., Letchford, A.N.: Non-convex mixed-integer nonlinear programming: a survey. Surv. Oper. Res. Manage. Sci. **17**, 97–106 (2012)
3. Floudas, C.A.: Generalized benders decomposition. In: Nonlinear and Mixed-Integer Optimization, pp. 114–143. Oxford University Press, Oxford (1995)
4. Le Thi, H.A., Pham Dinh, T., Le, D.M.: Exact penalty in D.C. programming. Vietnam J. Math. **27**(2), 169–178 (1999)
5. Le Thi, H.A., Pham Dinh, T.: The DC (difference of convex functions) programming and DCA revisited with DC models of real wourd nonconvex optimization problems. Ann. Oper. Res. **133**, 23–46 (2005)
6. Le Thi, H.A., Pham, D.T., Huynh, V.N.: Exact Penalty Techniques in DC Programming. Technical Report, LMI, INSA-Rouen, France, July 2007
7. Le Thi, H.A., Nguyen, Q.T.: A robust approach for nonlinear UAV task assignment problem under uncertainty. In: Transactions on Computational Collective Intelligence II, pp. 147–159 (2010)
8. Le Thi, H.A., Nguyen, Q.T., Phan, K.T., Pham Dinh, T.: DC programming and DCA based cross-layer optimization in multi-hop TDMA networks. In: Selamat, A., Nguyen, N.T., Haron, H. (eds.) Intelligent Information and Database Systems. ACIIDS. LNCS, vol. 7803. Springer, Heidelberg (2013)
9. Le Thi, H.A., Pham Dinh, T., Quynh, T.D.: A DC programming approach for a class of bilevel programming problems and its application in portfolio selection. NACO Numer. Algebra Control Optim. **2**(1), 167–185 (2012)
10. Liu, C., Fan, Y., Ordonez, F.: A two-stage stochastic programming model for transportation network protection. Comput. Oper. Res. **36**, 1582–1590 (2009)

11. Lu, J., Gupte, A., Huang, Y.: A mean-risk mixed integer nonlinear program for transportation network protection. Eur. J. Oper. Res. **265**(1), 277–289 (2018)
12. Nguyen, Q.T., Le Thi, H.A.: Solving an inventory routing problem in supply chain by DC programming and DCA. In: Nguyen, N.T., Kim, C.G., Janiak, A. (eds.) Intelligent Information and Database Systems. ACIIDS. LNCS, vol. 6592. Springer, Heidelberg (2011)
13. Pham Dinh, T., Le Thi, H.A.: Convex analysis approach to DC programming. theory, algorithms and applications. Acta Math. Vietnamica **22**(1), 289–357 (1997). Dedicated to Professor Hoang Tuy on the occassion of his 70th birthday
14. Pham, D.T., Le Thi, H.A.: DC optimization algorithms for solving the trust region subproblem. SIAM J. Optim. **8**, 476–505 (1998)
15. Traffic assignment manual for application with a large, high speed computer. Washington, U.S. Dept. of Commerce, Bureau of Public Roads, Office of Planning, Urban Planning Division (1964)

A Novel Approach for Travel Time Optimization in Single-Track Railway Networks

Nguyen Quang Thuan[1(✉)] and Nguyen Duc Anh[2]

[1] Vietnam National University Hanoi-International School, Hanoi, Vietnam
nguyenquangthuan@vnu.edu.vn
[2] Hanoi University of Science and Technology, Hanoi, Vietnam

Abstract. Train scheduling plays an important role in the operation of railways systems. This work focuses on a model of scheduling in which one minimizes the total travel time of trains in a single track railways network. The model can be written in the form of a mixed 0–1 linear program which has the worst case exponential complexity to calculate the optimal solution. In this paper, we propose a computationally efficient approach to solve the train scheduling problem. Our approach is based on a so-called Difference of Convex functions Algorithm (DCA) to provide good feasible solutions with finite convergence. The algorithm is tested on three different railway network topologies including one topology introduced in [18] and two practical topologies in Northern Vietnam. The numerical results are encouraging and demonstrate the efficiency of the approach.

Keywords: Train scheduling · Penalty function · DC Algorithm

1 Introduction

Railways have advantage over the roadways in that they can carry a large number of passengers and large or heavy freight loads to long distances. It becomes an essential pubic transport in most countries. Among many problems arising in operating a railway system, train scheduling is critical to reduce costs, increase profits or improve service quality. It generates train timetables to optimize total cost (time cost or financial cost) and satisfy some given conditions such as passenger demands, investment capital, time resource, etc. Train scheduling is usually classified into two groups [18]: line planning and scheduling generation. The former determines frequencies, routes, and scheduled times at each stop while the later finds the departure time and the arrival time of each train at sidings or stations.

Szpigel (1973) can be considered as a pioneer in studying train scheduling problems. The author modeled a problem of minimizing the travel time of trains on single track line to a job shop scheduling problem then used a branch-and-bound technique to solve it [31]. Afterwards, many researchers focused on

© Springer Nature Switzerland AG 2020
H. A. Le Thi et al. (Eds.): ICCSAMA 2019, AISC 1121, pp. 27–38, 2020.
https://doi.org/10.1007/978-3-030-38364-0_3

two aspects: modeling and solution methods for a given problem. For modeling, practical problems were normally formulated into the form of a mathematical optimization problem such as an integer program [2–6] a mixed-integer program [1,7,10–15,18,25,26], a multi-objective linear programming [8,11]. This kind of problems is NP-hard. Finding a solution method (exact or heuristic) for them is a challenging mission. Branch and bound techniques were usually used to get the optimal solution [14,16,31]. However, it takes much time to get the optimal solution in the case of large-scale and complex instances. Thus, heuristic approaches were often proposed to find feasible solutions, for instance, the priority-rule-based heuristics [1,9,19,28,28], backtracking search [1], look-ahead search [30], and meta-heuristic algorithms [14,15,17].

Each solution method above furnishes different feasible solutions. Its efficiency depends on the structure of formulations, the network topology, the size of instance, etc. It is worthy to have a new solution method that may find a good feasible solution. If this solution is not optimal, it is a good upper bound in the schema of branch-and-bound algorithms in order to accelerate the time of computing the optimal solution. Our contribution is to propose a method finding such a good feasible solution. It is based on DC (difference of two convex functions) programming and DCA (DC Algorithms) that has been efficiently applied to real world non convex programs in various fields [22,23].

Obviously, we study a typical model that was proposed by Higgins et al. [14]; and modified by Karoonsoontawong and Taptana [18]. It is formulated in the form of a mixed 0–1 linear programming (MILP). By employing the exact penalty method, we first show that the MILP can be equivalently recast as a concave minimization problem. Next, we reformulate the concave minimization problem in the form of a DC program then use DCA to solve. We test the proposed algorithm on three different railway network topologies including one topology introduced in [18] and two practical topologies in Northern Vietnam. The preliminary results demonstrate the efficiency of the proposed method.

The paper is structured as follows. After the introduction, we describe the problem in Sect. 2. Section 3 presents DC programming, DCA and show how to apply DCA to the problem. In Sect. 4, we provide some numerical experiments to evaluate the proposed approach. The last section is dedicated to some conclusions.

2   Problem Description

Among train scheduling problems in single-track railway line, the travel time optimization problem has specially attracted many researchers. Higgins et al. [14], Zhou and Zhong [32] introduced a mixed integer linear program for the aforementioned problem. Basing on the formulation in [14], Karoonsoontawong and Taptana in 2017 proposed a modified formulation [18]. This section presents the problem described in [18] with the following assumptions: Networks includes sidings or stations that divide railways into segments; Two trains or more are not allowed on any track segment; There are two tracks at sidings/stations and each segment has the most 2 tracks; and a pre-specified path is assigned for each train.

2.1   Notation

The sets, the parameters, the variables are denoted as follows.

Sets

$I = \{1, 2, ..., n_I\}$ - set of trains, $n_I = |I|, n_I \in \mathbb{N}$ is total number of trains in the railway system;

$P = \{1, 2, ..., n_P\}$ - set of rail segments, $n_P = |P|, n_P \in \mathbb{N}$ is total number of segments in the railway system;

$Q = \{1, 2, ..., n_Q\}$ - set of stations or sidings. $n_Q = |Q|, n_Q \in \mathbb{N}$ is the number of stations (or sidings) in the railway system;

$P(i)$ - ordered set of rail segments traversed by train $i \in I$;

$Q(i)$ - ordered set of stations (or sidings) traversed by train $i \in I$;

P_1 - set of single-track segments;

P_2 - set of double-track segments;

$P^{same\_d}(i, j)$ - set of common rail segments for trains i and j, which traverse in the same direction;

$P_1^{opp\_d}(i, j)$ - set of common single-track segments for trains i and j, which traverse in opposite directions;

D is set of segment directions: inbound or outbound.

Parameters

$q_1(p, d)$ - starting station (or siding) of segment p in direction d;

$q_2(p, d)$ - terminal station (or siding) of segment p in direction d;

$d_{i,p}$ - direction in which train $i \in I$ traverses segment $p \in P(i)$;

$h_{1,p}^{i,j}$ - minimum headway between trains $i, j \in I$ traversing in the same direction on $p \in P$;

$h_{2,p}^{i,j}$ - minimum headway between trains $i, j \in I$ traversing in opposite directions on $p \in P$;

l_p - length of segment $p \in P$;

Y_{dO}^i - earliest departure time of train $i \in I$;

$\underline{v}_p^i$ - minimum allowable average velocity of train $i \in I$ on segment $p \in P(i)$;

$\overline{v}_p^i$ - maximum achievable average velocity of train $i \in I$ on segment $p \in P(i)$;

W_i - weight showing the priority for train $i \in I$;

S_q^i - scheduled stop time for train $i \in I$ at station $q \in Q(i)$;

M - sufficiently big constant.

Decision Variables

A_{ijp} equals to 1 if train $i \in I$ traverses track segment $p \in P^{same\_d}(i, j)$ before train $j \in I$ when trains i and j traverse track segment p in the same direction; 0 otherwise.

B_{ijp} equals to 1 if train $i \in I$ traverses track segment $p \in P_1^{oppd}(i, j)$ before train $j \in I$ when trains i và j traverse track segment p in opposite directions; 0 otherwise.

$X_{a,q}^i$ is the arrival time of train $i \in I$ at station/siding $q \in Q(i)$.
$X_{d,q}^i$ is the departure time of train $i \in I$ from station/siding $q \in Q(i)$.
X_{dO}^i is the departure time of train $i \in I$ from its origin station.
X_{aD}^i is the arrival time of train $i \in I$ at its destination station.

2.2   Mathematical Model

The problem can be formulated as follows:

$$min\Big\{ z = \sum_{i \in I} W_i (X_{aD}^i - Y_{dO}^i) \Big\}, \tag{1}$$

subject to:

$$M * A_{ijp} + X_{d,q_1(p,d_{i,p})}^i \geq X_{a,q_2(p,d_{j,p})}^j + h_{1,p}^{i,j} \quad \forall i \neq j \in I, p \in P^{same\_d}(i,j) \tag{2}$$

$$M * (1 - A_{ijp}) + X_{d,q_1(p,d_{j,p})}^j \geq X_{a,q_2(p,d_{i,p})}^i + h_{1,p}^{i,j} \quad \forall i \neq j \in I, p \in P^{same\_d}(i,j) \tag{3}$$

$$M * B_{ijp} + X_{d,q_1(p,d_{i,p})}^i \geq X_{a,q_2(p,d_{j,p})}^j + h_{2,p}^{i,j} \quad \forall i \neq j \in I, p \in P_1^{opp\_d}(i,j) \tag{4}$$

$$M*(1-B_{ijp})+X_{d,q_1(p,d_{j,p})}^j \geq X_{a,q_2(p,d_{i,p})}^i + h_{2,p}^{i,j} \quad \forall i \neq j \in I, p \in P_1^{opp\_d}(i,j) \tag{5}$$

$$\frac{l_p}{v_p^i} \leq X_{a,q_2(p,d_{i,p})}^i - X_{d,q_1(p,d_{i,p})}^i \leq \frac{l_p}{v_p^i} \quad \forall i \in I, p \in P(i) \tag{6}$$

$$X_{dO}^i \geq Y_{dO}^i \quad \forall i \in I \tag{7}$$

$$X_{a,q}^i + S_q^i \leq X_{d,q}^i \quad \forall i \in I, q \in Q(i) \tag{8}$$

$$A_{ijp} \in \{0,1\} \quad i,j \in I, p \in P^{same\_d}(i,j) \tag{9}$$

$$B_{ijp} \in \{0,1\} \quad i,j \in I, p \in P_1^{opp\_d}(i,j). \tag{10}$$

The objective function (1) minimizes the weighted sum of total train travel times. Constraints (2) and (3) state that for any two trains i, j traversing the same segment p in same direction, A_{ijp} equals to zero if and only if train j traverses segment p before train i, and train j must leave segment p for the period of $h_{1,p}^{i,j}$ before train i can enter it. Constraints (4) and (5) also indicate that B_{ijp} equals to zero if and only if train j traverses segment p before train i with the time headway not less than the minimum safety headway. Constraints (6) ensure that the travel time of trains on any rail segment is in the range of the corresponding upper and lower limits. Constraints (7) allow the train departure time from its origin station to be bigger than or equal to its earliest departure time. Constraints (8) state that a train leaves a station siding after it arrives at this station and stops there for at least the scheduled stop time.

The problem above is a mixed 0–1 linear programming. Finding a suitable method for solving this kind of problems is always challenging. The challenge does not only come from the binary variables but also the size of problems. We propose here a method to solve the problem efficiently. For this, we first use the theory of exact penalization in DC programming [24] to reformulate the MILP as that of minimizing a DC function over a polyhedral convex set. The resulting problem is then handled by DCA which was introduced and extensively developed over the last decades [23]. The mentioned approach has been applied successfully in several large scale problems (see [20–23, 27] and reference therein). The details are provided in the following section.

3   Solution Method

3.1   DC Reformulation

By using an exact penalty result, we reformulate the MILP in the form of a concave minimization program. The exact penalty technique aims at tranforming the orignal MILP into a more tractable equivalent problem in the DC optimization framework. Let S be the feasible set of the problem MILP (1)–(10). For notational simplicity, we group all arrival and departure time variables in a column vector $U = [U_1^1, U_1^2, ..., U_1^{n_I}, U_2^1, U_2^2, ..., U_2^{n_I}, ..., U_{n_Q}^1, ..., U_{n_Q}^{n_I}, U_{n_Q+1}^1, ..., U_{2n_Q}^{n_I}]^T$, where $\mathcal{T}$ denotes the transpose operator; $U_q^i = X_{a,q}^i$ and $U_{n_Q+q}^i = X_{d,q}^i$ $\forall i \in I, q \in Q$. In the same way, we group all the binary variables (includes A_{ijp} and B_{ijp}) into a column vector $C = [c_{111}, c_{112}, ..., c_{n_I n_I n_P}, c_{(n_I+1)n_I n_P}, ..., c_{(2n_I)n_I n_P}]^T$, where $c_{ijp} = A_{ijp}$ and $c_{(n_I+i)jp} = B_{ijp}$ $\forall i, j \in I, p \in P$. We denote a new set $K := \{(U, c) \in S : c \in [0, 1]^{2n_I n_I n_P}\}$. Assume that K is a nonempty, bounded polyhedral convex set in $\mathbb{R}^{2n_I n_Q} \times \mathbb{R}^{2n_I n_I n_P}$.

Therefore, the problem (1)–(10) can be expressed in the general form

$$(U_{opt}, c_{opt}) = argmin\{z : (U, c) \in S, c \in \{0, 1\}^{2n_I n_I n_P}\}, \qquad (11)$$

where $z = \sum_{i \in I} W_i(X_{aD}^i - Y_{dO}^i)$.

Let us consider the function p defined by

$$p(U, c) := \sum_{i,j \in I; p \in P} min\{c_{ijp}, 1 - c_{ijp}\}. \qquad (12)$$

It is clear that p is concave and finite on K, $p(U, c) \geq 0$ $\forall (U, c) \in K$, and $\{(U, c) \in S : c \in \{0, 1\}^{2n_I n_I n_P}\} = \{(U, c) \in K : p \leq 0\}$. Hence, Problem (11) can be written as

$$(U_{opt}, c_{opt}) = argmin\{z : (U, c) \in K, p(U, c) \leq 0\}. \qquad (13)$$

The following theorem is in order.

Theorem 1. *Let K be a nonempty bounded polyhedral convex set, f be a finite concave function on K and p be a finite nonnegative concave function on K. Then there exists $\tilde{t}_0 \geq 0$ such that for $\tilde{t} \geq \tilde{t}_0$ the following problem has the same optimal value and the same optimal solution set:*

$$(P_t) \quad \alpha(t) = \{f(x) + \tilde{t}p(x) : x \in K\}$$
$$(P) \quad \alpha = min\{f(x) : x \in K, p(x) \leq 0\}$$

Furthermore,

- If the vertex set of K, denoted by $V(K)$, is contained in $x \in K : p(x) \leq 0$, then $\tilde{t}_0 = 0$.
- If $p(x) > 0$ for some $x \in V(K)$, then $t_0 = min\left\{\dfrac{f(x) - \alpha(0)}{S_0} : x \in K, p(x) \leq 0\right\}$, where $S_0 = min\{p(x) : x \in V(K), p(x) > 0\}$.

Proof. The proof for the general case can be found in [24].

From Theorem 1 we get, for a sufficiently large number $\tilde{t}$ ($\tilde{t} > \tilde{t}_0$), the equivalent concave minimization problem to (13)

$$min\left\{z + \tilde{t}p(U, c) : (U, c) \in K\right\} \tag{14}$$

which is a DC program of the form

$$min\left\{g(U, c) - h(U, c)\right\} \tag{15}$$

where $g(U, c) = \chi_K(U, c)$ and $h(U, c) = -z - \tilde{t}\sum_{i,j \in I; p \in P} min\{c_{ijp}, 1 - c_{ijp}\}$. $\chi_K(U, c) = 0$ if $(U, c) \in K$, otherwise $+\infty$ (the indicator function of K).

We have successfully tranformed an optimization problem with integer variables into its equivalent form with continuous variables.

3.2   DC Algorithm for (14)

Now, we investigate a DC programming approach for solving (14). A DC program is that of the form:

$$\alpha := min\left\{f(x) := g(x) - h(x) : x \in \mathbb{R}^n\right\} \tag{16}$$

with g, h being lower semi-continuous proper convex function on $\mathbb{R}^n$, and its dual problem is defined as

$$\alpha := min\left\{h^*(y) - g^*(y) : y \in \mathbb{R}^n\right\} \tag{17}$$

where $g^*(y) := max\left\{x^Ty - g(x) : x \in \mathbb{R}^n\right\}$ is the conjugate function of g.

Based on local optimality conditions and duality in DC programming, the DCA consists in the construction of two sequences $\{x^k\}$ and $\{y^k\}$, candidates to be optimal solutions of primal and dual programs respectively, in such a way that $\{g(x^k) - h(x^k)\}$ and $\{h^*(y^k) - g^*(y^k)\}$ are decreasing and their limits points satisfy the local optimality conditions. The idea of DCA is simple: each iteration of DCA approximates the concave part $-h$ by its affine majorization (that corresponds to taking $y^k \in \partial h(x^k)$) and minimizes the resulting convex function.

Generic DCA scheme:

Initialization   Let $x^0 \in \mathbb{R}^n$ be a good guess, $k \leftarrow 0$;

Repeat

      Calculate $y^k \in \partial h(x^k)$;

      Calculate $x^{k+1} \in argmin\left\{g(x) - h(x^k) - \langle x - x^k, y^k \rangle : x \in \mathbb{R}^n\right\}$ (P_k);

      $k + 1 \leftarrow k$;

Until Convergence of x^k.

The convergence properties of DCA and its theoretical basis can be found in [23], for instant it is important to mention that:

• DCA is a descent method without line search;

• If the optimal value of problem (16) is finite and the sequence $\{x^k\}$ is bounded then every limit point x^* of $\{x^k\}$ is a critical point of $g - h$;

• DCA has a linear convergence for general DC programs;

• DCA has a finite convergence for polyhedral DC programs ((16) is called polyhedral DC program if either g or h is polyhedral convex).

We now describe the DCA applied to the DC program (14). By definition, a sub-gradient $(v^k, d^k) \in \partial h(U^k, c^k)$ can be chosen as follows:

• $v_q^i = -W_i$ if q is destination station on travel path of train i, otherwise 0;

• $d_{ijp} = \tilde{t}$ if $c_{ijp} \geq 0.5$, otherwise $d_{ijp} = -\tilde{t}$ for all $i, j \in I, p \in P$.

By using (v^k, d^k), we then compute (U^{k+1}, c^{k+1}) by solving the linear program:

$$(U^{k+1}, c^{k+1}) = argmin\left\{\chi_K(U, c) - \langle (U, c) - (U^k, c^k), (v^k, d^k) \rangle : (U, c) \in K\right\} \tag{18}$$

or

$$(U^{k+1}, c^{k+1}) = argmin\left\{ - \langle (U, c), (v^k, d^k) \rangle : (U, c) \in K\right\} \tag{19}$$

Thus, the DCA applied to (14) is as follows:

Algorithm DCA:

      Let $k = 0$; $cr = 10^6$;

      Choose a sufficiently small positive number ϵ;

      Choose an initial point (U^k, c^k);

      while $er > \epsilon$ **do**

          Compute $(v^k, d^k) \in \partial h(U^k, c^k)$;

          Solve $\left\{ - \langle (U, c), (v^k, d^k) \rangle : (U, c) \in K\right\}$ to obtain (U^{k+1}, c^{k+1});

          Compute error $er =|| (U^{k+1}, c^{k+1}) - (U^k, c^k) ||$; $k = k + 1$;

      endwhile

Regarding the complexity of the proposed DCA, besides the computation of the sub-gradients which is trivial, the algorithm requires one linear program at each iteration and it has a finite convergence. The linear program has polynomial complexity. The convergence of Algorithm DCA can be summarized in the next theorem [29].

Theorem 2. *(i) Algorithm DCA generates a sequence* $\{(U^k, c^k)\}$ *contained in* $V(K)$ *such that the sequence* $\{g(U^k, c^k) - h(U^k, c^k)\}$ *is decreasing.*

(ii) If at iteration r, *we have* $c^r \in \{0,1\}^{2n_I n_I n_P}$, *then* $c^k \in \{0,1\}^{2n_I n_I n_P}$ *and* $f(U^{k+1}, c^{k+1}) < f(U^k, c^k) \quad \forall k > r.$

(iii) The sequence $\{(U^k, c^k)\}$ *converges to* $\{(U^*, c^*)\} \in V(K)$ *after a finite number of iterations. The point* (U^*, c^*) *is critical point of the problem (14). Moreover such an* (U^*, c^*) *is almost always a strict local minimum of problem (14).*

Basing on the second affirmation of the theorem above, we should choose a feasible solution as an initial point of DCA. This ensures that the solution obtained by DCA is feasible to the original problem although DCA works on the continuous domain. The way we choose an initial point is as follows: We arrange the trains in the order of the earliest departure time at the original station of the itinerary. With each train, we determine the schedule on the whole itinerary for that train. The generated schedule of the next train is based on the schedule which was created by the previous trains to ensure that no conflict occurs. The result is a feasible solution of the problem.

4   Computational Experiments

In this section, we provide preliminary computational results of our approach. We have coded the algorithm in C++ programming language and tested instances using PC Intel core i7 3770 3.4 GHz, 16 GB RAM. The solver CPLEX 12.6.1 is used to solve the linear program in each iteration of DCA and get the optimal value of MILP. We investigate the algorithm performance on three network topologies that are the topology shown in [18] (toy network) and two topologies in Northern Vietnam (HN-HP, HN-LC network). The toy network includes 3 trains, 5 segments, 6 stations and the total length of segments is 297 km. The HN-HP network is a single-track railway system connecting Hanoi capital and Hai Phong city, including 8 trains (4 inbounds, 4 outbounds), 7 segments and 8 stations. The total length of HN-HP network is 102 km. The HN-LC network is a line connecting Hanoi capital and Lao Cai province. It is more complex as it

Table 1. The size of testing networks

	Toy network	HN-HP network	HN-LC network
Trains	3	8	8
Segment	5	7	27
Stations/sidings	6	8	28
Continuous Variables	36	128	448
Binary variables	90	896	3456
Constraints	216	1920	7360

consists of 8 trains (4 inbounds and 4 outbounds), 27 segments, 28 stations and the total distance is 294 km. The size of the networks are shown in Table 1.

In each network, 10 instances are generated by modifying the train parameters composed of the scheduled stop time at stations, the earliest departure time and the weight of trains. The tested results for three networks are presented in Table 2, 3, and 4 where ValDCA is the objective value obtained by the algorithm DCA; CPU is the computing time; OptVal is the optimal value; and GAP

$$= \frac{\text{ValDCA} - \text{OptVal}}{\text{OptVal}} \times 100\%.$$

Table 2. The computation result for Toy network

Instance	ValDCA	CPU	OptVal	GAP
1	2347.022	0.011	2190.053	7.17 %
2	2319.022	0.018	2174.437	6.65 %
3	2199.714	0.016	2199.714	0.00 %
4	2175.438	0.021	2143.438	1.49 %
5	2281.022	0.019	2136.438	6.77 %
6	2175.715	0.014	2175.715	0.00 %
7	2164.438	0.022	2136.438	1.31 %
8	2223.715	0.016	2223.715	0.00 %
9	4420.691	0.018	4420.691	0.00 %
10	4370.568	0.021	4224.106	3.47 %
Average	**2667.735**	**0.018**	**2602.475**	**2.69%**

From the tables of results, we can see that:

- The computing time of DCA is good for all the instances. Even for the big size (HN-LC network with 3456 binary variables and 7360 constraints), the time to get solution is still small.
- The GAP numbers are very small. This means the solution quality is quite good. For the Toy network, there are 4 out of 10 instances in which GAP equals to zero, i.e. DCA furnishes the optimal solution for 40%.

Table 3. The computation result for HN-HP network

Instance	ValDCA	CPU	OptVal	GAP
1	6755	0.077	6745	0.15 %
2	6749	0.061	6739	0.15 %
3	5493	0.075	5455	0.70 %
4	5482	0.076	5442	0.74 %
5	5447	0.067	5445	0.04 %
6	5283	0.081	5273	0.19 %
7	5340	0.068	5301	0.74 %
8	17014	0.071	16744	1.61 %
9	16722	0.063	16659	0.38 %
10	10087	0.077	10015	0.72 %
Average	**8437.2**	**0.072**	**8381.8**	**0.54 %**

Table 4. The computation result for HN-LC network

Instance	ValDCA	CPU	OptVal	GAP
1	4546	0.385	4470	1.70 %
2	4543	0.381	4505	0.84 %
3	4618	0.342	4528	1.99 %
4	4956	0.357	4687	5.74 %
5	4996	0.386	4683	6.68 %
6	4546	0.356	4470	1.70 %
7	7311	0.335	7150	2.25 %
8	11322	0.379	11137	1.66 %
9	6008	0.373	5914	1.59 %
10	6054	0.335	5969	1.42 %
Average	**5890.00**	**0.363**	**5751.30**	**2.56 %**

5   Conclusion

In this paper, we have studied a model minimizing total travel time of trains in single track networks. We have shown that the aforementioned problem can be formulated as a mixed-integer linear program. Realizing the inherent difficulty in computing the optimal solution of MILPs, our main contribution was to propose a computationally efficient approach based on DCA. The considered combinatorial optimization problem has been beforehand reformulated as a DC program with a natural choice of DC decomposition, and the resulting DCA then consists in solving a finite sequence of linear programs. DCA is original because it gives an integer solution while it works in a continuous domain. Preliminary numerical results were encouraging and demonstrated the effectiveness of the proposed method. The short computing time and the capacity for handling the large-scale instances make the proposed method valuable. Moreover, notice that most problem formulations arising in train scheduling can be formulated as some sort of MILP problems, our proposed approach seems attractive and needs more investigation.

References

1. Adenso-Diaz, B., Gonzalez, M.O., Gonzalez-Torre, P.: On-line timetable re-scheduling in regional train services. Transp. Res. Part B **33**(6), 387–398 (1999)
2. Bussieck, M.R., Kreuzer, P., Zimmermann, U.T.: Optimal lines for railway systems. Eur. J. Oper. Res. **96**(1), 54–63 (1997)
3. Cacchiani, V., Caprara, A., Toth, P.: A column generation approach to train timetabling on a corridor. 4OR **6**(2), 125–142 (2008)
4. Cacchiani, V., Caprara, A., Toth, P.: Non-cyclic train timetabling and comparability graphs. Oper. Res. Lett. **38**(3), 179–184 (2010)
5. Caimi, G., Fuchsberger, M., Laumanns, M., Luthi, M.: A model predictive control approach for discrete time rescheduling in complex central railway station areas. Comput. Oper. Res. **39**(11), 2578–2593 (2012)
6. Caprara, A., Fischetti, M., Toth, P.: Modeling and solving the train timetabling problem. Oper. Res. **50**(5), 851–861 (2002)
7. Castillo, E., Gallego, I., Urena, J.M., Coronado, J.M.: Timetabling optimization of a mixed double and single track rail network. Appl. Math. Model. **35**(2), 859–878 (2011)
8. Chang, Y.H., Yeh, C.H., Shen, C.C.: A multiobjective model for passenger train services planning: application to Taiwan's high-speed rail line. Transp. Res. Part B **34**, 91–106 (2000)
9. Chen, B., Harker, P.T.: Two moments estimation of the delay on single-track rail lines with scheduled traffic. Transp. Sci. **24**(4), 261–275 (1990)
10. Corman, F., D'Ariano, A., Pacciarelli, D., Pranzo, M.: A tabu search algorithm for rerouting trains during rail operations. Transp. Res. B **44**(1), 175–192 (2010)
11. Corman, F., D'Ariano, A., Hansen, I.A., Pacciarelli, D.: Optimal multi-class rescheduling of railway traffic. J. Rail. Transp. Plan Manage. **1**(1), 14–24 (2011)
12. D'Ariano, A., Pacciarelli, D., Pranzo, M.: A branch and bound algorithm for scheduling trains in a railway network. Eur. J. Oper. Res. **183**, 643–657 (2007)
13. D'Ariano, A., Corman, F., Pacciarelli, D., Pranzo, M.: Reordering and local rerouting strategies to manage train traffic in real time. Transp. Sci. **42**(4), 405–419 (2008)
14. Higgins, A., Kozan, E., Ferreira, L.: Optimal scheduling of trains on a single-line track. Transp. Res. Part B: Methodol. **30**(2), 147–161 (1996)
15. Higgins, A., Kozan, E., Ferreira, L.: Heuristic techniques for single-line train scheduling. J. Heuristics **3**, 43–62 (1997)
16. Jovanovic, D., Harker, P.T.: Tactical scheduling of rail operations: the SCAN I system. Transp. Sci. **25**(1), 46–64 (1991)
17. Jong, J.-C., Chang, S., Lai, Y.-C.: Development of a two-stage hybrid method for solving high speed rail train scheduling problem. In: Proceedings of the Transportation Research Board 92nd Annual Meeting, Washington D.C., 13–17 January 2013

18. Karoonsoontawong, A., Taptana, A.: Branch-and-bound-based local search heuristics for train timetabling on single-track railway network. Netw. Spat. Econ. **17**(1), 1–39 (2017)
19. Kraay, D.R., Harker, P.T.: Real-time scheduling of freight railroads. Transp. Res. Part B **29**, 213–229 (1995)
20. Le Thi, H.A., Nguyen, Q.T., Tran, P.K., Pham Dinh, T.: DC programming and DCA based cross-layer optimization in multi-hop TDMA networks. In: ACIIDS 2013, LNCS, vol. 7803, pp. 398–408 (2013)
21. Le Thi, H.A., Nguyen, Q.T.: A robust approach for nonlinear UAV task assignment problem under uncertainty. Trans. Comput. Collective Intell. **2**, 147–159 (2010)
22. Le Thi, H.A., Nguyen, Q.T., Huynh, T.N., Pham, D.T.: Solving the earliness tardiness scheduling problem by DC programming and DCA. Math. Balkanica (N.S.) **23**(3-4), 271–288 (2009)
23. Le Thi, H.A., Pham, D.T.: The DC (difference of convex functions) programming and DCA revisited with DC models of real world non convex optimization problems. Ann. Oper. Res. **133**, 23–46 (2005)
24. Le Thi, H.A., Pham, D.T., Le, D.M.: Exact penalty in DC programming. Vietnam J. Math. **27**(2), 169–178 (1999)
25. Lee, Y., Chen, C.: A heuristic for the train pathing and timetabling problem. Transp. Res. Part B **43**(8–9), 837–851 (2009)
26. Mu, S., Dessouky, M.: Scheduling freight trains traveling on complex networks. Transp. Res. Part B **45**, 1103–1123 (2011)
27. Nguyen, Q.T., Le Thi, H.A.: Solving an inventory routing problem in supply chain by DC programming and DCA. In: ACIIDS 2011, LNCS, vol. 6592, pp. 432–441 (2011)
28. Petersen, E.R., Taylor, A.J., Martland, C.D.: An introduction to computer aided train dispatching. J. Adv. Transp. **20**, 63–72 (1986)
29. Pham Dinh, T., Le Thi, H.A.: Convex analysis approach to DC programming: theory, algorithms and applications. Acta Math. Vietnamica **22**, 289–355 (1997). Dedicated to Professor Hoang Tuy on the occasion of his 70th birthday
30. Sahin, I.: Railway traffic control and train scheduling based on inter-train conflict management. Transp. Res. Part B **33**(7), 511–534 (1999)
31. Szpigel, B.: Optimal train scheduling on a single track railway. Oper. Res. **72**, 343–352 (1973). North-Holland, Amsterdam, Netherlands
32. Zhou, X., Zhong, M.: Single-track train timetabling with guaranteed optimality, branch-and-bound algorithms with enhanced lower bounds. Transp. Res. Part B **41**, 320–341 (2007)

DCA with Successive DC Decomposition for Convex Piecewise-Linear Fitting

Vinh Thanh Ho$^{(\boxtimes)}$, Hoai An Le Thi, and Tao Pham Dinh

Institute for Research and Applications of Optimization (VinOptima), VinTech, Vingroup, 7 Bang Lang 1 Street, Long Bien District, Ha Noi, Vietnam
{v.thanhhv9,v.hoaianlethi,v.taophamdinh}@vinoptima.org

Abstract. We study an approach based on DC (Difference of Convex functions) programming and DCA (DC Algorithm) for the convex piecewise-linear fitting problem. The objective is to fit a given set of data points by a convex piecewise-linear function. The problem is formulated as minimizing the squared ℓ_2-norm fitting error, and then reformulated as a DC program for which a standard DCA scheme is applied. Furthermore, a modified DCA scheme with *successive* DC decomposition is proposed with the aim to improve DCA by updating the convex approximation of the fitting error function during DCA iterations. These DCAs consist in solving a sequence of convex quadratic programs. Moreover, the modified DCA still has the same convergence properties as the standard DCA. Numerical results on synthetic/real datasets show the efficiency of our methods when comparing with the existing approaches.

Keywords: DC programming · DCA · DCA with successive DC decomposition · Convex piecewise-linear fitting

1 Introduction

The problem of fitting a set of data points by a certain function has been studied extensively. Our work focuses on solving the problem of fitting a given set of m points $(\mathbf{x}_i, y_i) \in \mathbb{R}^n \times \mathbb{R}$ by a convex piecewise-linear continuous function $f : \mathbb{R}^n \to \mathbb{R}$ with K affine functions $(K > 1)$, of the general form $f(\mathbf{x}) = \max_{j=1,\ldots,K} \{\langle \mathbf{a}_j, \mathbf{x} \rangle + b_j\}$ with $(\mathbf{a}_j, b_j) \in \mathbb{R}^n \times \mathbb{R}$, $j = 1, \ldots, K$. Considering the least-squares criterion, this fitting problem aims to find the vectors $(\mathbf{a}_j, b_j)$ such that the mean-square error (MSE) $\frac{1}{m} \sum_{i=1}^{m} [f(\mathbf{x}_i) - y_i]^2$ is as small as possible (see, e.g., [13]). It can be formulated as the following squared ℓ_2-norm optimization formulation

$$\min F(\alpha) := \sum_{i=1}^{m} \left(\max_{j=1,\ldots,K} \{\langle \mathbf{a}_j, \mathbf{x}_i \rangle + b_j\} - y_i \right)^2 \tag{1}$$

$$\text{s.t. } \alpha = (\mathbf{a}_1, b_1, \mathbf{a}_2, b_2, \ldots, \mathbf{a}_K, b_K) \in \mathbb{R}^{K(n+1)}.$$

© Springer Nature Switzerland AG 2020
H. A. Le Thi et al. (Eds.): ICCSAMA 2019, AISC 1121, pp. 39–51, 2020.
https://doi.org/10.1007/978-3-030-38364-0_4

It is easy to see that the problem (1) is, in general, not convex. Hence globally solving this problem is difficult in large-scale setting.

This problem has many applications ranging from mathematical modeling/optimization, to econometrics, transportation and forecasting, to statistics, machine learning and data mining (see, e.g., [1,5,13,14,19,20] and references therein). In machine learning, developing regression techniques using piecewise-linear functions plays an important role for practical problems such as energy storage optimization with a solar source, beer brewery optimization, customer demand forecasting (see, e.g., [4,7,20] for more practical problems).

Related Works. Previous and relevant works for piecewise-linear fitting functions are given completely in, e.g., [1–4,7,13,18–20]. In most of these works, the attention is on constructing any piecewise-linear function in which the number of affine functions is updated increasingly to obtain a better fit. The recent work of Balázs [4] showed that for a class of sub-Gaussian fitting problems, there exists a near-optimal convex piecewise-linear fitting function with at most $\lceil m^{n/(n+4)} \rceil$ affine functions. On the contrary, our work focuses on searching among convex piecewise-linear functions with a *fixed* number K of affine functions. Works in the same direction can be found, for example, in [6,13,19]. Several works for special convex piecewise-linear functions with the ℓ_∞/ℓ_1-norm-based fitting criterion were mentioned in [13]. When $K = m$, Boyd and Vandenberghe [6] reformulated (1) as a convex quadratic program with $m(n+1)$ variables and $m(m-1)$ linear inequality constraints, which is impractical for medium/large datasets. Magnani and Boyd [13] proposed a fast, heuristic Gauss-Newton method, named Least-square partition algorithm (LSPA). Its idea is the same as the K-mean algorithm for clustering: it partitions the set $\{\mathbf{x}_i\}_{i=1,\ldots,m}$ into K subsets based on the chosen K centroids, then fits an affine function to each subset by solving a linear least-squares problem, next updates the K centroids, and repeats until there is no change in partitioning. LSPA depends on the starting points and does not ensure to converge even for the small datasets [7,13]. On the other hand, Toriello and Vielma [19] investigated an exact approach, named MIQPM, for (1). The authors used the big-M technique to reformulate (1) as a mixed-integer quadratic program containing $K(n+1)$ continuous variables, mK binary variables and more than $m(2K+1)$ linear constraints. Solving this mixed-integer quadratic program becomes computationally expensive due to a huge number of binary variables; moreover, how large value of big-M to trade off between the quality of fitting error and the rapidity is still in question [18].

Overcoming these disadvantages of both heuristic and exact approaches motivates us to develop efficient algorithms for solving the nonconvex, nonsmooth fitting problem. Our algorithms have the advantage of LSPA and MIQPM: they require solving a convex quadratic program at each iteration, like LSPA, and still always guarantee to converge (quite often to globally optimal solutions in practice), like MIQPM. The backbone of our approach is DC (Difference of Convex functions) programming and DCA (DC Algorithm) (see, e.g., [11,12,15,17] and the references in [9,12]) which are well-known as powerful nonsmooth, nonconvex optimization tools. DCA aims to solve a standard DC program that consists in minimizing a DC function $F = G - H$ (with G, H being convex functions) over a convex set or on the whole space. Here $G - H$ is called a DC decomposition of F, while G and H are DC components of F. The idea of standard DCA is, at each iteration k, approximating the second DC component H by its affine minorant H_k and then solving the resulting convex subproblem. In other words, one approximates the DC function F by the convex majorization $F^k := G - H_k$. It is clear that the closer to F the function F^k is, the better DCA could be. Le Thi and Pham Dinh [12] has recently introduced a modified version of the standard DCA scheme, named DCA with *successive* DC decomposition. In this version, the DC decomposition of F is successively updated during DCA iterations in order to better approximate the DC function F i.e. $F := G^k - H^k$ at each iteration k and its corresponding convex majorization is $F^k := G^k - H^k_k$. Our approach based on DCA with *successive* DC decomposition for this work is similar to the recent work [10] for reinforcement learning problems. In a related work with the use of DC optimization, Bagirov et al. [3] considered the continuous piecewise-linear fitting problem over the class of maxima of minima of linear functions (which includes convex piecewise-linear functions). The authors indicated that the objective function is a DC function without getting explicit DC components and designed an algorithm based on the subdifferentials of DC components. However, the computation of these subdifferentials stated in Proposition 2 in their work is not correct.

In this paper, we address the convex piecewise-linear fitting problem (1) by an approach based on DC programming and DCA. Particularly, we formulate (1) as a DC program for which a standard DCA scheme is developed. Its DC decomposition is designed by using the affine minorization of each convex fitting function at an arbitrary given point. Changing this point at each iteration of DCA leads to a modified DCA version with successive DC decomposition for the problem (1). The resulting subproblem in these DCAs can be equivalently reformulated as a convex quadratic program with $(2m + K(n+1))$ variables and $m(K + 1)$ linear inequality constraints. Especially, the modified DCA still has the convergence properties of the standard DCA: it is a *descent* algorithm with *global convergence* (i.e. it always converges from an arbitrary starting point). We provide several numerical experiments of our DCA-based algorithms on various synthetic/real datasets in comparison with the heuristic/exact algorithms LSPA, MIQPM for solving the problem (1).

The rest of the paper is organized as follows. How to apply DC programming and DCA to the considered problem is shown in Sect. 2. Section 3 presents the numerical results on benchmark datasets. Finally, Sect. 4 concludes the paper.

2   Solution Method by DC Programming and DCA

DC programming and DCA were introduced by Pham Dinh Tao in a preliminary form in 1985 and have been extensively developed by Le Thi Hoai An and Pham Dinh Tao since 1994. DCA is well-known as an efficient approach in the nonconvex programming framework. In recent years, numerous DCA-based algorithms have been developed for successfully solving large-scale nonsmooth/nonconvex programs in several application areas (see the list of references in [9,12]). For a comprehensible survey on thirty years of development of DCA, the reader is referred to the recent work [12].

The idea of DCA relies on the DC structure of the objective function F. The standard DCA scheme is described below.

Standard DCA scheme

Initialization: Let $\alpha^0 \in \mathbb{R}^p$ be a best guess. Set $k = 0$.

repeat

1. Calculate $\beta^k \in \partial H(\alpha^k)$.
2. Calculate $\alpha^{k+1} \in \operatorname{argmin}\{G(\alpha) - H(\alpha^k) - \langle \alpha - \alpha^k, \beta^k \rangle : \alpha \in \mathbb{R}^p\}$ (P_k).
3. $k = k + 1$.

until convergence of $\{\alpha^k\}$.

Convergence properties of the standard DCA are described completely in [11,15,16].

2.1   DCA for Solving the Problem (1)

The convex piecewise-linear fitting problem (1) can be rewritten as

$$\min\left\{ F(\alpha) = \sum_{i=1}^m [p_i(\alpha)]^2 : \alpha \in \mathbb{R}^{K(n+1)} \right\} \tag{2}$$

where for $i = 1, \ldots, m$, the function $p_i : \mathbb{R}^{K(n+1)} \to \mathbb{R}$,

$$p_i(\alpha) := \max_{j=1,\ldots,K} \left\{ \langle \alpha, \mathbf{z}^{(i,j)} \rangle - y_i \right\},$$

and for $i = 1, \ldots, m$, $j = 1, \ldots, K$, the vector $\mathbf{z}^{(i,j)} = (\mathbf{z}_1^{(i,j)}, \ldots, \mathbf{z}_K^{(i,j)}) \in \mathbb{R}^{K(n+1)}$ with $\mathbf{z}_r^{(i,j)} = (\mathbf{x}_i, 1)$ if $r = j$, $(\mathbf{0}, 0)$ if $r \neq j$. Obviously, p_i is a convex piecewise-linear function.

It is known from, e.g., [15] that if p_i is a DC function with a nonnegative DC decomposition then $[p_i]^2$ is DC too. Thus, using the convexity of the function p_i, we highlight a nonnegative DC decomposition of p_i (see [10]). In particular, we define the affine minorization of p_i at an arbitrary point $\overline{\alpha}$ $(i = 1, \ldots, m)$ as follows:

$$l_i(\alpha) := p_i(\overline{\alpha}) + \langle \alpha - \overline{\alpha}, \overline{\gamma}_i \rangle = \langle \alpha, \overline{\gamma}_i \rangle - y_i \text{ with } \overline{\gamma}_i \in \partial p_i(\overline{\alpha}).$$

Let $(l_i)^- := \max\{0, -l_i\}$. Obviously, the functions $p_i + (l_i)^-$ and $(l_i)^-$ are nonnegative and convex on $\mathbb{R}^{K(n+1)}$, and so are $[p_i + (l_i)^-]^2$ and $[(l_i)^-]^2$. Then we have the nonnegative DC decomposition of p_i as follows:

$$p_i = [p_i + (l_i)^-] - (l_i)^-.$$

It leads to a DC decomposition of $[p_i]^2$, that is,

$$[p_i]^2 = 2\left\{ [p_i + (l_i)^-]^2 + [(l_i)^-]^2 \right\} - [p_i + 2(l_i)^-]^2.$$

As a result, F is a DC function. We thus derive the DC formulation of (2) as follows:

$$\min\{F(\alpha) := G(\alpha) - H(\alpha) : \alpha \in \mathbb{R}^{K(n+1)}\} \qquad (3)$$

where $G = \sum_{i=1}^{m} 2\left\{[p_i + (l_i)^-]^2 + [(l_i)^-]^2\right\}$,

$H = \sum_{i=1}^{m} [p_i + 2(l_i)^-]^2$.

Applying the standard DCA scheme to (3) leads us, at the iteration k, to compute a subgradient $\beta^k \in \partial H(\alpha^k)$ and a solution α^{k+1} to the following convex program of the form

$$\min_{\alpha} \sum_{i=1}^{m} 2\left\{[p_i(\alpha) + (l_i)^-(\alpha)]^2 + [(l_i)^-(\alpha)]^2\right\} - \langle \beta^k, \alpha \rangle.$$

The last program can be reformulated as

$$\min \sum_{i=1}^{m} (2\tau_i^2) + \sum_{i=1}^{m} (2t_i^2) - \langle \beta^k, \alpha \rangle$$
$$\text{s.t. } t_i \geq (l_i)^-(\alpha), \tau_i \geq t_i + p_i(\alpha), i = 1, \ldots, m,$$

which is in fact a convex quadratic program of the form

$$\min_{\alpha,t,\tau} \sum_{i=1}^{m} (2\tau_i^2) + \sum_{i=1}^{m} (2t_i^2) - \langle \beta^k, \alpha \rangle, \qquad (4)$$
$$\text{s.t. } t_i \geq 0, \ t_i \geq y_i - \langle \alpha, \overline{\gamma}_i \rangle, \ i = 1, \ldots, m,$$
$$\tau_i \geq t_i + \langle \alpha, \mathbf{z}^{(i,j)} \rangle - y_i, \ i = 1, \ldots, m, j = 1, \ldots, K.$$

Compute the Subdifferential ∂H: We have

$$\partial H(\alpha) = \sum_{i=1}^{m} \partial \left[p_i(\alpha) + 2(l_i)^-(\alpha)\right]^2,$$

$$\partial (l_i)^-(\alpha) = \{\mathbf{0}\} \text{ if } l_i(\alpha) > 0, \ [\mathbf{0}, -\overline{\gamma}_i] \text{ if } l_i(\alpha) = 0,$$
$$\{-\overline{\gamma}_i\} \text{ otherwise,}$$

$$\partial p_i(\alpha) = \partial \left[\max_{j=1,\ldots,K} \left\{\langle \alpha, \mathbf{z}^{(i,j)} \rangle - y_i\right\}\right]$$
$$= \text{co}\{\mathbf{z}^{(i,j_i)} : j_i \in I_i(\alpha)\}, \text{ and}$$
$$I_i(\alpha) := \text{argmax}_{j=1,\ldots,K} \langle \alpha, \mathbf{z}^{(i,j)} \rangle.$$

Here $[\mathbf{0}, -\overline{\gamma}_i]$ is a line segment between $\mathbf{0}$ and $-\overline{\gamma}_i$, co denotes the convex hull of a set of points. Hence we take subgradients $\overline{\gamma}_i \in \partial p_i(\overline{\alpha})$, $\gamma_i^k \in \partial p_i(\alpha^k)$ and $\beta^k \in \partial H(\alpha^k)$ as follows: $\overline{j}_i \in I_i(\overline{\alpha})$, $j_i^k \in I_i(\alpha^k)$, $\overline{\gamma}_i = \mathbf{z}^{(i,\overline{j}_i)}$, $\gamma_i^k = \mathbf{z}^{(i,j_i^k)}$,

$$\beta^k = 2 \sum_{i=1}^{m} [p_i(\alpha^k) + 2(l_i)^-(\alpha^k)](\gamma_i^k - 2\overline{\gamma}_i \mathbb{1}_{\{l_i(\alpha)<0\}}(\alpha^k)). \qquad (5)$$

Here the function $\mathbb{1}_A(\alpha) = 1$ if $\alpha \in A$, 0 otherwise.

Finally, DCA applied to (3) is summarized in Algorithm 1.

Algorithm 1. Standard DCA for solving (3) (DCA)

Initialization: Let ε be a sufficiently small positive number. Let $\alpha^0 \in \mathbb{R}^{K(n+1)}$. Set $k = 0$.

repeat

1. Compute $\beta^k \in \partial H(\alpha^k)$ using (5).
2. Solve the convex quadratic program (4) to obtain $(\alpha^{k+1}, t^{k+1}, \tau^{k+1})$.
3. $k = k + 1$.

until Stopping criteria are satisfied.

Theorem 1. *Convergence properties of* DCA

(i) DCA *generates the sequence* $\{\alpha^k\}$ *such that the sequence* $\{F(\alpha^k)\}$ *is decreasing.*

(ii) *Every limit point* α^* *of the sequence* $\{\alpha^k\}$ *is a critical point of* $G - H$ *i.e.* $\partial G(\alpha^*) \cap \partial H(\alpha^*) \neq \emptyset$.

2.2 DCA with Successive DC Decomposition for Solving the Problem (1)

From DCA for solving (3), we see that the affine minorization l_i of p_i is computed easily with any point $\overline{\alpha}$, and the closer to p_i the function l_i is, the better DCA could be. Hence, we suggest a modified DCA with successive DC decomposition for the problem (1) (see [10]), i.e. l_i is updated during DCA iterations by choosing $\overline{\alpha} = \alpha^k$ at the iteration k. Particularly, at the iteration k, we set $\gamma_i^k = \mathbf{z}^{(i,j_i^k)} \in \partial p_i(\alpha^k)$,

$$l_i^k(\alpha) := p_i(\alpha^k) + \langle \alpha - \alpha^k, \gamma_i^k \rangle = \langle \alpha, \gamma_i^k \rangle - y_i.$$

The resulting DC formulation of (2) at the iteration k takes the form

$$\min\{F(\alpha) := G^k(\alpha) - H^k(\alpha) : \alpha \in \mathbb{R}^{K(n+1)}\} \tag{6}$$

where $G^k = \sum_{i=1}^m 2\left\{[p_i + (l_i^k)^-]^2 + [(l_i^k)^-]^2\right\}$,

$H^k = \sum_{i=1}^m [p_i + 2(l_i^k)^-]^2$.

Similarly, DCA with successive DC decomposition applied to (6), named DCA^k, need to compute two sequences $\{\beta^k\}$ and $\{\alpha^k\}$ such that $\beta^k \in \partial H^k(\alpha^k)$ is calculated as

$$\beta^k = 2\sum_{i=1}^m p_i(\alpha^k)\mathbf{z}^{(i,j_i^k)} \quad \text{where } j_i^k \in I_i(\alpha^k), \tag{7}$$

Algorithm 2. DCA with successive DC decomposition for solving (6) (DCA^k)

Initialization: Let ε be a sufficiently small positive number. Let $\alpha^0 \in \mathbb{R}^{K(n+1)}$.
Set $k = 0$.
repeat
   1. Compute $\beta^k \in \partial H^k(\alpha^k)$ using (7).
   2. Solve the convex quadratic program (8) to obtain $(\alpha^{k+1}, t^{k+1}, \tau^{k+1})$.
   3. $k = k + 1$.
until Stopping criteria are satisfied.

α^{k+1} is an optimal solution to the convex quadratic program

$$\min_{\alpha, t, \tau} \sum_{i=1}^{m} (2\tau_i^2) + \sum_{i=1}^{m} (2t_i^2) - \langle \beta^k, \alpha \rangle, \tag{8}$$

$$\text{s.t. } t_i \geq 0, \; t_i \geq y_i - \langle \alpha, \mathbf{z}^{(i, j_i^k)} \rangle, \; i = 1, \ldots, m,$$

$$\tau_i \geq t_i + \langle \alpha, \mathbf{z}^{(i,j)} \rangle - y_i, \; i = 1, \ldots, m, j = 1, \ldots, K.$$

Especially, the convergence properties of DCA is still valid for DCA^k as stated in Theorem 2. Its proof is similar to the proof of Theorem 2 in the recent work [10].

Theorem 2. *Convergence properties of* DCA^k

(i) *DCA^k generates the sequence $\{\alpha^k\}$ such that the sequence $\{F(\alpha^k)\}$ is decreasing.*
(ii) *Every limit point α^* of the sequence $\{\alpha^k\}$ is a critical point of $F = G^\infty - H^\infty$ where the functions*

$$G^\infty = \sum_{i=1}^{m} 2 \left\{ [p_i + (l_i^\infty)^-]^2 + [(l_i^\infty)^-]^2 \right\}, H^\infty = \sum_{i=1}^{m} [p_i + 2(l_i^\infty)^-]^2$$

and l_i^∞ is the affine minorization of p_i at α^*, $i = 1, \ldots, m$.

2.3 Starting Point for DCA

We suggest a deterministic strategy to compute the starting point for our DCAs that is quite similar to the random version in [13]. It consists of three main steps: (Step 1) choose K points from the set $\{\mathbf{x}_i\}_{i=1,\ldots,m}$ by the KKZ method [8] but the first point is set as $\bar{\mathbf{x}} = \frac{1}{m} \sum_{i=1}^{m} \mathbf{x}_i$; (Step 2) find $j_i^0 \in \{1, \ldots, K\}$ ($i = 1, \ldots, m$) such that the j_i^0-th point in these K points is closest to the point $\mathbf{x}_i$ (see Voronoi partitions in [13] for more details); (Step 3) compute the starting point $\alpha^0 = (\mathbf{a}_1^0, b_1^0, \ldots, \mathbf{a}_K^0, b_K^0)$ where for $j = 1, \ldots, K$, $(\mathbf{a}_j^0, b_j^0)$ is an optimal solution to the linear least-square problem $\min_{(\mathbf{a}_j, b_j)} \sum_{i \in \mathcal{J}(j)} (\langle \mathbf{a}_j, \mathbf{x}_i \rangle + b_j - y_i)^2$, where $\mathcal{J}(j) := \{i \in \{1, \ldots, m\} : j_i^0 = j\}$.

Table 1. Descriptions of six synthetic datasets: n is the dimension of the space; $\mathcal{X}$ is the set of points $\mathbf{x}_i$ with the size of m; $q(\mathbf{x})$ is the function used to generate $y_i = q(\mathbf{x}_i)$; $K := \lceil m^{n/(n+4)} \rceil$ is the number of affine functions.

Dataset	Description
Data1	$n = 3$, $m = 729$, $K = 17$, $\mathcal{X} = \{0,\ldots,8\}^3$, $q(\mathbf{x}) = 0.01(\exp(x_1) + 0.5x_2^2 + x_3)$
Data2 ([4])	$n \in \{4,6,8\}$, $m \in \{1000, 1500, 2000\}$, $K = \lceil m^{n/(n+4)} \rceil$,
	The points $\mathbf{x}_i$ are generated from the uniform distribution $\mathcal{U}([-2,2]^n)$,
	$y_i = q(\mathbf{x}_i) + \epsilon_i$, $\epsilon_i \sim \mathcal{N}(0,1)$, $q(\mathbf{x}) = \max_{j=1,\ldots,n+3}\{\langle \nabla w(\overline{\mathbf{x}}_j), \mathbf{x}\rangle + w(\overline{\mathbf{x}}_j)\}$,
	$w(\mathbf{x}) = (1/2)\mathbf{x}^\top \mathbf{H}\mathbf{x} + \langle \mathbf{f}, \mathbf{x}\rangle - n$, $\mathbf{f} = (1/n, 2/n, \ldots, 1) \in \mathbb{R}^n$,
	$\mathbf{H}_{i,j} = (1+2i)/(2i)$ if $i = j$, $(i+j)^{-1}$ otherwise,
	$\overline{\mathbf{x}}_j = \mathbf{e}_j - (k/d)\mathbf{1}$ for $j = 1,\ldots,n$; $\overline{\mathbf{x}}_{n+1} = \mathbf{0}$, $\overline{\mathbf{x}}_{n+2} = -\mathbf{1}$, $\overline{\mathbf{x}}_{n+3} = \mathbf{1}$,
	$\mathbf{e}_j$ is the j-th unit vector in $\mathbb{R}^n$, $\mathbf{1}$ is the vector of ones in $\mathbb{R}^n$
Data3	Data2 with $w(\mathbf{x}) = (1/2)[(\mathbf{x})^+]^\top \mathbf{H}(\mathbf{x})^+$
Data4	Data2 with $m = 2000$, $\epsilon = 0$
Data5	$n = 5$, $m = 1024$, $K = 47$, $\mathcal{X} = \{-2,-1,1,2\}^5$, $q(\mathbf{x}) = 1/2\mathbf{x}^\top \mathbf{H}\mathbf{x} + \langle \mathbf{f}, \mathbf{x}\rangle - n$
Data6	Data5 with $\mathcal{X} = \{0,1,2,3\}^5$, $q(\mathbf{x}) = 0.01(x_1 + 0.5x_2 + x_3)^2 - x_4 + (x_5)^2$

3   Numerical Experiments

In the section, we make a comparison between DCA, DCA^k and two heuristic/exact algorithms LSPA [13], MIQPM [18,19] on six synthetic datasets, similarly taken from the literature (see, e.g., [4,7,13]) and on six real, large regression datasets in different areas from UCI Machine Learning Repository[1] and LIBSVM website[2]. The detailed descriptions of synthetic/real datasets are summarized in Tables 1 and 3.

Comparison Criteria: We consider the following criteria: the root-mean-square (RMS) error defined as $\text{RMS}(\alpha) := (\frac{1}{m}\sum_{i=1}^{m}[p_i(\alpha)]^2)^{1/2}$, the CPU time (in seconds), and the number of the so-called *active* affine functions where a function $\langle \mathbf{a}_j^*, \mathbf{x}\rangle + b_j^*$ in the obtained function $f(\mathbf{x}) = \max_{j=1,\ldots,K}\{\langle \mathbf{a}_j^*, \mathbf{x}\rangle + b_j^*\}$ is active if it is the maximum at some point $\mathbf{x}_i \in \mathcal{X}$.

Set Up Experiments: All experiments were tested in MATLAB R2016b on a PC Intel(R) Xeon(R) CPU E5-2630 v2, @2.60 GHz of 32 GB RAM. We use CPLEX 12.8 for solving convex quadratic programs in DCA, DCA^k, and mixed-integer quadratic programs in MIQPM. The function lsqlin in MATLAB was used for solving linear least-squares problems in LSPA. The big-M of MIQPM is set to 1000 [1,18] and the maximum executing time for a call to CPLEX to 1000 s. As LSPA is fast but very sensitive to the starting point, and its convergence is not ensured [13], the final RMS error of LSPA is reported by the best result found in the N runs with random starting points and thus its CPU time is calculated as the total executing time of LSPA over these runs. The maximum number

[1] http://www.ics.uci.edu/~mlearn/MLRepository.html.
[2] https://www.csie.ntu.edu.tw/~cjlin/libsvmtools/datasets/.

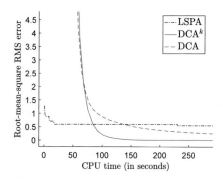

Fig. 1. The RMS error of DCA, DCA^k and LSPA versus the CPU time on Data4: $n = 8$, $m = 2000$.

of iterations of LSPA for each run and the value of N are set to 50 and 200 respectively (see [13]). As for DCAs, the starting point is described in Sect. 2.3 and the stopping criterion is $|\text{RMS}(\alpha^k) - \text{RMS}(\alpha^{k-1})| \leq \varepsilon(1 + \text{RMS}(\alpha^{k-1}))$. The default tolerance ε is set to 10^{-3} for Data1, Data6; 10^{-2} for the others. The point $\overline{\alpha}$ is set to α^0 in DCA. As for real datasets, we divide each dataset into two subsets: training set containing 75% of dataset, and test set containing 25% of dataset. First, we learn a PL model on the training set and collect its RMS error. Then, we use that PL model on the test set to compute the RMS error (named Test RMS).

Descriptions of Results' Table, Figures: The comparative results of all four algorithms on synthetic datasets in terms of the RMS error, the CPU time and the number of active affine functions are reported in Table 2. Moreover, we also perform and plot the RMS error (resp. the best RMS error) of DCA, DCA^k (resp. LSPA) versus the CPU time on Data4 ($n = 8$, $m = 2000$) in Fig. 1. As for real datasets, we report several results of DCA^k and LSPA in Table 3.

Comments on Computational Results

- In terms of the RMS error, both DCA and DCA^k are efficient and they are better than other comparative algorithms on all synthetic datasets. Indeed, the RMS error of the standard DCA is better than that of LSPA and MIQPM with the ratio of gain from 0.69% to 99.0% in 22/24 cases and from 27.0% to 98.9% in all cases, respectively. Meanwhile, DCA^k obtains the best fitting error on all datasets – the ratio of gain of DCA^k versus DCA, LSPA, and MIQPM varies from 1.01% to 66.7%, from 7.42% to 89.9%, and from 29.1% to 99.1%, respectively, for Data1, Data2, Data3, and Data5. Especially, as for Data4 and Data6, DCA^k seems to furnish the globally optimal solution to the problem (1) (with the RMS error less than 10^{-5} after about 200 s, see Fig. 1) and the ratio of gain is significantly increasing, it varies from 82.7% to 97.1%, from 85.3% to 99.8%, and around 99.8%, respectively. This turns out the advantage of updating the DC decomposition in DCA^k.

Table 2. Comparative results of DCA, DCA^k, LSPA, and MIQPM on six synthetic datasets in terms of the root-mean-square RMS error, the CPU time (in seconds) and the number of active affine functions. Bold values indicate the best results.

Dataset	n	m	K	RMS error				CPU time				Number of active affine functions			
				DCA	DCA^k	LSPA	MIQPM	DCA	DCA^k	LSPA	MIQPM	DCA	DCA^k	LSPA	MIQPM
Data1	3	729	17	0.018	**0.006**	0.011	0.706	15.9	**11.0**	15.8	1000	17	17	12	7
Data2	4	1000	32	0.986	**0.976**	1.169	2.497	6.37	**6.06**	33.2	1003	12	12	20	9
		1500	39	1.022	**1.006**	1.193	5.579	**12.4**	13.3	43.9	1003	17	16	25	1
		2000	45	1.031	**1.010**	1.324	5.665	**17.7**	20.9	54.9	1006	13	17	23	1
	6	1000	63	1.150	**1.003**	1.158	7.684	38.3	**27.3**	53.8	1006	20	20	27	1
		1500	80	1.141	**1.032**	1.279	–	70.7	**43.1**	74.5	–	17	16	25	–
		2000	96	1.213	**1.022**	1.390	–	111	**38.7**	103	–	17	10	31	–
	8	1000	100	1.081	**0.962**	1.219	–	93.8	**40.1**	78.0	–	23	18	37	–
		1500	131	1.132	**0.970**	1.195	–	86.4	**55.2**	125	–	17	18	50	–
		2000	159	1.097	**0.987**	1.193	–	265	**129**	170	–	25	18	43	–
Data3	4	1000	32	0.998	**0.969**	1.069	1.368	8.56	**7.49**	27.9	1003	17	16	19	12
		1500	39	1.029	**0.997**	1.077	5.644	14.4	**13.4**	41.4	1004	17	16	23	1
		2000	45	1.051	**1.025**	1.113	5.671	25.4	**20.4**	51.3	1006	13	13	18	1
	6	1000	63	1.176	**0.977**	1.158	7.866	**12.4**	13.6	52.2	1005	15	16	19	1
		1500	80	0.971	**0.960**	1.102	–	74.0	**67.9**	71.5	–	26	26	27	–
		2000	96	1.078	**1.007**	1.174	–	80.8	**58.0**	99.0	–	22	19	15	–
	8	1000	100	1.002	**0.942**	1.049	–	74.1	**55.3**	72.2	–	26	21	20	–
		1500	131	1.020	**0.956**	1.096	–	124	**98.9**	116	–	29	24	19	–
		2000	159	1.085	**1.007**	1.127	–	85.7	**81.4**	160	–	16	17	14	–
Data4	4	2000	45	0.151	**0.023**	0.157	–	73.2	**60.5**	61.5	–	19	22	22	–
	6	2000	96	0.210	**6.6e-3**	0.476	–	486	184	**111**	–	25	23	43	–
	8	2000	159	0.284	**8.1e-3**	0.589	–	231	**148**	180	–	18	16	48	–
Data5	5	1024	47	0.271	**0.144**	1.431	4.396	100	85.1	**65.5**	1005	47	47	40	1
Data6	5	1024	47	0.011	**1.9e-3**	1.116	1.000	37.1	31.4	**13.6**	1004	39	38	7	13

Here the sign "_" means that the results are not obtained by running out of memory.

A function $\langle a_j^*; x \rangle + b_j^*$ in the obtained function $f(\mathbf{x}) = \max_{j=1,\ldots,K} \left\{ \langle a_j^*, \mathbf{x} \rangle + b_j^* \right\}$ is active if it is the maximum at some point $\mathbf{x}_i \in \mathcal{X}$.)

Table 3. Comparative results of DCA^k and LSPA on six real datasets.

Real dataset	m	n	K	Test RMS		Training CPU	
				DCA^k	LSPA	DCA^k	LSPA
winequality-red	1599	11	224	**5.14e+00**	7.08e+00	**7.09e+01**	8.34e+01
space_ga	3107	8	125	**1.19e−01**	1.80e−01	**5.01e+01**	5.82e+01
abalone	4177	8	259	**2.17e+00**	2.51e+00	**8.39e+01**	2.02e+02
winequality-white	4898	11	508	**7.16e−01**	1.85e+01	**2.84e+02**	5.12e+02
CCPP	9568	4	98	**4.16e+00**	7.89e+00	1.58e+02	**1.35e+02**
cadata	20640	8	500	**2.76e+06**	3.00e+06	**8.40e+02**	1.32e+03

In our experiments, the best RMS error of LSPA is moderately improved (see Fig. 1) while MIQPM does not work in most of the medium/large datasets as running out of memory or exceeding the limited time (1000 s). Moreover, the smaller the stopping tolerance ε is, the slightly better the RMS error of our algorithms would be.

- Concerning the CPU time, DCA^k is the fastest in most cases – the ratio of gain of DCA^k versus DCA, LSPA, and MIQPM varies from 1.05 to 2.86 times in 21/24 cases, from 1.01 to 5.47 times in 21/24 cases, and from 11.8 to 165 times in all cases, respectively. DCA runs slower than LSPA from 1.00 to 4.37 times in 12/24 cases in particular for the large datasets. Updating the DC decomposition improves the rapidity of DCA. Note that the CPU time of the four algorithms depends on the fixed number K of affine functions. From Table 2, the number of active affine functions of DCA and DCA^k is smaller than the fixed value of K and that of LSPA in most cases. This observation leads us to a perspective: how to update the number of affine functions in our DCAs to improve the rapidity but still ensure the quality of fitting error.

- As for real datasets, we observe that the RMS of DCA^k is significantly better than that of LSPA in all test sets – the ratio of gain in terms of Test RMS varies from 8% to 96.1%, specially 96.1%, 47.28%, 8% on the large datasets with size of 4898, 9568, 20640, respectively. Meanwhile, the CPU time of DCA^k is reasonable (less than 900 s on large datasets) and faster than LSPA on 5/6 datasets.

4   Conclusions

We have investigated DC programming and DCA for solving the convex piecewise-linear fitting problem. Two DCA schemes, DCA and DCA^k, have been developed. Differing from the standard DCA, the DC decompositions in DCA^k are updated during DCA iterations to better approximate the DC objective function. Numerical results on six synthetic/real datasets have turned out that the standard DCA is efficient and DCA^k improves both the quality and the rapidity of DCA. DCA^k is then suggested as the best among related existing approaches for this problem. From these promising results, we plan in future works to update the number of affine functions in our DCA schemes, to explore other DC decompositions, and to develop our DC programming and DCA-based approach for more general fitting problems with DC fitting functions, and different fitting criteria as well.

References

1. Amaldi, E., Coniglio, S., Taccari, L.: Discrete optimization methods to fit piecewise affine models to data points. Comput. Oper. Res. **75**, 214–230 (2016)
2. Bagirov, A., Clausen, C., Kohler, M.: An algorithm for the estimation of a regression function by continuous piecewise linear functions. Comput. Optim. Appl. **45**(1), 159–179 (2010)
3. Bagirov, A., Taheri, S., Asadi, S.: A difference of convex optimization algorithm for piecewise linear regression. J. Ind. Manag. Optim. **15**(2), 909–932 (2019)
4. Balázs, G.: Convex regression: theory, practice, and applications. Ph.D. thesis, University of Alberta (2016)
5. Boyd, S., Kim, S.J., Vandenberghe, L., Hassibi, A.: A tutorial on geometric programming. Optim. Eng. **8**(1), 67–127 (2007)
6. Boyd, S., Vandenberghe, L.: Convex Optimization. Cambridge University Press, New York (2004)
7. Hannah, L.A., Dunson, D.B.: Multivariate convex regression with adaptive partitioning. J. Mach. Learn. Res. **14**, 3261–3294 (2013)
8. Katsavounidis, I., Kuo, C.C.J., Zhang, Z.: A new initialization technique for generalized lloyd iteration. IEEE Signal Process. Lett. **1**(10), 144–146 (1994)
9. Le Thi, H.A.: DC Programming and DCA. Homepage (2005). http://www.lita.univ-lorraine.fr/~lethi/index.php/en/research/dc-programming-and-dca.html
10. Le Thi, H.A., Ho, V.T., Pham Dinh, T.: A unified DC programming framework and efficient dca based approaches for large scale batch reinforcement learning. J. Global Optim. **73**(2), 279–310 (2019)
11. Le Thi, H.A., Pham Dinh, T.: The DC (difference of convex functions) programming and DCA revisited with DC models of real world nonconvex optimization problems. Ann. Oper. Res. **133**(1–4), 23–46 (2005)
12. Le Thi, H.A., Pham Dinh, T.: DC programming and DCA: thirty years of developments. Math. Program. **169**(1), 5–68 (2018)
13. Magnani, A., Boyd, S.P.: Convex piecewise-linear fitting. Optim. Eng. **10**(1), 1–17 (2009)
14. Martinez, N., Anahideh, H., Rosenberger, J.M., Martinez, D., Chen, V.C.P., Wang, B.P.: Global optimization of non-convex piecewise linear regression splines. J. Global Optim. **68**(3), 563–586 (2017)
15. Pham Dinh, T., Le Thi, H.A.: Convex analysis approach to DC programming: theory, algorithms and applications. Acta Mathematica Vietnamica **22**(1), 289–355 (1997)
16. Pham Dinh, T., Le Thi, H.A.: DC optimization algorithms for solving the trust region subproblem. SIAM J. Optim. **8**(2), 476–505 (1998)
17. Pham Dinh, T., Le Thi, H.A.: Recent advances in DC programming and DCA. In: Nguyen, N.T., Le Thi, H.A. (eds.) Transactions on Computational Intelligence XIII, vol. 8342, pp. 1–37. Springer, Heidelberg (2014)

18. Taccari, L.: Optimization methods for piecewise affine model fitting. Ph.D. thesis, Politecnico di Milano (2010)
19. Toriello, A., Vielma, J.P.: Fitting piecewise linear continuous functions. Eur. J. Oper. Res. **219**(1), 86–95 (2012)
20. Yang, L., Liu, S., Tsoka, S., Papageorgiou, L.G.: Mathematical programming for piecewise linear regression analysis. Expert Syst. Appl. **44**, 156–167 (2016)

Solving Efficient Target-Oriented Scheduling in Directional Sensor Networks by DCA

Anh Son Ta[(✉)], Hoai An Le Thi, and Tao Pham Dinh

Institute for Research and Applications of Optimization (VinOptima),
VinTech, Vingroup, Hanoi, Vietnam
v.sonta@vinoptima.org

Abstract. Unlike convectional omnidirectional sensors that consistently have an omniangle of detecting range, directional sensors may have a restricted point of detecting range because of specialized requirements or cost considerations. A directional sensor system comprises of various directional sensors, which can switch to several directions to broaden their detecting capacity to cover every one of the objectives in a given territory. Power preservation is still a significant issue in such directional sensor networks. In this paper, we consider the multiple directional cover sets (MDCS) problem of organizing the directions of sensors into a group of non-disjoint cover sets to extend the network lifetime. It is an NP-complete problem. Firstly, a new model of MDCS is introduced in the form of Mixed Binary Integer Linear Programming (MBILP). Secondly, we investigate a new method based on DC programming and DC algorithm (DCA) for solving MDCS. Numerical results are presented to demonstrate the performance of the algorithm.

Keywords: MDCS · Mixed Binary Integer Linear Programming (MBILP) · DC programming · DC algorithm (DCA)

1 Introduction

In promising stages for some applications, for example, ongoing years, sensor systems have risen as natural checking, combat zone reconnaissance, and medicinal services [2,3]. A sensor system may consists of an enormous number of sensor nodes that are composed of detecting, information handling, and communicating components. The ordinary research of sensor systems is constantly founded on the suspicion of omnidirectional sensors that have an omniangle of detecting range. Nonetheless, sensors may have a restricted edge of detecting range because of the specialized limitations or cost contemplations, which are meant by directional sensors.

Video sensors [4,5], ultrasonic sensors [6], and infrared sensors [3] are models of broadly utilized directional sensors. Note that the directional trademark we talk about in this paper is from the perspective of the detecting, but not from

© Springer Nature Switzerland AG 2020
H. A. Le Thi et al. (Eds.): ICCSAMA 2019, AISC 1121, pp. 52–63, 2020.
https://doi.org/10.1007/978-3-030-38364-0_5

the communicating activity of sensor nodes. There are a few different ways to expand the detecting capacity of directional sensors. One path is to put several directional sensors of a similar kind on one sensor node, every sensor has a different direction. Another way is to furnish the sensor node with a mobile device that permits the node to move around. The third path is to equip the sensor node with a gadget that empowers the sensor on the node to switch (or turn) to various directions, then a sensor can capture several directions. In this paper, we consider the scenario that every sensor node prepares precisely one sensor on it. A set of targets with given location are deployed in a two-dimensional Euclidean plane. Various directional sensors are arbitrarily dispersed near these targets. We assume that the detecting locale of each direction of a directional sensor is an area of the detecting disk centered at the sensor with a detecting span. Every sensor has a uniform detecting district and the detecting areas of different directions of a sensor do not overlap.

At the point when the sensors are randomly deployed, every sensor at first covers one of its directions. These sensors structure a directional sensor arrange with the goal that information can be assembled and moved to the sink, a central handling base station. If a directional sensor appearances to a direction, we state that the sensor works toward this path and the direction is the work direction of the sensor. When this sensor works toward a direction and a target is in the detecting locale of the sensor, we state that the direction of the sensor covers the target. Since a directional sensor has a small angle of detecting range than an omnidirectional sensor or even does not cover any target when it is sent, we have to plan sensors in the system capture specific directions to cover every one of the targets. We consider a subset of directions of the sensors wherein the directions cover every one of the targets as a cover set. Note that, in a cover set a sensor can have at most one direction. The issue of finding a cover set, called *directional spread set* (DCS) problem, is an NP-complete problem (see [1]).

Power protection is as yet a significant issue in directional sensor network because of the accompanying reasons. To start with, most sensors have restricted power sources and are non-rechargeable. Additionally, the batteries of the sensors are difficult to supplant due to antagonistic or blocked off conditions in numerous situations. We assume that every sensor is non-rechargeable and dies when it runs out its energy. To moderate vitality, we can leave necessary sensors in the active state and put necessary sensors into the sleep state, while keeping every one of the targets covered.

In this paper, our objective is to maximize the network lifetime of a directional sensor network, where the network lifetime is defined as the time duration when each target is covered by the work direction of at least one active sensor. The system is to organize the directions of sensors into non-disjoint subsets, each of which is a cover set, and designate the work time for each cover set.

Note that non-disjoint cover sets enable a direction or a sensor to take part in different cover sets. Only one cover set is active at any time and it is alternately. When one cover set is active, every sensor that has a direction in this cover set is in the active state and works toward this direction, while the rest of sensors are in the sleep state.

The problem of finding non-disjoint cover sets and allocating the work time for each of them to maximize the network lifetime is called multiple directional cover sets problem (MDCS). It also is an NP-complete problem (see [1]).

The main contributions of this paper are as follows. Section 2 introduces the problem statement and a new mathematical model of MDCS. The solution method based on DC programming and DCA, DCA combine with cutting plane method are presented in Sect. 3 while the numerical simulation and conclusion are reported in Sect. 4.

2   Problem Statement and Mathematical Modeling

In this section, the notations, the problem statement of the DCS and the MDCS are represented. The following notations are used in our this paper (Fig. 1).

- M: the number of targets,
- N: the number of sensors,
- W: the number of directions per sensor.
- a_m: the mth target, $1 \leq m \leq M$.
- s_i: the ith sensor, $1 \leq i \leq N$.
- d_{ij}: the jth direction of the ith sensor, $1 \leq i \leq N, 1 \leq j \leq W$.
- $\bar{d}_{ij} = \{a_m \mid a_m \in A, \ a_m \text{ is covered by } d_{ij}, \}$,
- $\bar{s}_i = \{d_{ij} \mid j = 1, \cdots, W\}$. So, if $a_m \in \bar{d}_{ij}$ then a_m is covered by d_{ij}.
- $A = \{a_1, a_2, \cdots, a_M\}$: the set of targets,
- $S = \{\bar{s}_1, \bar{s}_2, \cdots, \bar{s}_N\}$,
- $D = \{\bar{d}_{ij} \mid i = 1, \cdots, N, j = 1, \cdots, W\}$,
- L_i: the lifetime of a sensor s_i, which is the time duration when the sensor is in the active state all the time.

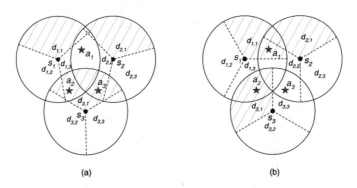

(a)   (b)

Fig. 1. An example of directional sensor network ([1]).

Problem statement is presented in [1] as follows.

Definition 1. *Cover Set: Given a collection D of subsets of a finite set A and a collection S of subsets of D, a cover set for A is a subset $D' \subseteq D$ such that every element in A belongs to at least one member of D' and every two elements in D' cannot belong to the same member of S.*

Definition 2. *DCS Problem: Given a collection D of subsets of a finite set A and a collection S of subsets of D, find a cover set for A.*

Definition 3. *MDCS Problem: Given a collection D of subsets of a finite set A and a collection S of subsets of D, find a family of K cover sets $D_1, D_2, \cdots, D_K \subseteq D$ for A, with nonnegative weights $t_1, t_2, \cdots, t_K$, such that $t_1 + t_2 + \cdots + t_k \geq p$ is maximized, and for each $s \in S$, $\sum_{i=1}^{K} |s \cap D_i|$, where L is a given positive number. a given positive number.*

Note that $|s \cap D_i|$ shows the quantity of the directions of s that are in D_i, where $|s \cap D_i| = 0$ or 1 since no more than one direction of a sensor can work in a cover set.

In [1] the MDCS is formulated as Mixed (0–1) Integer Programming with quadratic constraints. In our paper, a new model of MCDS is introduced in the form of Mixed Binary Integer Linear Programming (MBILP).

Consider a directional sensor network with a set of targets, a set sensors, and a set of directions. Each sensor i^{th} has W directions and an initial lifetime of L_i.

The directions is organize into K cover sets. The kth cover set is denoted by D_k, with the work time t_k. A direction d_{ij} can be joint in multiple cover sets.

Let us set a Boolean variable x_{ijk} as

$$x_{ijk} = \begin{cases} 1 & if \quad \bar{d}_{ij} \in D_k \\ 0 & otherwise \end{cases} \tag{1}$$

The MBILP problem formulated for the MDCS is as follows:

$$\max t_1 + t_2 + ... + t_K \tag{2}$$

subject to

$$\sum_{k=1}^{K} \sum_{j=1}^{W} y_{ijk} \leq L_i \ \forall \ \bar{s}_i \in S \tag{3}$$

$$y_{ijk} \leq M_0.x_{ijk} \tag{4}$$

$$y_{ijk} \leq t_k + M_0.(1 - x_{ijk}) \tag{5}$$

$$t_k - M_0.(1 - x_{ijk}) \leq y_{ijk} \tag{6}$$

$$\sum_{j=1}^{W} x_{ijk} \leq 1 \ \forall \bar{s}_i \in S, \ k = 1, ..., K \tag{7}$$

$$\sum_{\substack{a_m \in \bar{d}_{ij} \\ \bar{d}_{ij} \in D}} x_{ijk} \geq 1 \ \forall a_m \in A, \ k = 1, ..., K \tag{8}$$

$$x_{ijk} \in \{0,1\}, \quad t_k \geq 0, \quad y_{ijk} \geq 0, \tag{9}$$

where M_0 is positive large enough.

The objective function maximizes the total work time of all the K cover sets. The constraints (3) demonstrate the lifetime for every sensor. The W directions of any sensor work across all the cover sets underlying lifetime of the sensor. The constraints (7) show the selection among various directions of a single sensor, at most one direction of the sensor can work in a cover set. The constraints (8) indicate the coverage guarantee for each target. For each cover set, each target in an absolute necessity be covered by in at least one direction of this cover set. The constraints (9) speak to the limitations on the variables.

We rewrite the problem (2)–(9) in the form of BMILP:

$$\min\{-\langle c, x\rangle\} \text{ s.t. } x \in X,$$

where $X := \{x \in \mathbb{R}^q : Ax \leq b, \ x \geq 0, \ x_j \in \{0,1\} \forall j \in J\}$.

3   DC Programming and DCA

By using an exact penalty result, we can reformulate the problem (2)–(9) in the form of a concave minimization program. The exact penalty technique aims at transforming the original problem into a more tractable equivalent DC program. Let $K := \{x \in \mathbb{R}^q : Ax \leq b, \ x \geq 0, \ x_j \in [0,1] \ \forall j \in J\}$. The feasible set of the original problem is then $X = \{x : \ x \in K, \ x_j \in \{0,1\} \ \forall j \in J\}$. The original program is rewritten as problem (P),

$$\min\{-\langle c, x\rangle : \ x \in X\}. \quad (P) \tag{10}$$

Let us consider the function $p : \mathbb{R}^q \to \mathbb{R}$ defined by:

$$p(x) = \sum_{j \in J} \min\{x_j, 1 - x_j\}.$$

It is clear that $p(x)$ is concave and finite on K, $p(x) \geq 0 \ \forall x \in K$ and that:

$$\{x : \ x \in X\} = \{x : x \in K, p(x) \leq 0\}. \tag{11}$$

Hence the problem (P) can be rewritten as:

$$\min\left\{\langle c, x\rangle : x \in K, \ p(x) \leq 0\right\}. \tag{12}$$

The following theorem can then be formulated.

Theorem 1. *Let K be a nonempty bounded polyhedral convex set, f be a finite concave function on K and p be a finite nonnegative concave function on K. Then there exists $\eta_0 \geq 0$ such that for $\eta > \eta_0$ the following problems have the same optimal value and the same solution set:*

$$(P_\eta) \quad \alpha(\eta) = \min\{f(y) + \eta p(y) : y \in K\},$$
$$(P) \quad \alpha = \min\{f(y) : y \in K, p(y) \leq 0\}.$$

Furthermore

- *If the vertex set of K, denoted by $V(K)$, is contained in $x \in K : p(y) \leq 0$, then $\eta_0 = 0$.*
- *If $p(y) > 0$ for some y in $V(K)$, then $\eta_0 = \min\left\{ \frac{f(y) - \alpha(0)}{S_0} : y \in K, p(y) \leq 0 \right\}$, where $S_0 = \min\left\{ p(y) : y \in V(K), p(y) > 0 \right\} > 0$.*

Proof. The proof for the general case can be found in [7].

From Theorem 1 we get, for a sufficiently large number η ($\eta > \eta_0$), the equivalent concave minimization problem:

$$\min\{f_\eta(x) := \langle c, x \rangle + \eta p(x) : x \in K\}, \tag{13}$$

which is a DC program of the form:

$$\min\{g(x) - h(x) : x \in \mathbb{R}^q\}, \tag{14}$$

where: $g(x) = \chi_K(x)$ and $h(x) = -f_\eta(x) = -\langle c, x \rangle - \eta p(x)$.

We have successfully transformed an optimization problem with integer variables into its equivalent form with continuous variables. Notice that (14) is a polyhedral DC program where g is a polyhedral convex function (i.e., the pointwise supremum of a finite collection of affine functions). DCA applied to the DC program (14) consists of computing, at each iteration k, the two sequences $\{x^k\}$ and $\{y^k\}$ such that $y^k \in \partial h(x^k)$ and x^{k+1} solves the next linear program of the form (P_k).

$$\min \left\{ g(x) - \langle x - x^k, y^k \rangle : x \in \mathbb{R}^q \right\} \Leftrightarrow \min\{-\langle x, y^k \rangle : x \in K\}. \tag{15}$$

From the definition of h, a sub-gradient $y^k \in \partial h(x^k)$ can be computed as follows:

$$y^k = \begin{cases} -c_i - \eta & \text{if } x_i \geqslant 1/2 \\ -c_i + \eta & \text{if } x_i < 1/2 \end{cases} \tag{16}$$

The DCA scheme applied to (14) can be summarized as follows:

Algorithm 1

Initialization:
Choose an initial point x^0, set $k = 0$;
Let ϵ_1, ϵ_2 be sufficiently small positive numbers;
Repeat
Compute y^k via (16);
Solve the linear program (15) to obtain z^{k+1};
$k \leftarrow k + 1$;
Until either $\|x^{k+1} - x^k\| \leq \epsilon_1(\|x^k\| + 1)$ or $|f_\eta(x^{k+1}) - f_\eta(x^k)| \leq \epsilon_2(|f_\eta(x^k)| + 1)$.

Theorem 2. *(Convergence properties of Algorithm DCA)*

- DCA generates the sequence $\{x^k\}$ contained in $V(K)$ such that the sequence $\{f_\eta(x^k)\}$ is decreasing.
- The sequence $\{x^k\}$ converges to $x^* \in V(K)$ after a finite number of iterations.
- The point x^* is a critical point of Problem (14). Moreover if $x_i^* \neq \frac{1}{2}$ for all $i \in J$, then x^* is a local solution to (14).
- For a number η sufficiently large, if at iteration r we have $x_j^r \in \{0,1\}$ $\forall j \in J$, then $x_j^k \in \{0,1\}$ $j \in J$ for all $k \geq r$.

Proof. The proof can be found in [7].

To improve the solution we combine DCA and cutting plane method (see [8,9] and [10]).

4   DCA-Cut for Global Solution

Let us consider the following problem:

$$\min\{\chi_K(z) + c^T x + d^T y + tp(x) \; : \; z = (x,y) \in \mathbb{R}^n \times \mathbb{R}^p\}, \tag{17}$$

where

$$\chi_K(z) := \begin{cases} 0 & \text{si } z \in \mathrm{K}, \\ +\infty & \text{otherwise} \end{cases} \tag{18}$$

is the indicator function on K.

We set $g(z) := \chi_K(z)$ and

$$h(z) := -c^T x - d^T y + t(-p)(z) = -c^T x - d^T y + t \sum_{j=1}^n \max\{-x_j, x_j - 1\}. \tag{19}$$

Thus, the problem is equivalent with a DC program:

$$\min\{g(z) - h(z) \; : \; z = (x,y) \in \mathbb{R}^n \times \mathbb{R}^p\}. \tag{20}$$

We define a valid inequality for all point of X from a solution of penalty function p on K. Let $z^\star \in K$, we define:

$$I_0(z^\star) = \{j \in \{1,\ldots,n\} : x_j^\star \leq 1/2 \} \quad , \quad I_1(z^\star) = \{1,\ldots,n\} \setminus I_0(z^\star).$$

and

$$l_{z^\star}(z) \equiv l_{z^\star}(x) = \sum_{i \in I_0(z^\star)} x_i + \sum_{i \in I_1(z^\star)} (1 - x_i).$$

Lemma 1. *(see [8]) Let $z^\star \in K$, we have*

(i) $l_{z^\star}(x) \geq p(x) \; \forall x \in \mathbb{R}^n$.

(ii) $l_{z^*}(x) = p(x)$ *if and only if*

$$(x, y) \in R(z^*) := \{(x, y) \in K : x_i \leq 1/2, i \in I_0(x^*); x_i \geq 1/2, i \in I_1(x^*) \}.$$

Lemma 2. *(see [8]) Let $z^* = (x^*, y^*)$ be a local minimum of function p on K, then the inequality*

$$l_{z^*}(x) \geq l_{z^*}(x^*) \tag{21}$$

is valid for all $(x, y) \in K$.

Theorem 3. *(see [8,9]) There exists a finite number $t_1 \geq 0$ such that, for all $t > t_1$, if $z^* = (x^*, y^*) \in V(K) \setminus S$ is a local minimum of problem (20) then*

$$l_{z^*}(x) \geq l_{z^*}(x^*), \ \forall(x, y) \in K. \tag{22}$$

Construction of a Cut from an Infeasible Solution

Let z^* be a solution that is not a feasible point of X such that $l_{z^*}(z) \geq l_{z^*}(z^*) \ \forall z \in K$. In this case, there exists at least one index $j_0 \in \{1, \ldots, n\}$ such that $x_{j_0}^*$ is non binary. We consider two following cases:

Case 1: The value of $l_{z^*}(z^*)$ is not integer.
As $l_{z^*}(z)$ is integer for all $z \in S$, we have immediately:

$$\begin{cases} l_{z^*}(z) \geq \rho := \lfloor l_{z^*}(z^*) \rfloor + 1, \forall z \in S \\ l_{z^*}(z^*) \leq \rho. \end{cases} \tag{23}$$

In other words, the inequality

$$l_{z^*}(z) \geq \rho \tag{24}$$

is a strictly separate cut z^* of S.

Case 2: the value of $l_{z^*}(z^*)$ is integer.
It is possible that there are feasible points z' such that $l_{z^*}(z') = l_{z^*}(z^*)$. If such a point exists, we could update the best solution (PLM01) and also improve the upper bound of the optimal value.

Otherwise, for all $z \in S$, we have $l_{z^*}(z) > l_{z^*}(z^*)$. That is to say,

$$l_{z^*}(z) \geq l_{z^*}(z^*) + 1 \tag{25}$$

is a separate cut z^* of S.

We consider below a procedure (called *Procedure P*) is providing a cut, or a feasible point, or a potential point.

Let us set $I_F(z^*) := \{i \in I : x_i^* \notin \{0, 1\}\}$.

Let us set $I_{\overline{F}}(z^*) := \{i \in I : x_i^* \in \{0, 1\}\}$, then $I = I_{\overline{F}}(z^*) \cup I_F(z^*)$ and $I_{\overline{F}}(z^*) \cap I_F(z^*) = \emptyset$.

We observe that if we can find z^1 such that $l_{x^*}(x^1) = l_{x^*}(x^*)$ and there exists $i_1 \in I_{\overline{F}}$ and $x_{i_1}^1 = 1 - x_{i_1}^*$ (there are only two possibilities $x_{i_1}^1 = 1 - x_{i_1}^*$ or $x_{i_1}^1 = x_{i_1}^*$) then $z^1 \notin R(z^*)$.

By Lemma 1 we have:

$$p(x^\star) = l_{x^\star}(x^\star) = l_{x^\star}(x^1) > p(x^1).$$

Step 1. Let us set $K_1 = \{z = (x, y) \in K; \ x_i = x^\star_i \ \forall i \in I_{\overline{F}}\}$.

Step 2. Choose $i_s \in I_F(z^\star)$.

Step 2.1. If $i_s \in I_0(x^\star)$ then we solve the linear program:

$$(Pmax1) \quad \overline{x}_{i_s} = \max\{x_{i_s} \ : \ z = (x, y) \in K_1; \ l_{x^\star}(x) = l_{x^\star}(x^\star)\}. \tag{26}$$

- If $\overline{x}_{i_s} = 1$ then $\overline{z} \notin R(z^\star)$ and by Lemma 1 we have:

$$p(x^\star) = l_{x^\star}(x^\star) = l_{x^\star}(\overline{x}) > p(\overline{x}).$$

- If $\overline{x}_{i_s} < 1$ then $x_{i_s} = 0$ and we update the indices set $I_{\overline{F}}(z^\star) = I_{\overline{F}}(z^\star) \cup \{i_s\}$, $I_F(z^\star) = I_F(z^\star) \backslash \{i_s\}$ and

$$K_1 = K_1 \cap \{z = (x, y) \in K; \ x_{i_s} = 0\}.$$

- If Problem $(Pmax1)$ is infeasible then added a cut $l_{x^\star}(x) \geq l_{x^\star}(x^\star) + 1$ in our problem.

Step 2.2. If $i_s \in I_1(x^\star)$ then we solve the linear program:

$$(Pmin2) \quad \overline{x}_{i_s} = \min\{x_{i_s} \ : \ z = (x, y) \in K_1; \ l_{x^\star}(x) = l_{x^\star}(x^\star)\}. \tag{27}$$

- If $\overline{x}_{i_s} = 0$ then $\overline{z} \notin R(z^\star)$ and by Lemma 1 we have:

$$p(x^\star) = l_{x^\star}(x^\star) = l_{x^\star}(\overline{x}) > p(\overline{x}).$$

- If $\overline{x}_{i_s} > 0$ then $x_{i_s} = 1$ and we update the indices set $I_{\overline{F}}(z^\star) = I_{\overline{F}}(z^\star) \cup \{i_s\}$, $I_F(z^\star) = I_F(z^\star) \backslash \{i_s\}$ and

$$K_1 = K_1 \cap \{z = (x, y) \in K; \ x_{i_s} = 1\}.$$

- If Problem $(Pmin2)$ is infeasible then added a cut $l_{x^\star}(x) \geq l_{x^\star}(x^\star) + 1$ in our problem.

From a feasible point $x*$, we can be add cut $l_{x^\star}(x) \geq 1$ (see [9]) and restart DCA on new feasible region. The numerical results show the efficient of our proposed method.

5   Numerical Simulation and Conclusion

We evaluate the performance of our approach through simulations running on a computer with Core i7-8750 2.2 GHz CPU and 24 GB memory. The optimization algorithm is implement in Matlab and it is compared with CPLEX 12.7. The data is generated based on [1]. N sensors with detecting radius r = 100(m) and M targets are deployed uniformly in a region of 400×400 m^2. Each sensor has W directions. The maximal number of cover sets is equal to the number of sensors, i.e., K = N. The initial lifetime of each sensor is set as 1 and W is set as 3. The Gap1 and Gap2 are defined by the gap between the result of DCA, DCA-CUT with CPLEX 12.7, respectively (Table 1).

In Tables of results, Data, DCA, T-DCA, Gap1, DCACUT, T-CUT, Cplex, T-Cplex and Gap2 stand for name of data, objective value of DCA, running time by DCA, Gap between DCA and Cplex, objective value of DCACUT, running time by DCACUT, objective value of Cplex 12.7, running time of Cplex and objective value of Cplex 12.7 and Gap between DCA and Cplex (the global optimal value), respectively, where

$$Gap = \frac{\text{Upper bound - Lower bound}}{\text{Upper bound}}.$$

Table 1. Results of DCA, DCACUT and CPLEX 12.7.

	Data	DCA	T-DCA(s)	Gap1(%)	DCACUT	T-CUT	Cplex	T-Cplex(s)	Gap2(%)
M = 10 N = 80	Data1	–	–	–	77	0.6	77	5.8	0
	Data2	78	0.36	1.3	78	0.5	79	4.9	1.3
	Data3	77	0.4	0	77	1.6	77	4.1	0
	Data4	78	0.48	1.3	78	1.0	79	6.3	1.3
	Data5	–	–	–	77	0.4	78	6.1	1.3
M = 10 N = 90	Data6	87	0.5	2.2	87	1.5	89	2.7	2.2
	Data7	–	–	–	86	0.6	87	2.5	1.1
	Data8	88	0.6	1.1	88	1.13	89	9.1	1.1
	Data9	87	0.6	2.2	88	1.2	89	9.3	1.1
	Data10	89	0.3	0	89	9.5	89	9.7	0
	Data11	88	0.5	0	88	1.1	88	12.9	0
	Data12	87	0.4	2.2	88	1.1	89	9.8	1.1
M = 15 N = 100	Data13	97	0.5	1.0	98	0.9	98	16.8	0
	Data14	97	0.6	0	97	1.4	97	14	0
	Data15	–	–	–	96	17.6	96	15.6	0
	Data16	–	–	–	98	0.7	98	13.7	0
	Data17	97	0.8	1.0	98	1.5	98	6.2	0
M = 20 N = 150	Data18	–	–	–	147	7.1	148	8.2	0.6
	Data19	–	–	–	145	2.9	145	56.2	0
	Data20	147	2.1	0	147	3.5	147	54.4	0

From the numerical results, we observe that:

– In many cases DCA provides an integer solution after a few number of iterations.
– DCA-CUT always provides an integer solution.
– The GAPs are small. It means that the objective value obtained by DCA-CUT is rather close to the global optimal value, almost all GAP is less than 2.0%.

In this paper, we introduce a new model of MDCS problem in the form of the BMILP problem. An efficient approach based on DC algorithm (DCA) and Cutting plane method is proposed for solving this problem. The computational results obtained show that this approach is efficient and original as it can give integer solutions while working in a continuous domain. From the promising outcome, in a future work we plan to combine DCA, Branch-and-Bound and Cutting plane method to globally solve the general MDCS problem with high dimension and real data.

References

1. Cai, Y., Lou, W., Li, M., Li, X.Y.: Energy Efficient target-oriented scheduling in directional sensor networks. IEEE Trans. Comput. **58**(9), 1259–1274 (2009)
2. Akyildiz, I.F., Su, W., Sankarasubramaniam, Y., Cayirci, E.: A survey on sensor networks. ACM Trans. Multimedia Comput. Commun. Appl. **40**(8), 102–114 (2002)
3. Szewczyk, R., Mainwaring, A., Polastre, J., Anderson, J., Culler, D.: An analysis of a large scale habitat monitoring application. In: Proceedings ACM Conference, Embedded Networked Sensor Systems (SenSys) (2004)
4. Rahimi, M., Baer, R., Iroezi, O.I., Garcia, J.C., Warrior, J., Estrin, D., Srivastava, M.: Cyclops: in situ image sensing and interpretation in wireless sensor networks. In: Proceedings of ACM Conference on Embedded Networked Sensor Systems (SenSys) (2005)
5. Feng, W., Kaiser, E., Feng, W.C., Baillif, M.L.: Panoptes: scalable low-power video sensor networking technologies. ACM Trans. Multimedia Comput. Commun. Appl. **1**(2), 151–167 (2005)
6. Djugash, J., Singh, S., Kantor, G., Zhang, W.: Range-only slam for robots operating cooperatively with sensor networks. In: Proceedings of IEEE Conference, Robotics and Automation (2006)
7. Le Thi, H.A., Pham Dinh, T.: A continuous approach for globally solving linearly constrained quadratic zero - one programming problems. Optimization **50**, 93–120 (2001)
8. Nguyen Quang, T.: Approches locales et globales bases sur la programmation DC et DCA pour des problemes combinatoires en variables mixtes 0-1: applications a la planification operationnelle, These de doctorat dirigee par Le Thi H.A Informatique Metz (2010)
9. Nguyen, V.V.: Methodes exactes basees sur la programmation DC et nouvelles coupes pour l optimisation en variables mixtes zero-un, These de doctorat dirigee par Pham Dinh T. and Le Thi H.A , LMI, INSA Rouen (2006)

10. Ta, A.S., Le Thi, H.A., Ha, T.S.: Solving relaxation orienteering problem using DCA-CUT. In: Modelling, Computation and Optimization in Information and Management Sciences, Advances in Intelligent Systems and Computing AISC 359, pp. 191–202 (2015)

A Combination of CMAES-APOP Algorithm and Quasi-Newton Method

Duc Manh Nguyen[✉]

Institute for Research and Applications of Optimization,
VinTech, Vingroup, Hanoi, Vietnam
v.manhnd14@vinoptima.org

Abstract. In this paper, we present an approach for combining the CMAES-APOP with a local search in order to make a hybrid evolutionary algorithm. This combination is based on the information of population size in the evolution process of the CMAES-APOP algorithm while the local search is quasi-Newton line search algorithm. We will give some conditions to efficiently active the local search inside CMAES-APOP. Some numerical experiments on multi-modal optimization problems will show the efficiency of proposed approach.

Keywords: Evolutionary computation · Hybrid evolutionary algorithm · Global optimization · CMA-ES · CMAES-APOP

1 Introduction

A large number of real-world problems can be considered as multi-modal optimization problems. Solving multi-modal optimization problems is very important for the decision-making processes. However, finding global or even good local solution is a challenge since the objective function may have several global solution, or it may have too many local solutions. The CMAES-APOP [11–13] is a variant of the well-known CMA-ES (Covariance Matrix Adaptation Evolution Strategy) algorithm [8] which adapts the population size in the CMA-ES to deal with multi-modal functions. This approach is inspired from a natural desire when solving an optimization problem as well as one prospect when using larger population size to search: *we want to see the decrease of the objective function*. In this approach, the non-decrease of objective function in a slot of $S = 5$ successive iterations is tracked to adjust the population size for the next S successive iterations. This implies that in each slot of S iterations we change the population size for searching. Consequently, the variation of population size takes a staircase form in iterations.

In fact, adapting the population size in the CMA-ES seems to be a right way for solving multi-modal functions, since the default value of population size in the CMA-ES, say $\lambda := \lfloor 4 + 3 \ln(n) \rfloor$, is known to be insufficient [7]. In the literature there are well-known and successful strategies for adapting the

© Springer Nature Switzerland AG 2020
H. A. Le Thi et al. (Eds.): ICCSAMA 2019, AISC 1121, pp. 64–74, 2020.
https://doi.org/10.1007/978-3-030-38364-0_6

population size, such as IPOP-CMA-ES [2] in which the CMA-ES is restarted with increasing population size by a factor of two whenever one of the stopping criteria is met; and BIPOP-CMA-ES strategy [6] in which two restart regimes are defined: one with large populations (IPOP part), and another one with small populations. In each restart, BIPOP-CMA-ES selects the restart regime with less function evaluations used so far. Ahrari and Shariat-Panahi [1] introduced a population size adaptation method for the CMA-ES which uses a measure, the oscillation of objective value of x_{mean}, to quantify multi-modality of the region being explored. This quantity is iteratively updated based on the optimization history and eventually used to increase the population size when facing highly multi-modal regions and vice versa. In [15], Nishida and Akimoto have presented another population size adaptation strategy for the CMA-ES that is based the estimation accuracy of the natural gradient.

In this paper, we introduce a new method for combining the CMAES-APOP algorithm with a local search to make hybrid evolutionary algorithm. This method is based on an important property of CMAES-APOP: it tries to increase the population size when recognizing the roughness of the objective function in order to allocate the position of optimal/good-local solution; whenever a such solution is detected, the population size is gradually decreased until the algorithm converges. The local search which will be used is the quasi-Newton line search. The motivation for this work is that integrating local search algorithms into a population-based algorithm sometimes helps to improve its performance, for continuous/combinatorial optimization problems [3–5,9,10,14,16].

The rest of paper is organized as follows. In Sect. 2, we present briefly the CMAES-APOP algorithm and some motivations of the proposed method. The method of combining local search algorithm with the CMAES-APOP is given in Sect. 3. Numerical experiments are reported in Sect. 4 while some conclusions and perspectives are discussed in Sect. 5.

2   The CMAES-APOP Algorithm

The CMAES-APOP [13] is a variant of the CMA-ES which adapts the population size in the CMA-ES to deal with the multi-modal functions. In the following, we give a short presentation of the CMAES-APOP (for more details, see [11,13]) in Algorithm 1. Here, some notations were used:

- iter: number of iterations.
- S: number of iterations in each slot.
- $f^{\text{med}} := \text{median}(f(\mathbf{x}_{i:\lambda}), i = 1, ..., \mu)$: the median of objective function of μ elite solutions in each iteration; $f_{\text{prev}}^{\text{med}}$ and $f_{\text{cur}}^{\text{med}}$ denote the medians in the previous and current iteration respectively.
- n_{up}: the number of times "$f_{\text{cur}}^{\text{med}} - f_{\text{prev}}^{\text{med}} > 0$" occurs during a slot of S iterations.
- t_{up}: the history of n_{up} in each slot recorded.
- no_{up}: the number of most recent slots we do not see the non-decrease.

The lines 7–14 in Algorithm 1 gather information of n_{up} during $S = 5$ iterations in the CMA-ES. After each S iterations, the lines 21–37 adapt the population size to the next S iterations based on the information of n_{up} and its history.

Algorithm 1. CMAES-APOP algorithm

1: **Input:** $\mathbf{m} \in \mathbb{R}^n, \sigma \in \mathbb{R}_+$
2: **Initialize:** $\mathbf{C} = \mathbf{I}, \mathbf{p}_c = 0, \mathbf{p}_\sigma = 0, \lambda = k_n \times \lambda_{\text{default}}$
3: **Set:** $\mu = \lfloor \lambda/2 \rfloor, w_i = \log(\mu + 0.5) - \log i, i = 1, ..., \mu, \mu_w = \frac{1}{\sum_{i=1}^{\mu} w_i^2}, c_c =$
$\frac{4}{n+4}, c_\sigma = \frac{\mu_w + 2}{n + \mu_w + 3}, c_1 = \frac{2}{(n+1.3)^2 + \mu_w}, c_\mu = \frac{2(\mu_w - 2 + \frac{1}{\mu_w})}{(n+2)^2 + \mu_w}, c_1 + c_\mu \leq 1, d_\sigma =$
$1 + 2 \max\left(0, \sqrt{\frac{\mu_w - 1}{n+1}} - 1\right) + c_\sigma$, iter $= 0, S = 5, r_{\max} = 30, n_{\text{up}} = 0, t_{\text{up}} = [\ ]$.
4: **while not terminate**
5: $\quad$ iter $=$ iter $+ 1$;
6: $\quad \mathbf{x_i} = \mathbf{m} + \sigma \mathbf{y_i}, \mathbf{y_i} \sim \mathbf{N(0, C)}$, for $i = 1, ..., \lambda$
7: $\quad$ **if** iter $= 1$
8: $\quad\quad f_{\text{prev}}^{\text{med}} \leftarrow \text{median}(f(\mathbf{x}_{i:\lambda}), i = 1, ..., \mu)$
9: $\quad$ **else**
10: $\quad\quad f_{\text{cur}}^{\text{med}} \leftarrow \text{median}(f(\mathbf{x}_{i:\lambda}), i = 1, ..., \mu)$
11: $\quad\quad$ **if** $f_{\text{cur}}^{\text{med}} - f_{\text{prev}}^{\text{med}} > 0$ $\quad$ //Check if f^{med} increases
12: $\quad\quad\quad n_{\text{up}} = n_{\text{up}} + 1$;
13: $\quad\quad$ **end**
14: $\quad$ **end**
15: $\quad f_{\text{prev}}^{\text{med}} \leftarrow f_{\text{cur}}^{\text{med}}$ $\quad$ //Reset $f_{\text{prev}}^{\text{med}}$
16: $\quad \mathbf{m} \leftarrow \sum_{i=1}^{\mu} w_i \mathbf{x}_{i:\lambda} = \mathbf{m} + \sigma \mathbf{y}_w$, where $\mathbf{y}_w = \sum_{i=1}^{\mu} w_i \mathbf{y}_{i:\lambda}$
17: $\quad \mathbf{p}_c \leftarrow (1 - c_c)\mathbf{p}_c + \mathbf{1}_{||\mathbf{p}_\sigma < 1.5\sqrt{n}||} \sqrt{(1 - (1 - c_c)^2)} \sqrt{\mu_w} \mathbf{y}_w$
18: $\quad \mathbf{p}_\sigma \leftarrow (1 - c_\sigma)\mathbf{p}_\sigma + \sqrt{(1 - (1 - c_\sigma)^2)} \sqrt{\mu_w} \mathbf{C}^{-\frac{1}{2}} \mathbf{y}_w$
19: $\quad \mathbf{C} \leftarrow (1 - c_1 - c_\mu)\mathbf{C} + c_1 \mathbf{p}_c \mathbf{p}_c^T + c_\mu \sum_{i=1}^{\mu} w_i \mathbf{y}_{i:\lambda} \mathbf{y}_{i:\lambda}^T$
20: $\quad \sigma \leftarrow \sigma \times \exp\left(\frac{c_\sigma}{d_\sigma}\left(\frac{||\mathbf{p}_\sigma||}{E||N(0,1)||} - 1\right)\right)$
21: $\quad$ **if** $(\text{mod}(\text{iter}, S) = 1)$ & $(\text{iter} > 1)$ $\quad$ //Adapting the population size
22: $\quad\quad t_{\text{up}} = [t_{\text{up}}; n_{\text{up}}]$; $\quad$ //History of n_{up}
23: $\quad\quad$ **if** $n_{\text{up}} > 1$
24: $\quad\quad\quad \lambda \leftarrow \left\lfloor \min\left(\exp\left(\frac{n_{\text{up}} \cdot (4 + 3\log(n))}{S \cdot \sqrt{\lambda - \lambda_{\text{default}} + 1}}\right), r_{\max}\right) \times \lambda \right\rfloor$;
25: $\quad\quad\quad \sigma \leftarrow \sigma \times \exp\left(\frac{1}{n}\left(\frac{n_{\text{up}}}{S} - \frac{1}{5}\right)\right)$; //Enlarge σ a little bit
26: $\quad\quad$ **elseif** $n_{\text{up}} = 0$
27: $\quad\quad\quad \text{no}_{\text{up}} = \text{length}(t_{\text{up}}) - \max(\text{find}(t_{\text{up}} > 0))$;
28: $\quad\quad\quad$ **if** $\lambda > 2\lambda_{\text{default}}$
29: $\quad\quad\quad\quad \lambda \leftarrow \max\left(\lfloor \lambda \times \exp(-\text{no}_{\text{up}}/10)\rfloor, 2\lambda_{\text{default}}\right)$;
30: $\quad\quad\quad$ **end**
31: $\quad\quad$ **end**
32: $\quad\quad$ **if** λ is changed $\quad$ //Only when $n_{\text{up}} > 1$ or $n_{\text{up}} = 0$
33: $\quad\quad\quad$ Update $\mu, w_{i=1...\mu}, \mu_w$ w.r.t the new population size λ
34: $\quad\quad\quad$ Update the parameters $c_c, c_\sigma, c_1, c_\mu, d_\sigma$
35: $\quad\quad$ **end**
36: $\quad\quad n_{\text{up}} \leftarrow 0$ $\quad$ //Reset n_{up} back to 0
37: $\quad$ **end**

The population size will be enlarged if $n_{\mathrm{up}} > 1$, and reduced if $n_{\mathrm{up}} = 0$ (i.e., there is no "going up" during S iterations) while it is not changed if $n_{\mathrm{up}} = 1$. The threshold of 1 is required for n_{up} to decide about the increase of the population size because apart from the roughness of objective function, the (randomly) sampling process may also affect on the information of n_{up}. It is worth noting that the median used to measure the non-decrease signal makes the CMAES-APOP algorithm invariant to scaling and shifting operators on the objective function.

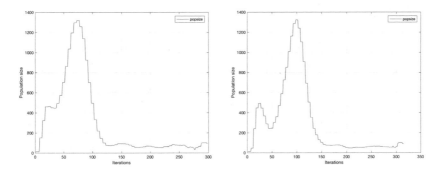

Fig. 1. Adapting the population size of CMAES-APOP in 20-D (Left: Rastrigin function, Right: Scale Rastrigin function).

When testing the performance of CMAES-APOP on some multi-modal functions with the conditions as in Sect. 4, and with the small initial population size $\lambda = \lambda_{\mathrm{default}}$ (i.e, set $k_n = 1$), and without the upper bound for the population size in the dimensions $n = 10, 20, 40$, we obtain high success rates (more than 80%) for all tests (see [11] for the details of experiment). Moreover, we found an important property of this algorithm: it tries to increase the population size to allocate the position of optimal/good-local solution; whenever an such solution is detected, the population size is gradually decreased until the algorithm converges. For instance, Fig. 1 shows the adaption of the population size in average over successful runs of the CMAES-APOP in the dimension 20.

This property and the high success rate in the experiment suggest us that it is possible to insert local searches during the process of CMAES-APOP when the algorithm locates the position of optimum to speed it up. In the next section, we will propose an approach to do that.

3   Combining Local Search with the CMAES-APOP

Using local search inside a global search algorithm is not too new. In the literature, there are several successful approaches using this technique for continuous/combinatorial optimization problems [3–5,9,10,14,16]. Nevertheless, how to use efficiently a local search strongly depends on the characteristics of global search algorithms, and we have to balance the exploration-exploitation trade-off in such frameworks. We normally have to answer the following questions:

- When/where should a local search be used?
- Which type of local search should be used? and what is the stopping condition for the local search?
- How many times should the local search be used?

In the Algorithm 2, we give a description of the combination of CMAES-APOP with a local search algorithm. It differs from the Algorithm 1 at only one point (lines 8–11 in Algorithm 2): it tries to use local search when some conditions are satisfied.

3.1   When/where Should a Local Search Be Used?

During the evolution process, we expect whenever the CMAES-APOP detects the position of optimal solution we will active a local search to get a quick convergence. Therefore, we will use a local search if the following conditions are all satisfied:

- The first one is based on the is_LS(hist_popsize) function (Algorithm 3) which can be described as follows: we use the information of population size in the last 40 iterations which corresponds to the array hist of 8 different population sizes (since the population size has a stair-case form in each 5 iterations). Then we fit the 8 data entries of hist with a quadratic function in the form $p1 \times t^2 + p2 \times t + p3$. The first condition holds if and only if $p1 < 0$, that is the population size is making a (locally) concave quadratic form.
- The second condition: the local search will be used after $\max(4 \times n, 40)$ iterations, where n is the problem dimension, since we need the population size to be quite large to spot the location of optimal solution.
- The third condition: at the current iteration the algorithm generates a better candidate than the best one recorded so far.

In this work, we limit the number of times using the local search: $\max_{LS} := 5$.

3.2   The Quasi-Newton Line Search Method

In this paper, we use the quasi-Newton line search (denoted by qN for short) as local search procedure. The qN is chosen since it has a super-linear convergence rate.

In optimization, the descent methods are classical techniques for exploring the neighborhood of the point $\mathbf{x}_{\text{best-so-far}}$. These approaches start from a point $\mathbf{x}_k$, construct a descent direction $\mathbf{d}_k$, then find an approximate solution for the line search problem:

$$\text{minimize}_{\alpha \geq 0} f(\mathbf{x}_k + \alpha \mathbf{d}_k). \tag{1}$$

When an approximate solution α^* of the problem (1) is found, we obtain a new point $\mathbf{x}_{k+1} = \mathbf{x}_k + \alpha^* \mathbf{d}_k$. Then, the previous procedure is repeated from that point. The gradient descent method uses the direction $\mathbf{d}_k = -\nabla f(\mathbf{x}_k)$, while the

---

Algorithm 2. CMAES-APOP algorithm using local search

1: **Input: m** $\in \mathbb{R}^n, \sigma \in \mathbb{R}_+$
2: **Initialize: C** $= \mathbf{I}, \mathbf{p}_c = 0, \mathbf{p}_\sigma = 0, \lambda = k_n \times \lambda_{\text{default}}$
3: **Set:** $\mu = \lfloor \lambda/2 \rfloor, w_i = \log(\mu + 0.5) - \log i, i = 1, ..., \mu, \mu_w = \frac{1}{\sum_{i=1}^{\mu} w_i^2}, c_c =$
   $\frac{4}{n+4}, c_\sigma = \frac{\mu_w + 2}{n + \mu_w + 3}, c_1 = \frac{2}{(n+1.3)^2 + \mu_w}, c_\mu = \frac{2(\mu_w - 2 + \frac{1}{\mu_w})}{(n+2)^2 + \mu_w}, c_1 + c_\mu \leq 1, d_\sigma = 1 +$
   $2 \max \left(0, \sqrt{\frac{\mu_w - 1}{n+1}} - 1 \right) + c_\sigma, \text{iter} = 0, S = 5, r_{\max} = 30, n_{\text{up}} = 0, t_{\text{up}} = [\ ], \text{hist\_popsize} =$
   $[\ ], n_{\text{LS}} = 0, \max_{\text{LS}} = 5.$
4: **while not terminate**
5:     iter = iter + 1;
6:     hist_popsize = [hist_popsize; λ];   //History of the population size
7:     $\mathbf{x_i} = \mathbf{m} + \sigma \mathbf{y_i}, \mathbf{y_i} \sim \mathbf{N(0, C)}$, for $i = 1, ..., \lambda$
8:     **if** (iter $>= \max(4 * n, 40)$) & ((mod(iter, S) == 1) || (mod(iter, S) == 3)) &
       is_LS(hist_popsize) & ($f(\mathbf{x}_{1:\lambda}) < f_{\text{best-so-far}}$) & ($n_{\text{LS}} < \max_{\text{LS}}$)
9:       • Do a local search with the starting point $\mathbf{x}_{1:\lambda}$. If the local search can find
         a better solution $\mathbf{x}_{\text{local}}^{\text{best}}$ than $\mathbf{x}_{1:\lambda}$, then we update $\mathbf{x}_{1:\lambda} \leftarrow \mathbf{x}_{\text{local}}^{\text{best}}$.
         Especially, if the local search finds an optimal solution, the whole
         algorithm is stopped.
10:      • $n_{\text{LS}} = n_{\text{LS}} + 1$;   //$n_{\text{LS}}$ is number of times using local search
11:    **end**
12:    **if** iter = 1
13:        $f_{\text{prev}}^{\text{med}} \leftarrow \text{median}(f(\mathbf{x}_{i:\lambda}), i = 1, ..., \mu)$
14:    **else**
15:        $f_{\text{cur}}^{\text{med}} \leftarrow \text{median}(f(\mathbf{x}_{i:\lambda}), i = 1, ..., \mu)$
16:        **if** $f_{\text{cur}}^{\text{med}} - f_{\text{prev}}^{\text{med}} > 0$ //Check if f^{med} increases
17:            $n_{\text{up}} = n_{\text{up}} + 1$;
18:        **end**
19:    **end**
20:    $f_{\text{prev}}^{\text{med}} \leftarrow f_{\text{cur}}^{\text{med}}$ //Reset $f_{\text{prev}}^{\text{med}}$
21:    $\mathbf{m} \leftarrow \sum_{i=1}^{\mu} w_i \mathbf{x}_{i:\lambda} = \mathbf{m} + \sigma \mathbf{y}_w$, where $\mathbf{y}_w = \sum_{i=1}^{\mu} w_i \mathbf{y}_{i:\lambda}$
22:    $\mathbf{p}_c \leftarrow (1 - c_c)\mathbf{p}_c + \mathbf{1}_{||\mathbf{p}_\sigma < 1.5\sqrt{n}||} \sqrt{(1 - (1 - c_c)^2)} \sqrt{\mu_w} \mathbf{y}_w$
23:    $\mathbf{p}_\sigma \leftarrow (1 - c_\sigma)\mathbf{p}_\sigma + \sqrt{(1 - (1 - c_\sigma)^2)} \sqrt{\mu_w} \mathbf{C}^{-\frac{1}{2}} \mathbf{y}_w$
24:    $\mathbf{C} \leftarrow (1 - c_1 - c_\mu)\mathbf{C} + c_1 \mathbf{p}_c \mathbf{p}_c^T + c_\mu \sum_{i=1}^{\mu} w_i \mathbf{y}_{i:\lambda} \mathbf{y}_{i:\lambda}^T$
25:    $\sigma \leftarrow \sigma \times \exp\left(\frac{c_\sigma}{d_\sigma} \left(\frac{||\mathbf{p}_\sigma||}{E||N(0,1)||} - 1 \right) \right)$
26:    **if** (mod(iter, S) = 1) & (iter > 1)
27:        $t_{\text{up}} = [t_{\text{up}}; n_{\text{up}}]$;   //History of n_{up}
28:        **if** $n_{\text{up}} > 1$
29:            $\lambda \leftarrow \left\lfloor \min\left(\exp\left(\frac{n_{\text{up}} \cdot (4 + 3\log(n))}{S \cdot \sqrt{\lambda - \lambda_{\text{default}} + 1}} \right), r_{\max} \right) \times \lambda \right\rfloor$;
30:            $\sigma \leftarrow \sigma \times \exp\left(\frac{1}{n} \left(\frac{n_{\text{up}}}{S} - \frac{1}{5} \right) \right)$; //Enlarge σ a little bit
31:        **elseif** $n_{\text{up}} = 0$
32:            $\text{no}_{\text{up}} = \text{length}(t_{\text{up}}) - \max(\text{find}(t_{\text{up}} > 0))$;
33:            **if** $\lambda > 2\lambda_{\text{default}}$
34:                $\lambda \leftarrow \max\left(\lfloor \lambda \times \exp(-\text{no}_{\text{up}}/10) \rfloor, 2\lambda_{\text{default}} \right)$;
35:            **end**
36:        **end**
37:        **if** λ is changed   //Only when $n_{\text{up}} > 1$ or $n_{\text{up}} = 0$
38:            Update $\mu, w_{i=1...\mu}, \mu_w$ w.r.t the new population size λ
39:            Update the parameters $c_c, c_\sigma, c_1, c_\mu, d_\sigma$
40:        **end**
41:        $n_{\text{up}} \leftarrow 0$ //Reset n_{up} back to 0
42:    **end**

Algorithm 3. The is_LS(hist_popsize) function

1: **function** y = is_LS(hist_popsize)
2:     y = 0;
3:     hist = hist_popsize((end-40):5:(end-1));
4:     popsize_model = fit((1:length(hist))', hist, 'poly2');
                //Fit 'hist' with a quadratic function: $p1 \times t^2 + p2 \times t + p3$
5:     **if** (popsize_model.p1 < 0) //popsize_model is concave quadratic
6:         y = 1;
7:     **end**
8: **end**

Newton's method takes $\mathbf{d}_k = -[\nabla^2 f(\mathbf{x}_k)]^{-1}\nabla f(\mathbf{x}_k)$, where $\nabla^2 f(\mathbf{x}_k)$ is the Hessian matrix of f at $\mathbf{x}_k$. Although Newton's method has a (locally) quadratic convergence rate, computing $\nabla^2 f(\mathbf{x}_k)$ requires a large amount of computation. The quasi-Newton methods avoid this disadvantage by using the observed behavior of $\mathbf{x}$ and $\nabla f(\mathbf{x})$ to make an approximation to the Hessian matrix. These methods normally have a super-linear convergence rate. Among many updating techniques, the Broyden-Fletcher-Goldfarb-Shanno (BFGS) formula is thought to be the most effective one:

$$H_{k+1} = H_k + \frac{\mathbf{q}_k\mathbf{q}_k^T}{\mathbf{q}_k^T\mathbf{s}_k} - \frac{H_k\mathbf{s}_k\mathbf{s}_k^T H_k^T}{\mathbf{s}_k^T H_k\mathbf{s}_k} \tag{2}$$

where T refers to the transpose and

$$\mathbf{s_k} = \mathbf{x}_{k+1} - \mathbf{x}_k, \tag{3}$$
$$\mathbf{q_k} = \nabla f(\mathbf{x}_{k+1}) - \nabla f(\mathbf{x}_k). \tag{4}$$

We can choose the starting matrix H_0 to be any symmetric positive definite matrix, for example, the identity matrix. The source code of the quasi-Newton line search is provided in the Matlab function fminunc. In fact, we will use this Matlab function with the option of quasi-Newton algorithm and with the default parameters.

4   Numerical Experiment

In this section, we test the performance of proposed algorithm on some test multimodal functions. We used the matlab implementation of CMA-ES[1], version 3.40.beta to make the proposed algorithm. The experiment is tested on a MacBook Air Intel(R) Core(TM) i5-5250U, CPU 1.60 GHz, RAM 8G. Table 1 summarizes the unconstrained multi-modal test problems while the initial parameters for the algorithms are given in Table 2.

All considered functions have a large number of local optima, are scalable in the problem dimension, and have a minimal function value of 0. The known

---

[1] https://www.lri.fr/~hansen/cmaes20091024.m.

Table 1. Test functions.

Name	Function
Rastrigin	$f_{\text{Rastrigin}}(\mathbf{x}) = 10n + \sum_{i=1}^{n}(x_i^2 - 10\cos(2\pi x_i))$
Scale Rastrigin	$f_{\text{RastScale}}(\mathbf{x}) = 10n + \sum_{i=1}^{n}((10^{\frac{i-1}{n-1}}x_i)^2 - 10\cos(2\pi 10^{\frac{i-1}{n-1}}x_i))$
Schaffer	$f_{\text{Schaffer}}(\mathbf{x}) = \sum_{i=1}^{n-1}(x_i^2 + x_{i+1}^2)^{0.25}[\sin^2(50(x_i^2 + x_{i+1}^2)^{0.1}) + 1]$
Griewank	$f_{\text{Griewank}}(\mathbf{x}) = \frac{1}{4000}\sum_{i=1}^{n}x_i^2 - \prod_{i=1}^{n}\cos\left(\frac{x_i}{\sqrt{i}}\right) + 1$
Ackley	$f_{\text{Ackley}}(\mathbf{x}) = 20 - 20 \cdot \exp\left(-0.2\sqrt{\frac{1}{n}\sum_{i=1}^{n}x_i^2}\right) + e - \exp\left(\frac{1}{n}\sum_{i=1}^{n}\cos(2\pi x_i)\right)$
Bohachevsky	$f_{\text{Bohachevsky}}(\mathbf{x}) = \sum_{i=1}^{n-1}(x_i^2 + 2x_{i+1}^2 - 0.3\cos(3\pi x_i) - 0.4\cos(4\pi x_{i+1}) + 0.7)$

global minimum is located at $\mathbf{x} = \mathbf{0}$. The performance of proposed algorithm is tested for dimensions $n = [10, 20, 40]$. The bound constraints for the Ackley function in $[-32.768, 32.768]^n$ are considered via quadratic penalty terms. That is $f_{\text{Ackley}}(\mathbf{x}) + \sum_{i=1}^{n}\theta(|x_i| - 32.768).(|x_i| - 32.768)^2$ will be minimized, where $\theta(x) = 1$ if $x > 0$ and $\theta(x) = 0$ if $x \leq 0$.

Table 2. Initial conditions.

Function	Initial point	σ
Rastrigin	$\mathbf{x}^0 = (5, ..., 5)$	2
Scale Rastrigin	$\mathbf{x}^0 = (5, ..., 5)$	22
Schaffer	$\mathbf{x}^0 = (55, ..., 55)$	220
Griewank	$\mathbf{x}^0 = (305, ..., 305)$	100
Ackley	$\mathbf{x}^0 = (15, ..., 15)$	25
Bohachevsky	$\mathbf{x}^0 = (8, ..., 8)$	23

There are 51 runs for each function in each dimension. Each run is stopped and regarded as successful if the objective value is smaller than $f_{\text{stop}} = 10^{-10}$ ($f_{\text{stop}} = 10^{-8}$ for the Schaffer function). We need some additional conditions for the Schaffer function: $\text{TolX} = 10^{-30}, \text{TolFun} = 10^{-20}, \text{TolHistFun} = 10^{-20}$. We denote CMAES-APOP using quasi-Newton method as a local search by CMAES-APOP-qN. The default parameters for fminunc function in Matlab is applied for the qN method. The performance of algorithms is measured by the average number of function evaluations to reach the target.

Table 3 shows the comparative results provided by the CMAES-APOP with its modified version combining the local search qN. We also compare these algorithms with the well-known IPOP-CMA-ES algorithm [2]. In this table, we take directly the results of CMAES-APOP and IPOP-CMA-ES from [11]. We observe that, in general the CMAES-APOP with local search does not always bring better performance. For example, on the Schaffer and Ackley functions, the integrated algorithm appears to be inferior. This is because the vicinity of optimal solution of this function is rugged; thus the proposed local search is not capable to find the optimal solution even though the CMAES-APOP has located its

Table 3. The comparative results between CMAES-APOP, CMAES-APOP-qN and IPOP-CMA-ES (n: problem dimension, SR: success rate, aRT (average Running Time) = number of function evaluations divided by the number of successful trials).

Function	n	CMAES-APOP		CMAES-APOP-qN		IPOP-CMA-ES
		SR	aRT	SR	aRT	aRT
Rastrigin	10	0.86	3.317e+04	0.90	2.328e+04	7.471e+04
	20	0.94	9.022e+04	0.94	7.338e+04	2.594e+05
	40	1.00	2.981e+05	1.00	2.709e+05	7.744e+05
Scale Rastrigin	10	0.80	3.527e+04	0.94	2.204e+04	8.498e+04
	20	0.80	1.111e+05	0.96	7.122e+04	2.794e+05
	40	0.90	3.427e+05	0.94	2.716e+05	9.179e+05
Schaffer	10	0.96	3.098e+04	0.96	3.298e+04	4.801e+04
	20	0.94	8.175e+04	0.96	9.189e+04	1.285e+05
	40	0.90	2.255e+05	0.90	2.591e+05	3.206e+05
Griewank	10	0.98	1.215e+04	1.00	7.659e+03	7.192e+03
	20	0.98	2.479e+04	0.98	1.752e+04	7.170e+03
	40	1.00	5.769e+04	1.00	3.795e+04	1.186e+04
Ackley	10	1.00	1.403e+04	1.00	1.471e+04	1.890e+04
	20	0.98	3.105e+04	1.00	3.252e+04	1.249e+05
	40	0.92	7.204e+04	0.94	7.379e+04	1.162e+06
Bohachevsky	10	1.00	1.002e+04	1.00	6.507e+03	5.947e+03
	20	1.00	2.397e+04	1.00	1.639e+04	1.813e+04
	40	0.98	5.536e+04	1.00	3.994e+04	4.537e+04

neighborhood. Nevertheless, we see that, on Rastrigin function, CMAES-APOP-qN runs significantly faster, about 1.25 times in average over 3 dimensions, than CMAES-APOP does. Especially on the Scale Rastrigin function, CMAES-APOP-qN runs significantly much faster, about 1.47 times, than CMAES-APOP does. In addition, CMAES-APOP-qN gives better performance than CMAES-APOP does, about 1.5 and 1.46 times on the Griewank, Bohachevsky functions respectively. Besides, CMAES-APOP-qN is slightly better than IPOP-CMA-ES on the Bohachevsky in the dimensions 20 and 40.

5   Conclusion

In this paper, we have presented an approach for using a local search, say the quasi-Newton line search method, inside the CMAES-APOP. The proposed algorithm is tested on some benchmark multi-modal functions. The numerical results show that this approach can improve the performance of CMAES-APOP in some cases. For the multi-modal functions having regular structure around the optimal solution, using qN as local search may bring a significant improvement. However, for the multi-modal function having rugged structure around the optimal solution, using qN as local search is insufficient and that may lead to the waste of function evaluations. Combining with another local search to overcome this drawback will be investigated in the future. Also, we will test our algorithm on other test problems.

References

1. Ahrari, A., Shariat-Panahi, M.: An improved evolution strategy with adaptive population size. Optimization **64**(12), 2567–2586 (2015)
2. Auger, A., Hansen, N.: A restart CMA evolution strategy with increasing population size. In: 2005 IEEE Congress on Evolutionary Computation, pp. 1769–1776 (2005)
3. Bagirov, A.-M., Rubinov, A.-M., Zhang, J.: Local optimization method with global multidimensional search. J. Glob. Optim. **32**(2), 161–179 (2005)
4. Chu, P.-C., Beasley, J.-E.: A genetic algorithm for the multidimensional knapsack problem. J. Heuristics 4(1), 63–86 (1998)
5. Cotta, C., Troya, J.-M.: A hybrid genetic algorithm for the 0–1 multiple knapsack problem. In: Artificial Neural Nets and Genetic Algorithms, pp. 250–254. Springer, Vienna (1998)
6. Hansen, N.: Benchmarking a bi-population CMA-ES on the BBOB-2009 function testbed. In: Proceedings of the 11th Annual Conference Companion on Genetic and Evolutionary Computation Conference: Late Breaking Papers, GECCO 2009, pp. 2389–2396 (2009)
7. Hansen, N., Kern, S.: Evaluating the CMA evolution strategy on multimodal test functions. In: Yao, X., et al. (eds.) Parallel Problem Solving from Nature - PPSN VIII, PPSN 2004. Lecture Notes in Computer Science, vol. 3242, pp. 282–291. Springer, Heidelberg (2004)
8. Hansen, N., Ostermeier, A.: Completely derandomized self-adaptation in evolution strategies. Evol. Comput. **9**(2), 159–195 (2001)
9. Lim, K. K., Ong, Y., Lim, M., Chen, X., Agarwal, A.: Hybrid ant colony algorithms for path planning in sparse graphs. Soft Comput. **12**(10), 981–994 (2008)
10. Merz, P., Freisleben, B.: Genetic local search for the TSP: new results. In: Proceedings of 1997 IEEE International Conference on Evolutionary Computation (ICEC 1997), pp. 159–164 (1997)
11. Nguyen, D.-M.: An adapting population size approach in the CMA-ES for multimodal functions. In: Proceedings of the Genetic and Evolutionary Computation Conference Companion (GECCO 2018), pp. 219–220. ACM, New York (2018)
12. Nguyen, D.-M.: Benchmarking avariant of the CMAES-APOP on the BBOB noiseless testbed. In: Proceedings of the Genetic and Evolutionary Computation Conference Companion (GECCO 2018), pp. 1521–1528. ACM, New York (2018)
13. Nguyen, D.-M., Hansen, N.: Benchmarking CMAES-APOP on the BBOB noiseless testbed. In: Proceedings of the Genetic and Evolutionary Computation Conference Companion (GECCO 2017), pp. 1756–1763. ACM, New York (2017)
14. Nguyen, D.-M., Le, T.-H.-A., Pham, D.-T.: Solving the multidimensional assignment problem by a cross-entropy method. J. Comb. Optim. **27**(4), 808–823 (2014)

15. Nishida, K., Akimoto, Y.: Population size adaptation for the CMA-ES based on the estimation accuracy of the natural gradient. In: Proceedings of the Genetic and Evolutionary Computation Conference 2016 (GECCO 2016), pp. 237–244. ACM, New York (2016)
16. Reeves, C.: Hybrid genetic algorithms for bin-packing and related problems. Ann. Oper. Res. **63**(3), 371–396 (1996)

A Triple Stabilized Bundle Method for Constrained Nonconvex Nonsmooth Optimization

André Dembélé[1], Babacar M. Ndiaye[2(✉)], Adam Ouorou[3], and Guy Degla[1]

[1] Institute of Mathematics and Physical Sciences, University of Abomey Calavi,
Porto-Novo, Republic of Benin
{andre.dembele,gdegla}@imsp-uac.org
[2] Laboratory of Mathematics of Decision and Numerical Analysis,
University of Cheikh Anta Diop - Dakar, 45087 Dakar-Fann, Senegal
babacarm.ndiaye@ucad.edu.sn
[3] Orange Labs Research, Avenue de la République, 92320 Chatillon, France
adam.ouorou@orange.com

Abstract. In this paper, we provide an exact reformulation of Non-smooth Constrained optimization Problems (NCP) using the Moreau-Yosida regularization. This reformulation allows the transformation of (NCP) to a sequence of convex programs of which solutions are feasible for (NCP). This sequence of solutions of auxiliary programs converges to a local solution of (NCP). Assuming Slater constraint qualification and basing on an exact penalization, our reformulation combined with a nonconvex proximal bundle method provides a local solution of (NCP). Our bundle method allows a strong update of the level set, may reduce significantly the number of null-steps and gives a new stopping criterion. Finally, numerical simulations are carried out.

Keywords: Proximal algorithm · Bundle method · Nonconvex optimization · Nonsmooth optimization · Reformulation

1 Introduction

In this paper, we consider nonsmooth Nonconvex Constrained Problems (NCP) which may be stated as

$$\textbf{(NCP)} \quad \begin{aligned} \min \quad & f^0(x) \\ \text{s.t.} \quad & f^j(x) \leq 0, \quad j \in J = \{1, ..., m\} \\ & x \in X. \end{aligned} \tag{1}$$

where: m, n are integers, $f^j : \mathbb{R}^n \to \mathbb{R}$; for $j \in J \cup \{0\}$, are lower-$\mathcal{C}^2$; and X is a bounded polyhedron of $\mathbb{R}^n$.

The classical bundle method is inspired by Kelley's cutting-plane method proposed in 1960s [1] and can be applied to both convex and nonconvex optimization problems. For more details, we refer the reader to [2,3].

© Springer Nature Switzerland AG 2020
H. A. Le Thi et al. (Eds.): ICCSAMA 2019, AISC 1121, pp. 75–87, 2020.
https://doi.org/10.1007/978-3-030-38364-0_7

Two major practical issues in the bundle methods are the question of limiting the bundle size and reducing the number of iterations of null-steps.

We introduce a new reformulation that allows designing a proximal point algorithm for our problem (1). Consider the following assumptions.

Assumption 1. The objective function and constraint functions of (1) are *lower-*$\mathcal{C}^2$ on the same open and bounded set $\mathcal{O}$ which contains X.

Assumption 1 ensures the existence of η such that functions $f^j + \eta|.|$, for $j = 0, ..., m$ are strictly convex and thus allows to make a convex reformulation for (1).

Assumption 2. The initial point is not a local solution for (**NCP**).

When the point initial is a local maximum, the proximal point method may fail to find a descent direction. In the nonsmooth nonconvex case, it may not be easy to show that $0 \in \partial f(x)$. And, to avoid being tempted to check that the considered point is a local maximum (i.e. 0 is a subgradient), it is better to start from an infeasible point.

Assumption 2 ensures global convergence (not depending on the initial point). Moreover, since (NCP) is nonconvex, we assume that:

Assumption 3. The feasible set of problem (1) has a non-empty interior.

From the lower-$\mathcal{C}^2$ property, this Assumption allows having a descent direction from any feasible solution (which is not a critical point of (1)). Therefore, it ensures to have a KKT point of the primal problem (NCP).

The method proposed in this work is a new proximal point method for constrained nonsmooth nonconvex optimization. The new method is combined with a bundle method to design a new triple stabilized bundle method.

This type of problem has been recently studied by Yang et al. in [3] and a nonconvex bundle method is proposed. Thanks to an exact penalty function, in [3], the problem is transformed into a parametric unconstrained problem whose local convexity requires a more stringent special Slater constraint qualification. In the nonconvex bundle method, this local convexity is necessary to have a local solution.

In this paper, we present a new reformulation of the problem that guarantees convergence and then, takes advantage of the known results (such as those of bundle methods) on unconstrained convex problems to solve (NCP). In fact, (NCP) will be converted to a parametric unconstrained optimization problem that is convex for sufficiently large parameter tactically detected by a nonconvex bundle method.

The paper is organized as follows. Under certain assumptions, a bundle method is described in Sect. 2. In Sect. 3, we provide numerical results for the proposed algorithms applied to a problem in wireless cellular networks. We end the paper with our conclusion and some discussions in Sect. 4.

2   A Nonconvex Nundle Method

This section presents a proximal bundle method for our problem of interest (1). All results that will follow, will be under the assumptions (1), (2) and (3) presented in Sect. 1.

2.1   Work Model

For a point $\bar{x} \in \mathbb{R}^n$ and the scalars $c > 0$ and $\eta > 0$, let us define, for the optimization problem $(CP_{\bar{x}})$, the exact penalty function by $\psi_{c,\eta,\bar{x}} : \mathbb{R}^n \longrightarrow \mathbb{R}$,

$$\psi_{c,\eta,\bar{x}}(x) = f^0(x) + \frac{\eta}{2}|x - \bar{x}|^2 + cF_{\bar{x},\eta}(x)^+ \tag{2}$$

where: $F_{\bar{x},\eta_k}(x)^+ := \max\left\{0, f^1(x) + \frac{\eta_k}{2}|x - x^k|^2, ..., f^m(x) + \frac{\eta_k}{2}|x - \bar{x}|^2\right\}$

Remark 1. This penalty function is better than the one given by

$$f^0(x) + c\max\left\{0, f^1(x) + \frac{\eta_k}{2}, ..., f^m(x)\right\} + \eta\frac{c+1}{2}|\cdot - \bar{x}|^2$$

which penalizes too much the steps of descents.

Suppose we have generated the points y^i, $i \in I_k = \{1, ..., k\}$, where $x^k = y^k$ is the last obtained and that $\psi_{c,\eta,\bar{x}}(y^i)$ and $v^i \in \partial\psi_{c,\eta,\bar{x}}(y^i)$ have already been computed. We consider the following piecewise linear function (or model)

$$\check{\psi}_{c_k,\eta_k,x^k}(x) = \max_{i \in I_k}\{\psi_{c_k,\eta_k,\bar{x}}(y^i) + \langle v^i, x - y^i\rangle\} \tag{3}$$

where c_k grows and η_k is updated such that $\check{\psi}_{c_k,\eta_k,x^k}$ is convex on $\{y^1, ..., y^k\}$. This model can be reritten as:

$$\check{\psi}_{c_k,\eta_k,x^k}(x) = \psi_{c_k,\eta_k,\bar{x}}(x^k)+$$
$$\max_{i \in I_k}\left\{\psi_{c_k,\eta_k,\bar{x}}(y^i) + \langle v^i, x^k - y^i\rangle - \psi_{c_k,\eta_k,\bar{x}}(x^k) + \langle v^i, x - x^k\rangle\right\} \tag{4}$$

Thus,

$$\check{\psi}_{c_k,\eta_k,x^k}(x) - \psi_{c_k,\eta_k,\bar{x}}(y^k) = \max_{i \in I_k}\left\{-\alpha_{c_k,\eta_k}(x^k, y^i) + \langle v^i, x - x^k\rangle\right\}, \tag{5}$$

where: $\alpha_{c_k,\eta_k}(x^k, y^i) = \psi_{c_k,\eta_k,\bar{x}}(x^k) - [\psi_{c_k,\eta_k,\bar{x}}(y^i) + \langle v^i, x^k - y^i\rangle]$ is the linearization error between y^i et x^k. Since $\psi_{c_k,\eta_k,\bar{x}}$ is convex on $\{y^1, ..., y^k\}$ and $v^i \in \partial\psi_{c_k,\eta_k,\bar{x}}(y^i)$, for $i \in I_k$, then we have

$$\psi_{c_k,\eta_k,\bar{x}}(x) \geq \psi_{c_k,\eta_k,\bar{x}}(y^i) + \langle v^i, x - y^i\rangle, \quad \forall\, x \in \{y^1, ..., y^k\}. \tag{6}$$

In particular for $x = x^k$ in (6), (2.1) gives us

$$\alpha_{c_k,\eta_k}(x^k, y^i) \geq 0 \text{ and } -\alpha_{c_k,\eta_k}(x^k, y^i) = \check{\psi}_{c_k,\eta_k,x^k}(x^k) - \psi_{c_k,\eta_k,\bar{x}}(y^k) \leq 0 \tag{7}$$

Now, we consider the function

$$\hat{\psi}_{c_k,\eta_k,\bar{x}}(x) = \max_{i\in I_k}\big\{-\alpha_{c_k,\eta_k}(x^k,y^i) + \langle v^i, x - x^k\rangle\big\} \tag{8}$$

and at iteration k, the next point generated is the solution of quadratic program $(\mathcal{QP}_k)$ defined by

$$(\mathcal{QP_k})\ \text{s.t.}\ \begin{array}{c}\min\ \hat{\psi}_{c_k,\eta_k,\bar{x}}(x) + \frac{\mu_k}{2}||x - x^k||^2\\ \breve{\psi}_{c_k,\eta_k,\bar{x}}(x) \leq l_k\\ x \in X\end{array} \tag{9}$$

Let us write more explicitly the terms in $(\mathcal{QP_k})$. For $i \in I_k$, let

$$t^i \in \partial\big(\max\{f^1(y^i),...,f^m(y^i)\}\big),\quad g^i \in \partial f(y^i),\quad \delta_i^k := \begin{cases} 1 & \text{if } F_k(y^i)^+ \geq 0 \\ 0 & \text{otherwise}\end{cases}$$

$$d_i^k = ||y^i - x^k||^2/2 \qquad \text{and} \qquad \Delta_i^k = y^i - x^k$$

For f^0 and $F_{k\,+} = \max\{f^1, f^2, ..., f^m\}$, the respective linearization errors $e_{f,i}^i$ and $E_{F_k,i}^k$ between x^k and y^i are defined as

$$e_{f^0,i}^i = f^0(x^k) - f^0(y^i) - \langle g^i, x^k - y^i\rangle,\quad E_{F_k,i}^k = F_k(x^k)_+ - F_k(y^i)_+ \tag{10}$$
$$+ \langle \delta_i^k t^i{}^+, x^k - y^i\rangle$$

It is known that if the function f^0 is convex on the line segment joining x^k and y^i, then $e_{f^0,i}^i$ is positive and $g^i \in \partial_{e_{f,i}^i} f^0(x^k)$. The functions f^0 and f^j are nonconvex but $f^0 + \frac{\eta_k}{2}|\cdot - x^k|^2$ and $f^j + \frac{\eta_k}{2}|\cdot - x^k|^2$ are locally convex so that the previous relationship can be applied. We point out that all functions involved in our problem of interest 1, i.e., f^0 and f^j, are regular. Thus, one can check that, for all $i \in I_k,\ \delta_i^k t^i \in \partial F_{\eta_k,x^k}(y^i)_+ \quad \forall t^i \in \partial\big(\max\{f^1(y^i),...,f^m(y^i)\}\big),\ s_k^i = g^i + \eta_k \Delta_i^k + c_k \delta_i^k(t^i + \eta_k \Delta_i^k) \in \partial \breve{\psi}_{c_k,\eta_k,x^k}(y^i),\ \alpha_{c_k,\eta_k}(x^k,y^i) = e_{f,i}^k + c_k E_{F_k,i}^k + \eta_k(1 + c_k \delta_i^k)d_i^k$. Therefore, we can be rewriten $(\mathcal{QP_k})$ as

$$\min_x\ \nu + \frac{\mu_k}{2}||x - x^k||^2$$
$$(\mathcal{QP_k})\ \text{s.t.}\ -\big(e_{f,i}^k + c_k E_{F_k,i}^k\big) - \eta_k\big(1 + c_k \delta_i^k\big)d_i^k$$
$$+ \langle g^i + \eta_k(y^i - x^k) + c_k.\delta_i^k(t^i + \eta_k \Delta_i^k), x - y^i\rangle \leq \nu,\ i \in I_k \tag{11}$$
$$\nu \leq l_k,\ \nu \in \mathbb{R},\ x \in X. \tag{12}$$

We have this analogous result of Proposition in [4,5]; with an appropriate choise of l_k. Let's recall that:

1. X is a polyhedron of $\mathbb{R}^n$, so as in [4] we can determine the real numbers α_i^k and β_k as a solution to a minimization of a quadratic function on the unit simplex of $\mathbb{R}^{|I_k|+1+p}$ where p is the number of faces of X;

2. $\psi_{c,\eta_k,\bar{x}}$ becomes convex whenever the convexification parameter $\eta_k \geq \eta = \max\{\bar{\eta}_0, ..., \bar{\eta}_m\}$.

Let:

$$I_k^{act} = \{i \in I_k \ : \ \alpha_i^k > 0\}, \quad e_i^k = e_{f,i}^k + c_k E_{F_k,i}^k, \quad e^k = \sum_{i \in I_k^{act}} \alpha_i^k e_i^k, \qquad (13)$$

$$s^k = \sum_{i \in I_k^{act}} \alpha_i^k \big(g^i + \eta_k \Delta_i^k + c_k \delta_i^k(t^i + \eta_k \Delta_i^k)\big) = \sum_{i \in I_k^{act}} \alpha_i^k s_k^i, \qquad (14)$$

$$d^k = \sum_{i \in I_k^{act}} \alpha_i^k d_i^k, \quad d_{strict}^k = \sum_{i \in I_k^{act}} \alpha_i^k \delta_i^k d_i^k, \quad \Delta^k = \sum_{i \in I_k^{act}} \alpha_i^k \Delta_i^k, \qquad (15)$$

$$\varepsilon_k = e^k + \eta_k(d^k + c_k d_{strict}^k). \qquad (16)$$

Define the aggregate piece

$$\bar{\psi}_{c_k,\eta_k,x^k}(x) := \check{\psi}_{c_k,\eta_k,x^k}(y^{k+1}) + \langle s^k + v^k, x - y^{k+1} \rangle \qquad \forall x \in \mathbb{R}^n \qquad (17)$$

and the aggregate error $\bar{\varepsilon}_k = \psi_{c_k,\eta_k,x^k}(x^k) - \check{\psi}_{c_k,\eta_k,x^k}(x^k)$. Let us set

$\varepsilon_k^\psi = \psi_{c_k,\eta_k,x^k}(x^k) - \psi_{c_k,\eta_k,x^k}(y^{k+1})$, the real progress

$\varepsilon_k^y = \psi_{c_k,\eta_k,\bar{x}}(x^k) - \check{\psi}_{c_k,\eta_k,x^k}(y^{k+1})$, the expected or predicted progress
(related to the solution)

$\varepsilon_k^l = \hat{\psi}_{c_k,\eta_k,x^k}(x^k) - l_k$, the expected progress related to level parameter l_k

Proposition 1. *If $\eta_k \geq \eta$ then,*

1. $s^k + v^k \in \partial \hat{\psi}_{c_k,\eta_k,x^k}(y^{k+1})$ *and* $s^k + v^k \in \partial \check{\psi}_{c_k,\eta_k,x^k}(y^{k+1})$
2. $s^k + v^k \in \partial_{\bar{\varepsilon}_k} \psi_{c_k,\eta_k,x^k}(x^k)$, *where:* $\bar{\varepsilon}_k = \varepsilon_k^y - \frac{1}{\mu_k}\|s^k + v^k\|^2 \geq 0$
3.

$$\begin{aligned} \psi_{c_k,\eta_k,x^k}(x) &\geq \bar{\psi}_{c_k,\eta_k,x^k}(x) \\ &= \hat{\psi}_{c_k,\eta_k,x^k}(y^{k+1}) + \langle s^k + d^k, x - y^{k+1} \rangle \qquad (18) \\ &= \psi_{c_k,\eta_k,x^k}(x^k) + \langle s^k + v^k, x - x^k \rangle - \bar{\varepsilon}_k, \ \forall x \in \mathbb{R}^n. \end{aligned}$$

Points 2 and 3 are analogous to Proposition 2.3 from [5].

In the nonconvex case, the linearization error may be negative. A convex cut can remove one portion of a feasible set (containing an optimal solution) and also lead to the emptiness of subproblem's feasible set. To correct this, some methods use the absolute value of the error to generate cuts. Here we make as in the "classical" proximal bundle method, and make sure to update correctly the convexification parameter η_k. In [3], the linearization error is $\tilde{e}_i^k + \eta d_i^k$, here it is $e_i^k + \eta_k(1 + c_k \delta_i^k) d_i^k$ (e_i^k instead of $\tilde{e}_i^k$ because our penalty functions are different). We have:

$$g^i + \eta_k \Delta_i^k + c_k \delta_i^k (t^i + \eta_k \Delta_i^k) \in \partial_{e_i^k + \eta_k(1+c_k\delta_i^k)d_i^k} \hat{P}_k(x^k)$$

$$\text{if} \quad e_i^k + \eta_k(1 + c_k\delta_i^k)d_i^k \geq 0 \qquad \forall i \in I_k. \quad (19)$$

We can ensure that the error is nonnegative by checking that $\eta_k \geq \eta_k^{min}$ where

$$\eta_k^{min} = \max_{\substack{i \in I_k,\ d_i^k \neq 0, \\ e_i^k + \eta_k(1+c_k\delta_i^k)d_i^k \leq 0}} \frac{-e_i^k}{(1+c_k\delta_i^k)d_i^k} = \max_{\substack{i \in I_k,\ d_i^k \neq 0, \\ e_i^k + \eta_k(1+c_k\delta_i^k)d_i^k \leq 0}} \frac{-\left(e_{f,i}^k + c_k E_{F_k,i}^k\right)}{(1+c_k\delta_i^k)d_i^k}$$

$$(20)$$

At the $k-1$ th iteration, the past information to update the model ψ_{c_k,η_k,x^k} is stored in $\mathcal{B}f_k \cup \mathcal{B}q_k$ defined from

$$\mathcal{B}f_k = \left\{ f(y^i),\ F_{\eta_k,x^k}(y^i)^+,\ g^i \in \partial f(y^i),\ \delta_i^k t^i \in \partial F_{\eta_k,x^k}(x^k)^+ \ :\ i \in I_k \right\} \quad (21)$$

$$\mathcal{B}q_k = \left\{ d_i^k = ||y^i - x^k||^2/2,\ \Delta_i^k = y^i - x^k \ :\ i \in I_k \right\}. \quad (22)$$

Proposition 2. *1.* $d_i^{k+1} = d_i^k + ||x^{k+1} - x^k||^2/2 - \langle \Delta_i^k, x^{k+1} - x^k \rangle$
2. $\Delta_i^{k+1} = \Delta_i^k + x^k - x^{k+1}$
3. $e_{f,i}^{k+1} = e_{f,i}^k + f(x^{k+1}) - f(x^k) - \langle g^i, x^{k+1} - x^k \rangle$
4. $E_{F_{k+1},i}^{k+1} = E_{F_k,i}^k + E_{F_{k+1},k}^{k+1} + \bar{E}_{F_k,i}^k$ *where:*

$$\bar{E}_{F_k,i}^k = K_i^k \max \left\{ f^1(x^k), ..., f^m(x^k) \right\}$$
$$- \left(K_i^k \max \left\{ f^1(y^i), ..., f^m(y^i) \right\} + \langle K_i^k t^i, x^k - y^i \rangle \right)$$

and K_i^k *is given at by (23).*

$$K_i^k = \begin{cases} 1 \text{ if } F_k(y^i)^+ < 0 \text{ and } F_{k+1}(y^i)^+ \geq 0 \\ 0 \text{ otherwise,} \end{cases} \quad (23)$$

2.2   Convergence

From a certain iteration k_0, the convexification parameter η_k does not change. We can, therefore, analyze Algorithm 1 as if it deals with a convex optimization problem. The convergence of proximal bundle methods requires that $\{\mu_k\}$ be bounded below (see [4,6]). And, through the update rule in step 12. of Algorithm 1, these requirements are met. Let $\bar{\eta}$ be the smallest scalar such that $f^j + \frac{\eta_k}{2}|\cdot - x|^2$, for $j \in \{0, 1, \ldots, m\}$, is strictly convex on a bounded set that contains $\mathcal{F}((NCP))$ and the sequence of optimal solutions generated by Algorithm 1. Denote by $\lfloor \cdot \rfloor$ the integer part of a scalar.

Proposition 3. *From an initial value* η_0 *of* η_k *with* $\eta_0 < \bar{\eta}$, *no more than* $\left\lfloor \dfrac{\log(\bar{\eta}) - \log(\eta_0)}{\log(\Gamma_\eta)} \right\rfloor + 1$ *iterations of convexification should be required (see 24).*

The relations: $s^k + v^k \in \partial_{\bar{\varepsilon}_k} \psi_{c_k, \eta_k, x^k}(x^k)$ *and* $s^k + v^k \in \partial \hat{\psi}_{c_k, \eta_k, x^k}(y^{k+1})$ $(**)$ (see statement 1. in Proposition 1) allow getting the convergence result. The limit point $x^\star$ of the sequence x^k satisfies $0 \in \partial \hat{\psi}_{c_k, \eta_k, x^\star}(x^\star)$ and, in view of $(**)$, $0 \in \partial \psi_{c_k, \eta_k, x^\star}(x^\star)$. Since the penalty parameter c_k tends to $+\infty$, then a local solution to the original problem (1) is obtained.

Algorithm 1

1. Select m_l, $m_f \in]0, 1[$, $\Gamma_c > 1$, $\Gamma_\eta > 1$, the tolerances tol_Δ, tol_g, tol_R, $tol_{\bar{\varepsilon}}$, and the penalty parameter c_0.
   Set $k = 1$, $I_k = \{1\}$, an initial point $x^1 = y^1$, compute $g^1 \in \partial f^0(y^1)$, $\delta_1^1 t^1 \in \partial F_{\eta_1, x^1}(y^1)_+$.
   Set $d_1^1 = 0$, $\Delta_1^1 = 0$, select η_1 and μ_1, compute $f^0(x^1)$ and $F_{\eta_1, x^1}(x^1)^+$ update $\mathcal{B}f_k \cup \mathcal{B}q_k$ (given in (21) and (22)).
2. Set $\psi_k^{low} = -\infty$, select $\varepsilon_1^l \geq 0$
3. Set $\varepsilon_k^\Delta = \psi_{c_1, \eta_1, x^1}(x^k) - \psi_k^{low}$
4. If $\epsilon_k^\Delta \leq tol_\Delta$ and $F_{\eta_k, x^k}(x^k)^+ \leq tol_R$, stop ($x^k$ is a solution).
5. Set $l_k = -\varepsilon_k^l$.
6. If $(\mathcal{QP}_k)$ has no solution, set

$$\psi_k^{low} = \psi_{c_k, \eta_k, x^k}(x^k) + l_k, \ \varepsilon_k^l = -(1 - m_l)l_k, \text{ go to step 3.}$$

7. Solve the quadratic subproblem $(\mathcal{QP}_k)$ (11) to get (y^{k+1}, ν_{k+1}) and the multiplier β_k associated to level constraint, set $\sum_{i \in I_k^{act}} \alpha_i^k = \beta_k + 1$, $\varepsilon_k^y = -\nu_{k+1}$
   $(||s^k + v^k||^2 = \mu_k^2 ||y^{k+1} - x^k||^2)$, $\bar{\varepsilon}_k = \varepsilon_k^y - 1/\mu_k ||s^k + v^k||^2$.
8. Update the convexification parameter, set

$$\begin{cases} \eta_{k+1} = \eta_k & \text{if } \eta_k \geq \eta_k^{min} \text{ (given in (20))} \\ \eta_{k+1} = \Gamma_\eta \eta_k & \text{otherwise and go to step7.} \end{cases} \qquad (24)$$

9. If $\bar{\varepsilon}_k \leq tol_{\bar{\varepsilon}}$, $||s^k + v^k||^2 \leq tol_g$ ($s^k + v^k$ is an ε-subgradient of ψ_{c_k, η_k, x^k} at y^{k+1} (see part 2 of Proposition 1)) and $F_{\eta_k, y^{k+1}}(y^{k+1}) \leq tol_R$, stop ($y^{k+1}$ is a solution).
10. Set $c_{k+1} = \Gamma_c c_k$.
11. Compute $\psi_{c_k, \eta_k, x^k}(y^{k+1})$, set $\psi_{k+1}^{low} = \psi_k^{low}$.
12. If $\psi_{c_k, \eta_k, y^{k+1}}(y^{k+1}) \leq \psi_{c_k, \eta_k, x^k}(x^k) - m_f \varepsilon_k^y$, set $\mu_{k+1} = \mu_k(\beta_k + 1)$.

$$x^{k+1} = y^{k+1}, \quad \varepsilon_{k+1}^l = \min\left\{\varepsilon_k^l, (1 - m_l)\left(\psi_{c_k, \eta_k, x^k}(x^{k+1}) - \psi_{k+1}^{low}\right)\right\} \qquad (25)$$

Otherwise, set $x^{k+1} = x^k$, $\mu_{k+1} = \mu_k$, $\eta_{k+1} = \eta_k$, $\varepsilon_{k+1}^l = \varepsilon_k^l$
13. Set $\mathcal{B}f_{k+1} \cup \mathcal{B}q_{k+1}$: compute $g^{k+1} \in \partial f^0(y^{k+1})$, $\delta_i^{k+1} t^i \in \partial F_{\eta_{k+1}, x^{k+1}}(y^{k+1})$ $e_{f,i}^{k+1}$, $E_{F_{k+1},i}^{k+1}$, d_i^{k+1}, Δ_i^{k+1} (given in 1 - 4 of Proposition 2).
14. Update the model, set $\hat{\psi}_{c_{k+1}, \eta_{k+1}, x^{k+1}}(x) \geq \max\{\hat{\psi}_{c_k, \eta_k, x^k}(x), \bar{\psi}_{c_k, \eta_k, x^k}(x)\}$ (see (17)), increase k by 1, go to step 3.

3   Numerical Experiments

The bundle method is implemented in *python* [9] using *CPLEX 12.5* [8]. The obtained solutions are compared with those obtained with *Couenne* [7] coupled with *Pyomo* [10]. Through this implementation, we compare different versions of the algorithm (named *proximal-level*) according to the rules for updating μ_k (named *proximal*) and rules for a_k and b_k (named *level*).

For *proximal*, we have *special (sp)*: corresponds to the rule of μ_k, based on *Lagrangian formulation* of *Hestenes* and *Powell*, *natural (na)*: the rule where μ_k is multiplied by a constant $\Gamma_\mu > 1$ each time we have a *null step*, *constant (co)*: μ_k remains constant (tightening the bounds a_k and b_k is enough to improve the proximity with the last *serious-step*), *classical (cl)*:μ_k is set as the maximum of the dual variables associated to the constraints (except the level constraint), *without (wt)*:$\mu_k = 0$.

For *level*, we have *classical (cl)*: the model includes the update rule of a_k and b_k, *without (wt)*: a_k does not change.

The *without-without* and *constant-without* are not considered. The update *classical* of μ_k, used in Algorithm 1, is done when we get a serious-step. For the *special*, *natural* and *constant* rules, there is an update when we get a *null step*. This allows, when μ_k increases (in *special* and *natural*), to decrease the descent step from the next iteration. *'without-classical'* is a *Kelley* cutting plane method which includes the update.

For the different versions of the bundle method ($\mu_k \neq 0$), a condition will be added to the one given in Algorithm 1 in step 8., it's: $b_k \geq -\epsilon_k$.

In the case where *level = classical* and $\mu_k \neq 0$, we add to the stop condition in step 8., the condition: $b_k - a_k \leq \epsilon_k$ and $b_k \geq -\epsilon_k$.

And when $\mu_k = 0$ (i.e. *proximal = without*), we consider only the next stop condition: $b_k - a_k \leq \epsilon_k$ and $b_k \geq -\epsilon_k$.

The *proximal-level* versions of Algorithm 1, where *proximal ≠ without*, use the same parameters. For versions *proximal-level* where ($\mu_k = 0$), except $\bar{\epsilon}_k$ update, we always consider the same parameters. When $\mu_k \neq 0$, $\bar{\epsilon}_k$ becomes

$$\bar{\epsilon}_k = \epsilon_k^y - \frac{1}{\mu_k}||s^k + v^k||^2 \geq 0 \quad \text{(see Proposition 1)}.$$

For the case where $\mu_k = 0$ (ie *without-classical*) which corresponds to our version of the *Kelley* cut, we take $\bar{\epsilon}_k = b_k - a_k$.

The parameters are initialized as follows: $m_f = 0.5$, $m_l = 0.2$, $c_0 = 1.2$ and $\Gamma_c = 1.5$, $\eta_0 = 1.5$ and $\Gamma_\eta = 1.2$, $\mu_0 = 0.1$, $\Gamma_\mu = 1.5$, $\mu_{\min} = 10^{-6}$, $tol_\Delta = 10^{-6}$, $tol_g = 10^{-6}$, $tol_R = 10^{-6}$, $tol_{\bar{\epsilon}} = 10^{-6}$, $\bar{\epsilon}_0 = 10^6$, $a_0 = -10^3$ (when $\mu_0 \neq 0$, $a_0 = \nu_0$, the value of ν after the first iteration), $b_0 = -40$, $\epsilon_l = 10^{-3}$ and $\bar{\epsilon}_l = 10^{-4}$, the initial point $x^0 = [-100, ..., -100]$.

3.1   Test Problems

The problems used for these tests are instances of channel allocation and user association problem in wireless cellular networks which is described in [11]. The quadratic reformulation of this problem, denoted by (**sinr**), is expressed below in the following page; where, with $n, m, k, q \in \mathbb{N}$. We have:

$$N = \{1, ..., n\}, M = \{1, ..., m+1\}, K = \{1, ..., \kappa\} \text{ and } Q = \{1, ..., q\}.$$

The nonconvex quadratic constraints are:

$$\gamma_{ijk} N_0 + \sum_{j'=0, j' \neq j}^{m} P_{j'k} \gamma_{ijk} G_{ij'k} = P_{jk} G_{ijk}, \quad i \in N, j \in M, k \in K.$$

$$\min \sum_{i=1}^{n} \sum_{j=1}^{m} \beta_{ij}$$

$$\text{s.t. } \sum_{j=1}^{m} x_{ij} = 1, \qquad\qquad i \in N$$

$$w_{ijk} \leq \phi_q y_{jk}, \qquad\qquad i \in N, j \in M, k \in K$$

$$\sum_{s=1}^{q} \nu_{sijk} \phi_s - \phi_q(1 - y_{jk}) \leq w_{ijk}, \qquad\qquad i \in N, j \in M, k \in K$$

$$w_{ijk} \leq \sum_{s=1}^{q} \nu_{sijk} \phi_s, \qquad\qquad i \in N, j \in M, k \in K$$

$$\xi_{ijk} \leq \kappa \phi_q x_{lj}, \qquad\qquad i, l \in N, j \in M$$

$$\sum_{l=1}^{n} \xi_{ijl} \leq \kappa \phi_q x_{ij}, \qquad\qquad i \in N, j \in M$$

$$\beta_{ij} - \kappa \phi_q(1 - x_{ij}) \leq \xi_{ijl} \leq \beta_{ij}, \qquad\qquad i, l \in N, j \in M$$

$$\sum_{k=1}^{\kappa} w_{ijk} - \kappa \phi_q(1 - x_{ij}) \leq \sum_{l=1}^{n} \xi_{ijl}, \qquad\qquad i \in N, j \in M$$

$$\sum_{l=1}^{n} \xi_{ijl} \leq \sum_{k=1}^{\kappa} w_{ijk}, \qquad\qquad i \in N, j \in M$$

$$\gamma_{ijk} N_0 + \sum_{j'=0, j' \neq j}^{m} P_{j'k} \gamma_{ijk} G_{ij'k} = P_{jk} G_{ijk}, \quad i \in N, j \in M, k \in K$$

$$\sum_{s \in Q} \nu_{sijk} \hat{\gamma}_s \leq \gamma_{ijk} \leq \sum_{s \in Q} \nu_{sijk} \hat{\gamma}_{s+1}, \qquad\qquad i \in M, j \in M, k \in K$$

$$\sum_{s \in Q} \nu_{sijk} = 1, \qquad\qquad i \in N, j \in M, k \in K$$

$$\sum_{k=1}^{\kappa} P_{jk} = \bar{P} u_j, \qquad\qquad j \in M$$

$$\sum_{k=1}^{\kappa} y_{jk} \leq \kappa \sum_{i=1}^{n} x_{ij}, \qquad\qquad j \in M$$

$$u_j \leq \sum_{k \in K} y_{jk}, \qquad\qquad j \in M$$

$$\sum_{k \in K} y_{jk} \leq \kappa u_j, \qquad\qquad j \in M$$

$$P_{jk} \leq \bar{P}_j y_{jk}, \qquad\qquad j \in M, k \in K$$

$$P_j - \bar{P}_j(1 - y_{jk}) \leq P_{jk} \leq P_j, \qquad j \in M, k \in K$$

$$x_{ij}, \, y_{jk}, \, \nu_{sijk}, \, u_j \in \{0,1\}, \qquad s \in Q, i \in N, j \in M, k \in K$$

$$0 \leq w_{ijk}, \, \xi_{ijl}, \, P_{jk}, \, \beta_{ij}, \, P_j, \, \gamma_{ijk}, \qquad i, l \in N, j \in M, k \in K$$

All integrality constraints 0–1, $z \in \{0,1\}$, are reformulated as $z^2 - z = 0$. Instances for (**sinr**) are defined according to the values of n, m, k, q. We denote by: F_0: the largest value of constraints at the starting point x^0, F_k: the largest value at the last obtained point, $v^\star$: the optimal value provided by *Couenne*, v: the optimal value provided by our algorithm, v_0: the value of the objective function at the starting point x^0, $\#x^k$: the number of serious steps, % gab: for a variable *var* indexed by $I(var)$, at optimum, var_{cou} denotes its value for *Couenne* and var_{bund} its value for the version *bund* of our algorithm (Table 3).

Then let:

– $A(\gamma) = \big\{ \min\{var_{bund}(s), 1 - var_{bund}(s)\} \leq \gamma, \; s \in I(var)\big\}$ if *var* is a 0–1 vector
– $A(\gamma) = \big\{ |var_{cou}(s) - var_{bund}(s)| \leq \gamma, \; s \in I(var)\big\}$ otherwise

and

$$\%\text{gab}(\gamma) = 100\frac{\#A(\gamma)}{\#var} \text{ and } \text{gab}_{\max} = \max A(\gamma).$$

Table 1. Description of instances

Instance (name)	Variables (var.)		Constraintes (cntr.)		F_0	$v^\star$
	Nb. var.	Nb. var. 0–1	Nb.	Nb. quad.		
2-2-2-2	103	39	247	24	10100.0	9.7032E-17
3-2-2-2	154	54	386	36	10100.0	1.4612E-16
4-2-2-2	211	69	549	48	10100.0	2.0597E-16
5-2-3-2	355	117	913	90	10100.0	3.1135E-16
6-2-5-5	901	486	1577	180	10100.0	3.1378E-17

3.2 Comments

As shown in Table 2 (*all-without* vs *all-classical* and *without-classical*) and as expected, the update rule can substantially reduce the number of *null step*.

Table 2. Iterations (1000) from Algorithm 1 ($all = na, cl, sp, co$) and $level = wt, sl$.

Instance (name)	all-without ($all \neq without$ i.e. $\mu_k \neq 0$)				all-classical ($all{:}\mu_k \neq 0$) ($all \neq without$ i.e. $\mu_k \neq 0$)				without-classical
	cl-wt	co-wt	na-wt	sp-wt	cl-cl	co-cl	na-cl	sp-cl	wt-cl
2-2-2-2	110	110	110	110	115	115	111	111	2
3-2-2-2	110	110	110	110	115	115	111	111	111
4-2-2-2	110	110	110	110	115	115	111	111	111
5-2-3-2	110	110	110	110	115	115	111	111	111
6-2-5-5	110	110	110	110	114	114	111	108	111

Bundle methods provide a better exploration of the feasible set of the auxiliary problem since they allow to generate cuts from interior points, unlike cutting plane methods which generate these cuts only from extreme points. However, in many cases, as shown by the results in tables 3, *without-classical* is better than the versions *all-level* ($all \neq without$) regarding the number of iterations and the quality of the solutions obtained. The cuts made in *without-classical* are deeper than those made with *all-level* ($all \neq without$).

Table 3. Results obtained with some instances ($all = natural, classical, special, constant$ and $level = without, classical$)

Instance name	all-level ($all{:} \mu_k \neq 0$)				without-classical			
	$v_0 - v^\star$	$v - v^\star$	F_k	$\sharp x^k$	$v_0 - v^\star$	$v - v^\star$	F_k	$\sharp x^k$
2-2-2-2	600.0	−9.7032E-17	0.2222	2	600.0	−9.7032E-17	0.0	2
3-2-2-2	900.0	−1.4612E-16	0.2222	2	900.0	−1.4612E-16	0.222	2
4-2-2-2	12000.0	−2.0597E-16	0.2222	2	12000.0	−2.0597E-16	0.25	2
5-2-3-2	15000.0	−3.1135E-16	0.2222	2	15000.0	−3.1135E-16	0.239	2
6-2-5-5	18000.0	−3.1378E-17	0.2222	2	18000.0	−3.1378E-17	0.222	2

4   Conclusion and Extensions

We have proposed a convexification scheme that is used in the proximal bundle method. A Kelley cutting plane version of this algorithm that includes a level update rule appears efficient to solve the considered instances. Several work models can be designed from the proposed convexification scheme. The polyhedral model is built from the penalty function

$$\psi_{c_k, \eta_k, x^k}(x) = f^0(x) + \frac{\eta_k}{2} ||x - x^k||^2 + c_k F_k(x)^+ \qquad \forall x \in \mathbb{R}^n$$

with $F_{\eta_k,x^k}(x)^+ = \max\left\{0, f^1(x)+\frac{\eta_k}{2}||x-x^k||^2, ..., f^m(x)+\frac{\eta_k}{2}||x-x^k||^2\right\}, c_k > 0$ (penalty parameter), $\eta_k > 0$ (convexification parameter). When the next point x^{k+1} is not feasible, then we have

$$\psi_{c_k,\eta_k,x^k}(x^{k+1}) = f^0(x^{k+1}) + c_k \max\{f^1(x^{k+1}), ..., f^m(x^{k+1})\}$$
$$+ \underbrace{\frac{c_k\eta_k+1}{2}||x^{k+1}-x^k||^2}_{b}.$$

The convexification term (b) proceeds as a proximity control term. Since c_k tends to become very large, descent step size from x^k is more and more penalized. We can also consider the case for which we would look for the proximal point of f^0 at x^k in a region *"centered"* at x^{k+1}.

The presence of a penalty term that must take a very large value leads to very small descent steps and numerical difficulties. To correct this, combined with our new reformulation, the following well-known function can be used

$$h_x(y) = \max_{y\in\mathbb{R}^n}\left\{f^0(y) - f^0(x), F(y)\right\}$$

where $F(y) = \max\{0, f^1(y), ..., f^m(y)\}$.

Acknowledgement. The first author was supported by the German Academic Exchange Service (DAAD). Gratitude is expressed to the projects CEA-SMA and NLAGA for the support of this work. The authors thank the anonymous referees for useful comments and suggestions.

References

1. Kelley Jr., J.E.: The cutting-plane method for solving convex programs. J. Soc. Ind. Appl. Math. **8**, 703–712 (1960)
2. Kiwiel, K.C.: An exact penalty function algorithm for non-smooth convex constrained minimization problems. IMA J. Numer. Anal. **5**(1), 111–119 (1985). https://academic.oup.com/imajna/article-lookup/doi/10.1093/imanum/5.1.111
3. Yang, Y., Pang, L., Ma, X., Shen, J.: Constrained nonconvex nonsmooth optimization via proximal bundle method. J. Optim. Theory Appl. **163**(3), 900–925 (2014). https://doi.org/10.1007/s10957-014-0523-9
4. Ouorou, A.: A proximal cutting plane method using Chebychev center for nonsmooth convex optimization. Math. Program. **119**(2), 239–271 (2009). https://doi.org/10.1007/s10107-008-0209-x
5. De Oliveira, W., Solodov, M.: A doubly stabilized bundle method for nonsmooth convex optimization. Math. Program. **156**(1–2), 125–159 (2016). https://doi.org/10.1007/s10107-015-0873-6
6. Nemirovski, A.: Efficient Methods in Convex Programming (2007). https://www.semanticscholar.org/paper/Efficient-Methods-in-Convex-Programming-Nemirovski/a2e65acd1dc3642ffd91aadc9c420573b611694d
7. Couenne, a solver for non-convex MINLP problems. https://www.coin-or.org/Couenne

8. us-en_analytics_SP_cplex-optimizer   —   IBM Analytics. https://www.ibm.com/
   analytics/cplex-optimizer
9. Welcome to Python.org. https://www.python.org/
10. Pyomo. http://www.pyomo.org/
11. Fooladivanda, D., Al Daoud, A., Rosenberg, C.: Joint channel allocation and user
    association for heterogeneous wireless cellular networks. In: 2011 IEEE 22nd Inter-
    national Symposium on Personal, Indoor and Mobile Radio Communications, pp.
    384–390. IEEE (2011). http://ieeexplore.ieee.org/document/6139988/

An Adapted Derivative-Free Optimization Method for an Optimal Design Application with Mixed Binary and Continuous Variables

Thi-Thoi Tran[1(✉)], Delphine Sinoquet[1], Sébastien Da Veiga[2], and Marcel Mongeau[3]

[1] IFPEN, Rueil-Malmaison, France
{thi-thoi.tran,delphine.sinoquet}@ifpen.fr
[2] Safran, Magny-Les-Hameaux, France
sebastien.da-veiga@safran.fr
[3] ENAC, Université Toulouse III, Toulouse, France
mongeau@recherche.enac.fr

Abstract. Numerous optimal design applications are black-box mixed integer nonlinear optimization problems: with objective function and constraints that are outputs of a black-box simulator involving mixed continuous and integer (discrete) variables. In this paper, we address an optimal design application for bladed disks of turbo-machines in aircraft. We discuss the formulation of an appropriate distance with respect to discrete variables which can deal with the cyclic symmetry property of the system under study. The necklace concept is introduced to characterize similar blade configurations and an adapted distance is proposed for discrete space exploration of a derivative-free optimization method. The results obtained with this method on a simplified industrial application are compared with results of state-of-the-art black-box optimization methods.

Keywords: Mixed integer non-linear programming · Black-box simulation · Derivative free trust region method · Necklace distance

1 Motivation

Air traffic is one of the most important means of transportation, especially in Europe. Besides, it is connected with very high costs of fuel [1,2], and also with high costs of maintenance and of manufacturing. Thus, reducing fuel consumption by increasing engine efficiency and maintenance savings by decreasing vibrations are two major concerns of the aviation industry.

There are several ways to optimize costs in aircraft: optimal trajectories, optimal seat arrangement designs, optimal cargo arrangements ... In our case, we want to optimize the design of turbo-machines, precisely, by maximizing the efficiency of compressor and minimizing the vibrations.

© Springer Nature Switzerland AG 2020
H. A. Le Thi et al. (Eds.): ICCSAMA 2019, AISC 1121, pp. 88–98, 2020.
https://doi.org/10.1007/978-3-030-38364-0_8

In the concrete application proposed by SAFRAN, the optimization variables are of 2 types: continuous shape parameters, x, e.g. thickness, length of blades and binary variables y associated to each blade, with the value 0 for a reference shape and 1 for the other predefined shape (mistuning shape). Binary variables are used to locate these reference blade shapes on the disk. This parameterization provides the distribution of the two shapes around the turbine disk.

There is a strong symmetry in this problem that should be taken into account. Two bladed disks that differ only by a rotation of the blade pattern around the disk will lead to the same simulation outputs, such arrangements are considered as the same solution or called redundant solution: e.g. 001101001101 and 010011010011. Note that the number of identical solutions increases rapidly with the number of blades (see Table 1).

In practice, engineers try to overcome this difficulty in practical optimization by using Reduce Order Modeling Technique (ROMT), detailed in [6–8,15]. Briefly, grouping blades by sectors of 2 patterns - 00 or 11 and 10 or 01 - tends to reduce the number of discrete variables from n to $n/2$.

In general these mixed problems are NP-hard and difficult to solve. Especially, for real applications, the difficulty lies in evaluating computationally expensive cost function: it can take several hours or even days to compute the functions to be optimized. Moreover, the derivatives are often not available. Most of algorithms for solving black-box Mixed Integer Nonlinear Problem are based on genetic algorithms which require a lot of function evaluations which are particularly costly in our application. In the next section, we motivate the use of derivative-free trust-region method.

Table 1. Number of distinct and total arrangements for a given number of blades

Total number of blades on the disk (N)	Number of distinct arrangements ($\sim 2^N/N$)	Total number of arrangements (2^N)
2	3	4
3	4	8
5	8	32
10	108	1024
12	352	4096
20	52488	1048576

2 Derivative Free Trust-Region Method

Among Derivative-free optimization (DFO) methods, one distinguishes direct search methods (e.g., directional or Nelder Mead simplex) and trust region methods based on simple interpolation or regression models (linear or quadratic). The direct search methods require a large number of simulations, whereas the second type of methods is generally more efficient to converge to a local solution. Convergence results to local minima are proved for the later methods but it requires adaptations to hope to converge to a global optimal solution (e.g., a multi-start approach with several initial points).

We focus on surrogate DFO methods [4,10]. These methods aim to explore the optimization variable space by replacing the costly-to-evaluate functions by response surfaces, with a choice of the points to simulate based on a compromise between exploration (points far from points simulated at previous iterations) and exploitation of information captured by the response surface (points around potential optima). Most popular response surfaces for surrogate optimization are Radial Basis Functions (RBF) and Gaussian processes [11,14,17].

Besides, derivative-free trust-region algorithm is based on local quadratic models defined inside a trust region

$$m(z) = \alpha + g^T z + \frac{1}{2} z^T H z,$$

whose coefficients are determined as the solution of minimal Frobenius norm problem

$$\begin{cases} \min_{\alpha, g, H=H^T} \frac{1}{2} \|H\|_F^2 \\ m(z_i) = f_i, i = 0, \dots, p. \end{cases}$$

where $Z = (z_0, \dots, z_p)$ is the interpolation set for both continuous and binary variables $z = (x, y)$, f_i the associated objective function evaluations and $\|.\|_F$ the Frobenius norm. The interpolation set is assumed to be poised, see details in [4,10].

The brief idea of the algorithm is to replace the initial problem, which involves the expensive simulations with quadratic optimization problem, simple to optimize. At the first step, we fix the binary variables to the best current solution and solve the quadratic sub-problem within the trust region with respect to continuous variables only. When a better point is obtained in this first step (a smaller objective function), the second step consists in minimizing the quadratic model with respect to both continuous and binary variables, thus solving a mixed binary continuous quadratic problem.

The model is built in order to efficiently approximate the function and fulfills the fully-linear or fully-quadratic model properties to ensure the local convergence of the algorithm. Thus, model improvement steps are performed in order to optimize the geometry of the interpolation set, see details in [4,9,10].

The introduction of binary (or integer) variables requires an adapted trust region definition. In [9], the authors introduce a l_1-norm trust region for continuous variables and the Hamming distance trust region for binary variables defined as

$$\sum_{j:y_{c_j}=0} y_j + \sum_{j:y_{c_j}=1} (1 - y_j) \leq \Delta_y, \tag{1}$$

with Δ_y, the size of the trust region for binary variables.

When the algorithm reaches a "local solution" (no further improvement), an exploration phase is necessary in binary variable domain. It is performed by adding a "no-good-cut" constraint to escape from this local solution y^*, defined as

$$\sum_{j:y_j^*=0} y_j + \sum_{j:y_j^*=1} (1 - y_j) > \Delta_y^*, \qquad (2)$$

where (y^*, Δ_y^*) are respectively the local solution and the current radius of the "sufficiently explored" area around this solution. Note that this approach leads to a bunch of constraints, one for each explored local solution.

In the next section, we propose an adapted distance that will be used for the trust region associated with binary variables for our application.

3   Adapted Distance for Blade Design Application

To avoid the solution redundancy (illustrated in Table 1), engineers use ROMT which reduces the optimization problem size but has the disadvantage of removing a large number of feasible solutions. Safran's application with 12 blades has 352 distinct arrangements but only 28 distinct arrangements for the two subproblems of ROMT, limiting a lot the explored configuration set with a high probability to remove "good" candidates.

Therefore, we attempt to define a new distance which can avoid the redundant solutions without arbitrary removal of configurations. The new distance should lead to simple constraints as the Hamming distance which leads to linear constraints (1).

In order to detect similar blade arrangements, we introduce the concept of "necklace" [12,13]. In combinatorics, a k-ary necklace of length n is an equivalence class of n-character strings over an alphabet $\sum^k = \{a_1, \ldots, a_k\}$ of size k, taking all rotations as equivalent. It represents a structure with n circularly connected beads which have k available colors. Our blade design application can be seen as a 2-color necklace optimization with a fixed number of beads.

Using the concept of necklace gives an exact formula of the number of distinct arrangements which is the number of necklace for given n beads: $1/n \sum_{d|n} \phi(d) 2^{n/d}$, where ϕ is Euler's totient function, the summation is taken over all divisors d of n.

There are numerous applications based on "necklace" concept and various distances, e.g. geometry distance for music, swap distance [22–24], Hamming distance with shift [16], and necklace alignment distance (NAD) [5].

Following the idea of NAD, we propose a new distance, that we call in the following the necklace distance,

$$d_{neck}(y, y') = \min_r d_H(y, Rot^r(y')), \qquad (3)$$

where d_H denotes the Hamming distance, $Rot^r(y)$ is the rotation of r positions from y. It is clear that this distance satisfies

- the non-negativity property: $d_{neck}(y, y') \geq 0$,
- the reflexivity property: $d_{neck}(y, y) = 0$,
- the commutativity property: $d_{neck}(y, y') = d_{neck}(y', y)$,
- the triangle inequality property: $d_{neck}(y, y'') \leq d_{neck}(y, y') + d_{neck}(y', y'')$.

Besides these metric properties, d_{neck} has one important property

$$d_{neck}(y, y') = 0 \iff y \in Rot(y') \tag{4}$$

which ensures to detect all the necklaces in the set. The definition of the necklace distance is not linear because of "min" operator. We propose in the next section an adaptation of this distance for the trust region and no-good-cut constraints used in the trust region algorithm.

Reformulating the Distance for Trust Region DFO Algorithm

The mixed binary continuous sub-problem associated with the minimization of the quadratic model is written as

$$\begin{cases} \min_{x,y} m(x,y) \\ \min(g_1(y, y_c), \ldots, g_n(y, y_c)) \leq \Delta_y \\ \|x - x_c\|_{l_1} \leq \Delta_x, \\ x \in \mathbb{R}^n, y \in \{0, 1\}^n, \end{cases} \tag{5}$$

with x_c and y_c the current centers of the two trust regions in continuous and binary variable space with Δ_x and Δ_y, the size of the trust region for respectively the continuous and binary variables and $g_i(y, y_c) = d_H(y, Rot^i(y_c)), i = 1, \ldots, n$.

To deal with "min" operator in the constraints, we add slack variables t, integer, and $y_{n+1}, y_{n+2}, \ldots, y_{n+n}$, binaries, and propose an exact reformulation,

$$\begin{cases} \min_{x,y,t} m_\mu(x, y, t) = \min_{x,y,t} m(x,y) + \mu t \\ t \geq g_1(y, y_c) - M y_{n+1} \\ \ldots \\ t \geq g_n(y, y_c) - M y_{n+n} \\ \sum_{i=1}^{n} y_{n_i} = n - 1 \\ t \leq \Delta_y \\ \|x - x_c\|_{l_1} \leq \Delta_x, \\ x \in \mathbb{R}^n, y \in \{0, 1\}^{2n}, t \in \mathbb{Z}_+, \end{cases} \tag{6}$$

with a real parameter $\mu > 0$ and an integer parameter $M > n$.

As explained before, the exploration phase uses "no-good-cut" constraints (2) to enforce to explore new values for binary variables. The maximum number of constraints is $2^n - 1$. If we apply the same trick as in (6) for exploration phase, we highly increase the dimension of the problem due to the additional slack variables, $n + 1$ for each "no-good-cut" constraint. We use instead the equivalence

$$\min(g_1, g_2, \ldots, g_n) \geq \Delta_y^* \iff g_i \geq \Delta_y^*, i = 1, \ldots, n,$$

and thus replace no-good-cut constraints by at most n linear constraints

$$d_H(y, Rot(y^*)) \geq \Delta_y^*, \dots, d_H(y, Rot^n(y^*)) \geq \Delta_y^*.$$

Note that, in practice, we use $\Delta_y^* = 1$.

4 Toy Problem

The problem of determining the best-case (or worst-case) mistuning pattern and maximizing the forced response vibration amplification by the addition of a small mistuning to a perfectly cyclical bladed disk is mentioned in literature [6–8, 15, 19–21]. We use a single degree-of-freedom (DOF) per blade disk model (see Fig. 1).

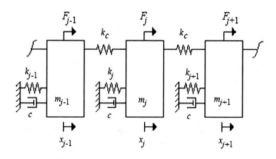

Fig. 1. Single DOF per blade disk model [7]

The DOF problem is formulated as

$$\begin{aligned} \underset{\omega, y}{\text{minimize}} \quad & \|A\|_\infty \\ \text{subject to} \quad & \omega \in [\omega_{min}, \omega_{max}], \\ & y_i \in \{0, 1\}, \end{aligned} \tag{7}$$

where $A = T^{-1}\bar{\mathcal{F}}, \bar{F}_i = F_0 e^{j\phi_i}, T = \begin{pmatrix} T_0 & -k_c & 0 & \dots & 0 & -k_c \\ -k_c & T_1 & -k_c & \dots & 0 & 0 \\ \vdots & \vdots & \vdots & \ddots & \vdots & \vdots \\ -k_c & 0 & 0 & \dots & -k_c & T_{N-1} \end{pmatrix},$

$T_i = -m_i\omega^2 + j\omega c + 2k_c + (1+\delta_i)k_b$. The nomenclature is detailed in [7].

5 Preliminary Numerical Results

We present some preliminary results of our trust-region method, called DFOb in the following, applied to the toy problem and to a simplified application provided by SAFRAN.

Results for the Toy Problem

In general, optimization researchers ideally seek for an algorithm which provides the smallest objective function value comparing to alternative algorithms. But, for derivative free optimization methods dealing with expensive-to-evaluate simulations, the efficiency of an optimization algorithm is generally measured by comparing the objective function value for a given budget of simulations (fixed number of simulations).

On this toy example, we run 10 times our method DFOb with different initial random points with two different distances for binary variables: Hamming distance and the necklace distance proposed in previous section (Eq. 3). The new implementation allows to reach a better point than the Hamming distance implementation as shown in Fig. 2. The number of simulations to reach the same objective value, '9.68272$e-04$, is larger with Hamming distance (90 simulations) than with the necklace distance (60 simulations).

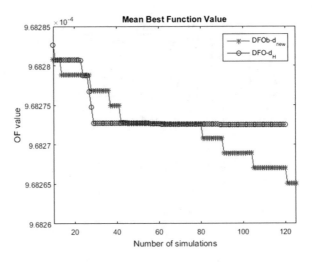

Fig. 2. Mean best average objective function obtained with 10 random initial datasets with Hamming distance and with the proposed distance.

Results on SAFRAN Application

We apply our method DFOb on the two sub-problems of ROMT and compare the results with NOMAD, [3,18] and RBFopt [11] optimization methods with a fixed budget of 100 simulations:

1. ROMT sub-problem 1 with patterns $00 - 11$,
2. ROMT sub-problem 2 with patterns $10 - 01$.

DFOb and RBFOpt share the same initial points (random choice from RBFOpt method). NOMAD's initial point is chosen as the best point of the initial set with regard to objective function value. Some constraints are added in order to avoid trivial solutions with only one type of patterns: only zero or one values

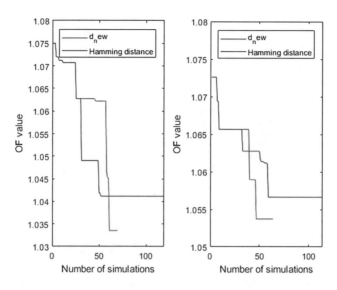

Fig. 3. Comparing results obtained with Hamming and necklace distances for the two ROMT sub-problems with patterns $00 - 11$ (left) and $10 - 01$ (right) of Safran application

in vector of binaries y. Figure 3 illustrates the efficiency of DFOb coupled with the necklace distance compared to Hamming distance for binary variables for the two sub-problesm of ROMT. For ROMT sub-problem with $00 - 11$ patterns, DFOb with the necklace distance provides an objective function of 1.033509 after 57 simulations while the same algorithm with the Hamming distance does not reach this value after more than 100 simulations. For the first ROMT sub-problem (see Fig. 4), DFOb method coupled with necklace distance has very good performance compared to RBFopt and NOMAD. NOMAD and RBFopt provide infeasible points during the exploration phasis (with small objective function values), the constraints being handled as soft constraints. In DFOb, the constraints are taken into account explicitely.

For the second ROMT sub-problem (see Fig. 5), DFOb finds good points rapidly after less than 10 simulations, nevertheless, it does not reach the best objective function value obtained by NOMAD and RBFopt within the fixed budget of simulations (100). With a different initial set, DFOb is able to find the global solution within 100 simulations. A future study will focus on the sensitivity of the results of our method to the initial set and of its size.

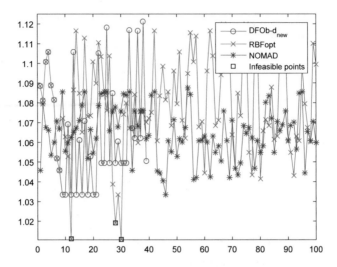

Fig. 4. Compare DFOb, Nomad and RBFOpt results obtained for sub-problem with patterns 00 − 11 of Safran application. Square black symbols indicate infeasible points.

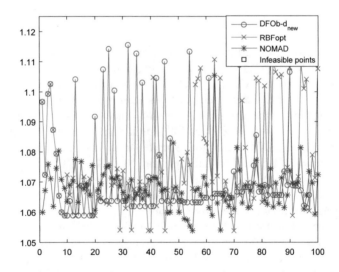

Fig. 5. Compare DFOb, Nomad and RBFOpt results obtained for sub-problem with patterns 10 − 01 of Safran application.

6   Conclusions

In this study we address black-box optimization problems with costly-to-evaluate objective functions. A trust region derivative free optimization method adapted to mixed binary and continuous variables is presented. In order to improve the exploration phase of this algorithm for a blade design application, we introduce a distance in binary variable domain, that takes into account the symmetry of the problem. Preliminary results illustrate the performances of this new distance compared to classical Hamming distance. The comparison of the proposed method with two state-of-the-art methods NOMAD and RBFopt shows some encouraging results.

References

1. Air Transport Action Group (ATAG). http://www.atag.org/facts-and-figures.html
2. International Air Transport Association, IATA price analysis. http://www.iata.org/publications/economics/fuel-monitor/Pages/price-analysis.aspx
3. Abramson, M., Audet, C., Couture, G., Dennis, J., Le Digabel, S., Tribes, C.: The NOMAD project. https://www.gerad.ca/nomad/
4. Audet, C., Hare, W.: Derivative-Free and Blackbox Optimization. Springer Series in Operations Research and Financial Engineering. Springer International Publishing, Cham, Switzerland (2017). https://doi.org/10.1007/978-3-319-68913-5
5. Bremner, D., Chan, T.M., Demaine, E.D., Erickson, J., Hurtado, F., Iacono, J., Langerman, S., Patrascu, M., Taslakian, P.: Necklaces, convolutions, and X+Y. CoRR abs/1212.4771 (2012). http://arxiv.org/abs/1212.4771
6. Choi, B., Lentz, J., Rivas-Guerra, A., Mignolet, M.: Optimization of intentional mistuning patterns for the reduction of the forced response effects of unintentional mistuning: formulation and assessment. J. Eng. Gas Turbines Power **125**(1), 131–140 (2003). https://doi.org/10.1115/1.1498270
7. Choi, B.: Pattern optimization of intentional blade mistuning for the reduction of the forced response using genetic algorithm. KSME Int. J. **17**(7), 966–977 (2003). https://doi.org/10.1007/BF02982981
8. Choi, B., Eun, K.H., Jung, K.H., Haneol, J., DongSik, G., Kwan, K.M.: Optimization of intentional mistuning for bladed disk: intentional mistuning intensity effect. In: Mathew, J., Kennedy, J., Ma, L., Tan, A., Anderson, D. (eds.) Engineering Asset Management, pp. 1024–1029. Springer, London (2006)
9. Conn, R., D'Ambrosio, C., Liberti, L., Sinoquet, D.: A trust region method for solving grey-box mixed integer nonlinear problems with industrial applications. https://modc2016.scicncccconf.org/filc/223761
10. Conn, R., Scheinberg, K., Vicente, L.: Introduction to Derivative-Free Optimization. Society for Industrial and Applied Mathematics (2009). https://doi.org/10.1137/1.9780898718768. https://epubs.siam.org/doi/abs/10.1137/1.9780898718768
11. Costa, A., Nannicini, G.: RBFOPT: an open-source library for black-box optimization with costly function evaluations. Math. Program. Comput. **10**(4), 597–629 (2018). https://doi.org/10.1007/s12532-018-0144-7
12. Fredricksen, H., Kessler, I.J.: An algorithm for generating necklaces of beads in two colors. Discrete Math. **61**(2), 181–188 (1986). https://doi.org/10.1016/0012-365X(86)90089-0. http://www.sciencedirect.com/ 0012365X86900890
13. Gabric, D., Sawada, J.: Constructing de Bruijn sequences by concatenating smaller universal cycles. Theor. Comput. Sci. **743**, 12–22 (2018). https://doi.org/10.1016/j.tcs.2018.06.039. http://www.sciencedirect.com/science/article/pii/S0304397518304559
14. Gutmann, H.M.: A radial basis function method for global optimization. J. Global Optim. **19**(3), 201–227 (2001). https://doi.org/10.1023/A:1011255519438

15. Han, Y., Murthy, R., Mignolet, M.P., Lentz, J.: Optimization of intentional mistuning patterns for the mitigation of the effects of random mistuning. J. Eng. Turbines Power **136**(6) (2014). https://doi.org/10.1115/1.4026141. 062505

16. Jiang, M.: On the sum of distances along a circle. Discrete Math. **308**(10), 2038–2045 (2008). https://doi.org/10.1016/j.disc.2007.04.025. http://www.sciencedirect.com/science/article/pii/S0012365X07002555

17. Jones, D.R., Schonlau, M., Welch, W.J.: Efficient global optimization of expensive black-box functions. J. Global Optim. **13**(4), 455–492 (1998). https://doi.org/10.1023/A:1008306431147

18. Le Digabel, S.: Algorithm 909: NOMAD: nonlinear optimization with the MADS algorithm. ACM Trans. Math. Softw. **37**(4), 1–15 (2011)

19. Liao, H., Wang, J., Yao, J., Li, Q.: Mistuning forced response characteristics analysis of mistuned bladed disks. J. Eng. Gas Turbines Power **132**(12) (2010). https://doi.org/10.1115/1.4001054. 122501

20. Óttarsson, G.: Dynamic modeling and vibration analysis of mistuned bladed disks. Thesis, University of Michigan, May 1994. https://tel.archives-ouvertes.fr/tel-00598068

21. Schonlau, M.: Conception robuste en vibration et aéroélasticité des roues aubagées de turbomachines. Ph.D. thesis, Université Paris-Est Marne la vallée (2009). https://tel.archives-ouvertes.fr/tel-00529002v2/document

22. Toussaint, G.: A mathematical analysis of African, Brazilian, and Cuban clave rhythms, pp. 157–168. Townson University (2002)

23. Toussaint, G.: The geometry of musical rhythm. In: Akiyama, J., Kano, M., Tan, X. (eds.) Discrete and Computational Geometry, pp. 198–212. Springer, Berlin (2005)

24. Toussaint, G.: Computational geometric aspects of rhythm, melody, and voice-leading. Comput. Geom. **43**(1), 2–22 (2010). https://doi.org/10.1016/j.comgeo.2007.01.003. http://www.sciencedirect.com/science/article/pii/S0925772110900042X, Special Issue on the 14th Annual Fall Workshop

Numerical Technologies for Investigating Optimal Control Problems with Free Right-Hand End of Trajectories

Tatiana Zarodnyuk$^{(\boxtimes)}$ [iD], Alexander Gornov[iD], Anton Anikin[iD], and Pavel Sorokovikov[iD]

Matrosov Institute for System Dynamics and Control Theory of SB RAS, Irkutsk, Russia
tzarodnyuk@gmail.com

Abstract. The paper considers numerical approaches for solving optimal control problems with free trajectories at the end of the time interval. A modification of the algorithm of the conjugate gradient for studying the controlled dynamic problem is presented. The proposed technique has been tested by using the test optimal control problems. We describe the results of solving an applied problem of nanophysics. It is considered two cells of a quantum computer which are based on four tunnel-coupled semiconductor quantum dots. The multistage series of computations for investigation of system dependence from changes of the model parameters values are carried out and allowed to demonstrate the effectiveness of the proposed approach.

Keywords: Numerical algorithm · Optimal control · Applied problem

1 Introduction

Optimization problems, both finite-dimensional and infinite-dimensional, quite often meet in various scientific and technical fields. The first extremal problem appeared in the framework of the calculus of variations. The main attention was paid to the analysis of smooth functionals that were defined in the whole space or limited by a smooth set. The development of computing technologies has led to the emergence in practice of new problems where control have changed in some closed set. A wide class of such problems was investigated in the works of L.S. Pontryagin, et al. [1], who received the necessary condition for an extremum (Pontryagin's maximum principle).

The first optimal control problems are considered to be the problems in which the search for the control law that ensures the minimum transition time is carried out. The problem of time-optimal control (performance problem) was developed by R. Bellman, N.N. Krasovsky, R.V. Gamkrelidze, M. Atans, P. Falb,

Supported by Russian Foundation of Basic Research, project number 18-07-00587.

© Springer Nature Switzerland AG 2020
H. A. Le Thi et al. (Eds.): ICCSAMA 2019, AISC 1121, pp. 99–105, 2020.
https://doi.org/10.1007/978-3-030-38364-0_9

which presented the results on the existence, uniqueness, and basic properties of a speed-optimal control (see, for example, [1–3]).

In the course of time, the new problems with functionals of a more complex structure and nonlinear systems described the dynamical processes under study appeared.

Today, the optimal control problems (OCP) with free trajectories at the end of the time interval for a differential system with an objective functional is a relevant topic in control theory. The optimal control problems can be considered as an auxiliary problem for the study of more complex problems, for example, with restrictions of the terminal and phase type. The solution of these problems is required to be performed repeatedly on the numerical methods iterations. In our opinion, this problem is a start stage for solving the actual problems of various classes.

Many theoretical studies focus only on a local search for an extremum in the optimal control problem. Nevertheless, it is necessary to study the global optimization problems for dynamic systems within a framework of the present control theory and applications. The investigation of the non-convex extremal OCP is an important to be a important scientific problem.

2   The Optimal Control Problems with Free Trajectories at the End of the Time Interval

The following problem statement is investigated: the controlled dynamical process is described by the differential equations system

$$\dot{x}(t) = f(x, u, t), \ x(t_0) = x^0, \ t \in T = [t_0, t_1]. \tag{1}$$

The initial conditions for the trajectories and control conditions are given

$$u \in U = \left\{ u \in E^r : ul_i \leq u \leq ug_i, \ i = \overline{1, r}. \right\} \tag{2}$$

We need to minimize the functional of a terminal type

$$I_0(u) = \varphi_0(x(t_1)). \tag{3}$$

It is necessary to find the optimal functional value (3) obtained using the admissible control from the convex set U. $f(x, u, t)$ and φ_0 are continuously differentiable in all arguments.

We can solve the problems of another class, reducing to the considerable statement (1)–(2) with the functional (3): in addition with the direct control constraints, also with the terminal and phase constraints, optimization the integral functional, optimization of the control-constants, as well as solving the time minimization problems and others.

3   Numerical Technologies for Studying OCP

The paper considers the numerical technique for investigating OCP with the free trajectories in is the end of the time interval. An idea of a multiple search for a minimum functional is one of the most popular and informative approach. The multi-start technique is based on multiple extreme search with various admissible controls. This method allows us to find the global extremum, construct the reachable set for the controlled system of the differential equations, and, alsow, evaluate the region of attraction for different extremal points [4].

Another effectiveness numerical methods for search the extremal value of the objective functional are Pontryagin's method [1] mofidication, based on the non-local maximum principle; method for convexifying of the reachable set, based on the solving the extended OCP [5,6], and methods oriented to the sequential search of local extremes in order to find the best solution (for example, technique of the curvilinear search [8] and algorithm of the tunnel type [9]).

Among the other approaches, it is necessary to mention the enumeration method of local extrema (for instance, heavy ball method, Branin's method), methods of integral representations (Chichinadze's method), and evolutionary algorithms [12–15]. Among the approaches developed in the optimal control theory, the methods of solving Bellman equation (Krotov-Bellman, Hamilton-Jacobi-Bellman methods) take a special place [16,17]. They are, in particular, characteristics method and semi-discretization method. These methods either impose strict restrictions on the structure of the problem or are focused on searching only for the local extremum of the target functional.

For nonlocal nonlinear OCP with controls of the relay type, we construct stochastic algorithms based on finite-dimensional optimization methods: the genetic algorithms of the extremal search, the random coverings algorithms, Shepard's algorithms and others. We realized the techniques of the postopti-mization analysis and used specialized visualizations for evaluating the quality of the obtained results.

As an example, we present one of the implemented algorithms of gradient type for investigating OCP with the free right-hand end of the trajectory. The conjugate gradient method is one of the most famous and well-studied methods. Nevertheless, this method is effective for considerable problems and continues to be one of the best in the class of methods that do not use quadratic memory:

Step 1. Select $u^0(t)$, $t \in T$.

Step 2. Set the update frequency K.

At the k-th iteration:

Step 3. Calculate the conjugate coefficient

$$\beta^k = \frac{\|\nabla I_0(u^k(t))\|^2}{\|\nabla I_0(u^{k-1}(t))\|^2}, \text{ if } k \neq K,$$
$$\beta^k = 0, \text{ if } k = K.$$

Step 4. Calculate the descent direction $q^k(t) = -\nabla I_0(u^k(t)) + \beta^k q^{k-1}(t)$, $t \in T$.

Step 5. Find $u^{k+1}(t) = \arg\min \left\{ I_0(u^k(t) + \alpha q^k(t), 0 \leq \alpha < +\infty \right\}$.

The iteration is complete.

To investigate the presented technique, it is developed the collection of optimal control problems [7], including up to now more than 180 model examples.

4   The Optimal Control Problem in the System of Four Semiconductor Quantum Dots

Computing technologies developed by the authors were tested by using test OCPs with the free right-hand end. We present the results of solving an applied problem from the field of nanophysics using the proposed approaches.

We consider two cells of a quantum computer which are based on four tunnel-coupled semiconductor quantum dots (QD) in the Ge-Si system. The size and the density of QD are determined to the existence of sufficient tunnel couple in the top layer for implementation of quantum logical operations. The tunnel couple in the lower layer does not exist. A voltage impulse applies to metal gates, which are above QD, lead to tunneling of electrons from lower to top dots. The exchange operation of information (SWAP) implements due to movement particles to a top layer. The problem consists of finding optimal form and length of the voltage impulse for the realization of SWAP operation.

Nonstationary Schrodinger equation is employed for the description of controlled electrons movement in the system of vertically combined QD. The mathematical model of the analyzed subject is formulated as a system of controlled differential equations:

$$\dot{x} = A_0(u)x, \qquad (4)$$

where the matrix dimension $A_0(u)$ is 32×32 [10]. The structure of this matrix is presented in the first initial statement [11]. It is necessary to transfer the system from $x(t_0) = x^0$ to $x(t_1) = x^1$ with minimum of the expended energy. The energy required for the movement of the electrons between the different QD is the control. It is formulated the objective functional which contains two parts: the control variation in all time interval and the derivation of the system trajectory from the terminal conditions

$$I(u) = s_1 \int_0^T u^2 dt + s_2 \sum_{j=1}^{32} (x_j(t_1) - x_j^1)^2 \to \min. \qquad (5)$$

We deal with variants of this problem with the different values of the coefficients s_1, s_2 and the functional of the response speed ($I(u) = t_1 \to \min$).

Numerical solutions to the problems of this series are implemented using the developed approach.

The first trouble of numerical solving of this problem is a slow convergence of algorithms in all-time interval. We use well-known method of parameter continuation for overcoming of this trouble, the system of differential equations is supplemented with a new parameter p: $\dot{x} = p_1 f(x, u, t)$. A numerical experiment in a small time interval allows us to get a close approximation for carrying out the next computation. It is found the strategy of parameter continuation which permits to pass from solving simple problems to initial one with accounting results from previous stages.

The most complicated and laborious problems turn out a response speed problem. Time estimation of response speed was given by experts and was equal to 1500 units. We can make better this estimation and find an optimal value proved to be 606 conventional units. It is found the optimal form of controlled voltage impulse and trajectories of the probability of electron's staying in different QD by developed algorithms for solving OCP. It is detected that the existence of a tunnel couple between QD in vertical line leads to appearance additional errors for implementation of SWAP operation for electrons in the lower layer.

To study the quantum logical operation $\sqrt{SWAP}$ we use the following functional [10]

$$I(u) = \sum_{i=1}^{32} x_i(t) + 4 - 2\sqrt{D^2 + E^2} \to \min, \tag{6}$$

here D and E are defined as $D = x_1 + 0.5((x_{10} + x_{11}) + (x_{14} - x_{15}) + (x_{18} + x_{19}) + (x_{23} - x_{22})) + x_{28}$ and $E = x_5 + 0.5((x_{11} - x_{10}) + (x_{14} + x_{15}) + (x_{18} - x_{19}) + (x_{22} + x_{23}))$. The structure of the control is given $\delta B(t) = A\cos(\omega t + \varphi) + C$, where $A \in [0, 3]$ is the amplitude of the oscillations, $\omega \in [0, 200]$ is the frequency, $\varphi \in [0, 2\pi]$ is the initial phase, $C \in [-3, 3]$ is the constant displacement.

The series of the computational experiments for the quantum logical operation $\sqrt{SWAP}$ made it possible to evaluate the accuracy and reduce its execution time. The structure of the obtained optimal control is presented in Fig. 1 (on the left side). The analysis of the dependence of the error functional on the change in the constant bias value (parameter C) for the operation is performed. The control structure was complicated: $\delta B(t) = A\cos(\omega t + \varphi) \cdot \sin(\pi \cdot t/0.1766) + C$.

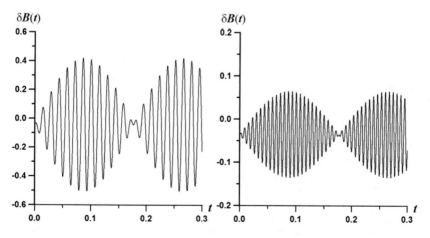

Fig. 1. Optimal control for quantum logical operation $\sqrt{SWAP}$.

We selected other intervals for changing the controlled parameters $A \in [0, 10]$, $\omega \in [0, 1000]$, $\varphi \in [0, 2\pi]$, $C \in [-3, 3]$. This made it possible to obtain the control of the different structure (Fig. 1, in the right side) and increase the accuracy in the functional (its value decreased from $4.77796 \cdot 10^{-3}$ to $4.23987 \cdot 10^{-3}$). The optimal parameter values for this control are as follows: $A = 0.100$, $\omega = 887.689$, $\varphi = 4.985$, $C = -0.036$. The harmonic dependence of used magnetic field $\delta B(t)$ provides the universal set of the logical operations necessary for the quantum computing.

5 Conclusion

The various control structures for the system of the quantum dots are investigated, the obtained solutions are presented for different cases. The results of study of the electron states in the model with two layers of the QD show the possibility of using these states for quantum computing.

Thus, the proposed numerical technology can be used to solve applied OCP from various scientific and technical fields.

References

1. Pontryagin, L.S., Boltyansky, V.G., Gamkrelidze, R.V., Mishchenko, E.V.: The Mathematical Theory of Optimal Processes. Fizmatgiz, Moscow (1961)
2. Atans, M., Falb, P.: Optimal control. Mechanical Engineering, Moscow (1968)
3. Krasovsky, N.N.: To the theory of optimal regulation. Autom. Telemech. **18**(11), 960–970 (1957)
4. Gornov, A.Y., Zarodnyuk, T.S., Finkelshtein, E.A., Anikin, A.S.: The method of uniform monotonous approximation of the reachable set border for a controllable system. J. Glob. Optim. **66**(1), 53–64 (2016)
5. Gamkrelidze, R.V.: Fundamentals of Optimal Control. Publishing house of the Tbilisi University (1977)
6. Tolstonogov, A.A.: Differential inclusions in a Banach space. Mathematics and its Applications. Kluwer Academic Publishers, Dordrecht (2000)
7. Gornov, A.Y., Zarodnyuk, T.S., Madzhara, T.I., Daneyeva, A.V., Veyalko, I.A.: A collection of test multiextremal optimal control problems. In: Optimization, Simulation, and Control. Springer Optimization and Its Applications, vol. 76, pp. 257–274 (2013)
8. Gornov, A.Y., Zarodnyuk, T.S.: The curvilinear search method of global extremum in optimal control problems. Mod. Tech. Syst. Anal. Simul. **3**, 19–26 (2009). (in Russian)
9. Gornov, A.Y., Zarodnyuk, T.S.: Tunneling algorithm for solving nonconvex optimal control problems. In: Optimization, Simulation, and Control. Springer Optimization and Its Applications, vol. 76, pp. 289–299 (2013)
10. Zinovieva, A.F., Nenashev, A.V., Koshkarev, A.A., Zarodnyuk, T.S., Gornov, A.Y., Dvurechenskii, A.V.: Quantum gates with spin states in continuous microwave field. Russ. Microelectron. **47**(4), 268–278 (2018)
11. Gornov, A.Y., Dvurechenskii, A.V., Zarodnyuk, T.S., Zinov'eva, A.F., Nenashev, A.V.: Problem of optimal control in the system of semiconductor quantum points. Autom. Remote Control **72**(6), 1242–1247 (2011)

12. Brent, R.: On the Davidenko-Branin methods for solving simultaneous nonlinear equations. IBM J. Res. Dev. **16**, 434–436 (1972)
13. Kasbah, A.B., Astolfi, A., Laila, D.S.: A modified Branin method. In: Proceedings of the 17th International Symposium on Mathematical Theory of Networks and Systems, Kyoto, Japan (2006)
14. Karpenko, A.P.: Population algorithms of global search searching optimization. Review of new and little-known algorithms. Inf. Technol. **7**, 1–32 (2012)
15. Diveev, A.I., Konstantinov, S.V.: Evolutionary algorithms for solving optimal control problem. Bull. Peoples' Friendsh. Univ. Russ. Ser.: Eng. Stud. **18**(2), 254–265 (2017)
16. Krotov, V.F.: Global Methods in Optimal Control Theory. Marcel Dekker Inc., New York (1996)
17. Dykhta, V.A.: Weakly monotone solutions of the Hamilton-Jacobi inequality and optimality conditions with positional controls. Autom. Remote Control **75**(5), 829–844 (2014)

A Genetic Algorithm Approach for Scheduling Trains Maintenance Under Uncertainty

Hanyu Gu and Hue Chi Lam$^{(\boxtimes)}$

School of Mathematical and Physical Sciences, University of Technology Sydney,
Sydney, NSW, Australia
hanyu.gu@uts.edu.au, hue.lam@student.uts.edu.au

Abstract. This paper investigates the overhaul maintenance scheduling problem in which the maintenance duration is uncertain at the time of planning. This problem involves specifying the dates of trains' arrival at the maintenance centre while taking into consideration the due windows, the desired number of trains in service, and the capacity of the maintenance centre. The cycle time of each type of trains is random with a known probability distribution. The objective is to minimise a weighted sum of two components: (i) the deviation of the assigned arrival dates from the due windows and (ii) the penalty for violating the resources' constraints. A combined genetic algorithm with sample average approximation solution approach is developed to solve this problem. The solution approach consists of a genetic algorithm for global search and an exact method to determine the arrival dates of train-sets when a sequence of train-sets is known. The results with data provided by one of the leading Australian maintenance center show that the proposed method can produce good solution within acceptable computation time.

Keywords: Genetic algorithm · Stochastic cycle time · Quadratic earliness/tardiness · Sample average approximation

1 Introduction

This paper deals with the scheduling of overhaul maintenance of trains, arising in the realm of passenger rail services. In the case of overhaul maintenance, the trains are withdrawn from service and are sent to a specialised maintenance centre where they will stay for at least one month for the entire maintenance process. A typical objective of this problem is to minimise the total cost of earliness and tardiness subject to the centre capacity. According to [8], this problem is known as the Overhaul Maintenance Scheduling Problem (OMSP). This paper addresses the stochastic version of OMSP, where the dwell time of the trains at the maintenance centre is uncertain at the time of planning. We refer to this problem as Stochastic Overhaul Maintenance Scheduling Problem (SOMSP).

© Springer Nature Switzerland AG 2020
H. A. Le Thi et al. (Eds.): ICCSAMA 2019, AISC 1121, pp. 106–118, 2020.
https://doi.org/10.1007/978-3-030-38364-0_10

In SOMSP, we consider the perspective of the maintenance centre's planning manager who must determine an arrival plan specifying the arrival dates for all trains for one year or for a longer period, where each train has a distinct due window and a random dwell time[1] that follows a known probability distribution associated with the type of train. During the planning phase, one must take into consideration the distinct due windows which are the desired arrival windows of the trains. It is noted that the definition of due window in this paper is different from those in most scheduling literature whereby it is referred to as the desired completion window of a job.

The train arrives at the maintenance centre in groups. Each group is comprised of several cars coupled together and is referred to as a set or a train-set (see for example [4]). All cars in a train-set undergo the maintenance in the maintenance centre simultaneously. A feasible arrival plan is one that satisfies a number of constraints. Firstly, a team of technicians and engineers with a broad range of skills is needed to perform the maintenance tasks. To complete the maintenance operations, the team must use various equipment and materials. These renewable and nonrenewable resources can be quantified as centre capacity. The centre capacity imposes a restriction on the number of train-sets which can undergo maintenance simultaneously.

Secondly, on the arrival date, the train-set is completely withdrawn from service and must stay at the maintenance centre for at least one month. This long cycle time directly impacts the number of train-sets available in active service. Therefore, a permissible number of train-sets that can be taken out of service simultaneously are specified for each type of train-sets.

Solving the SOMSP is challenging for various reasons: (i) the single machine scheduling problem with earliness and tardiness objectives, which is similar to the OMSP, is known to be NP-hard [11]. Hence, the considered SOMSP with stochastic dwell time is even harder to solve; and (ii) due to the non-uniform distribution of the desired arrival window, there is a trade-off between respecting the time window and satisfying the centre capacity and operational requirement [6].

Due to the complexity of SOMSP, a Genetic Algorithm (GA) is proposed to solve the considered problem. Based on previous study in [2], it is noted that the decoding procedure, where a chromosome is transformed into a feasible arrival plan, is a crucial step of GA. The decoding procedure used in [2] is a simple greedy heuristic which does not generate good solution. To improve the performance, we develop a new decoding procedure based on Sample Average Approximation (SAA) method in this paper.

The remainder of this paper is organised as follows. Section 2 presents the relevant literature. Section 3 gives the mathematical formulation of the considered problem. Section 4 shows the model formulation using SAA approach. Section 5 describes the proposed genetic algorithm in detail. Section 6 reports the results of the computational experiments. Finally, Sect. 7 concludes the study and gives directions for future research.

---

[1] In the context of this paper, dwell time and cycle time have the same meaning. The two terms are used interchangeably.

2   Literature Review

Papers that consider the scheduling and planning of rolling stock's maintenance within the railway industry can be categorised into two groups. The first focuses on the running maintenance which is performed during the connection between two consecutive paths or overnight at the rolling stock depot. These studies generally consider the perspective of the railway provider who must design a rolling stock rostering plan while taking into consideration various maintenance requirements. The design objectives often include minimising a weighted sum of maintenance cost, deadhead cost due to unexpected failure, and substitution cost due to the use of undesired train-sets [4]; and minimising the sum of deadhead costs and number of train units [9].

The second focuses on the overhaul maintenance which is performed at a specialised workshop. These studies generally consider the perspective of the workshop's planning manager who must determine the dates on which the trains must be sent to the workshop for maintenance. As an early work, [8] proposes a genetic algorithm for solving the OMSP. Doganay and Bohlin [1] formulates the OMSP as a mixed integer linear programming model and solves it by exact method. Lin et al. [6] formulates the high-level maintenance scheduling for high speed trains as a mixed integer linear programming model and proposes a simulated annealing algorithm for solving large-scale instances. From these studies, we can observe that the objective of minimising the total cost of earliness and tardiness is a common design objective of OMSP.

The proposed SOMSP is similar to a Stochastic Multiple Resource Constrained Scheduling Problem (SMRCSP), in which one need to decide the start times of jobs requiring different types of resources under uncertain processing time [3]. This kind of problem has application in appointment sequencing and scheduling where jobs correspond to operation appointments and resources correspond to doctors, nurses and operating rooms [7]. However, the key differences between the proposed SOMSP and the aforementioned problems involve: (i) each type of train-sets has a known processing time on the first operation line where preemptions are not allowed. That is, the scheduling on the first operation line can be treated as a single machine scheduling problem; and (ii) for train-sets of the same type, there exists a noticeable sequence among them. That is, train-set with earlier due window should always arrive for maintenance earlier than the others of the same type in order to minimise the objective function value.

3   Mathematical Formulation

Consider a set $N := \{1, \cdots, n\}$ of train-sets and a planning period of T days. The planning horizon is discretised into calendar days which are indexed from 0 to $T - 1$.

Each train-set $j \in N$ has a due window $[e_j, l_j]$, where e_j is the earliest desired arrival date and l_j is the latest desired arrival date. The earliness and tardiness of train-set j if it arrives on day t are defined as $E_{jt} = \max\{0, e_j - t\}$, and

$T_{jt} = \max\{0, t - l_j\}$, respectively. Let c_{jt} be the cost for assigning train-set j to arrive on day t, as indicated in (1).

$$c_{jt} = \lambda_1 E_{jt}^2 + \lambda_2 T_{jt}^2,\tag{1}$$

where λ_1 and λ_2 are the earliness and tardiness cost factors, respectively.

There are m types of train-sets and the set of all train-sets is partitioned into m families. Each train-set belongs to a train family $F^k, k \in K = \{1, \cdots, m\}$. Each type of train-sets requires a minimum duration on the first operation line, during which the maintenance operations must be performed without interruption. For each train family F^k, let p_k be the minimum duration on the first operation line. The minimum duration is assumed to be deterministic.

For each day t, let C_t be the centre capacity (i.e. the number of train-sets which can undergo maintenance simultaneously). Violation of this restriction is permitted for a penalty cost. Let δ be the unit penalty for violating the centre capacity. In practice, the penalty is associated with each additional train-set to account for overtime, outsourcing, and hiring contractors.

For each train family F^k and each day t, let C_{kt} be the permissible number of out-of-service train-sets. Violation of this restriction is permitted since the shortage of some types of train-sets can be substituted by a different train type. Let δ_{kt} be the unit penalty for violating the limit C_{kt}. In practice, a penalty is associated with each substitution to account for the passengers' dissatisfaction due to the differences in their configurations.

The cycle time of each train-set j is a random variable D_j. Train-sets in a train family follows the same probability distribution. It is assumed that the probability distribution for the cycle time of each type of train-sets is known. It is further assumed that all D_j are independent.

We introduce the time indexed binary variables $x_{jt} \in \{0,1\}$ which is equal to 1 if train-set $j \in N$ arrives at the maintenance centre on day t, and is equal to 0 otherwise. Then, for each train-set $j \in N$ its arrival day is defined as

$$s_j = \sum_{t=0}^{T-1} t x_{jt}, \ \forall j \in N\tag{2}$$

Accordingly, we can define an arrival plan $s = (s_1, \cdots, s_n)$. For any arrival plan s and any integers $1 \le k \le m$ and $0 \le t < T$, the number of trains of family k that present at the centre on day t is

$$Z_{kt}^s = \sum_{j \in F^k} B(s_j, t),\tag{3}$$

where

$$B(s_j, t) = \begin{cases} 1 & \text{if } s_j \le t \text{ and } s_j + D_j \ge t+1 \\ 0 & \text{otherwise} \end{cases}.\tag{4}$$

Then, the total number of trains present at the centre on day t is

$$Z_t^s = \sum_{k \in K} Z_{kt}^s.\tag{5}$$

If it is clear which arrival plan is considered, the superscript s can be dropped and the notation Z_t (Z_{kt}) can be used instead of Z_t^s (Z_{kt}^s).

Formulation of SOMSP

$$\text{Min } \alpha \sum_{j \in N} \sum_{t=0}^{T-1} c_{jt} x_{jt}$$

$$+ \beta \sum_{t=0}^{T-1} \left(\delta \mathbb{E} \left[(Z_t - C_t)^+ \right] + \sum_{k \in K} \delta_{kt} \mathbb{E} \left[(Z_{kt} - C_{kt})^+ \right] \right) \tag{6}$$

subject to

$$\sum_{t=0}^{T-1} x_{jt} = 1, \quad \forall j \in N \tag{7}$$

$$\sum_{k \in K} \sum_{j \in F^k} \sum_{s=\max(t-p_k+1,0)}^{t} x_{js} \leq 1, \quad \forall t \in [0, T-1] \tag{8}$$

$$(3), (4), (5)$$

$$x_{jt} \in \{0, 1\}, \quad \forall j \in N, \quad \forall t \in [0, T-1] \tag{9}$$

Where $(a)^+ := \max(a, 0)$; $E[.]$ denotes the expectation operator; α and β are weights reflecting the relative importance of the two components of the objective function. The objective function (6) minimises the weighted sum of two components: the cost incurred if train-set j arrives for maintenance on day t (i.e. the quadratic earliness and tardiness costs); and the expected penalties for violating the limits C_t and C_{kt}. Constraint (7) guarantees that each train-set is scheduled for maintenance on a particular day t within the planning horizon. Constraint (8) depicts the restriction on the first operation line. If a train-set j occupies the first operation line in a given time interval, other train-sets are not allowed to arrive during this period. Constraint (9) states that the decision variables are binary.

4   SAA Model

Using the sample average approximation approach, the SOMSP formulation, presented in Sect. 3, can be rewritten as a mixed integer programming model.

Consider a set $\Omega := \{1, \cdots, \omega\}$ of scenarios. Each scenario comprises a vector of realisations of cycle time which are drawn independently from the distributions corresponding to each train family. Let ξ_j^ω be the cycle time of train-set j in scenario ω.

Formulation of SAA Model

$$\text{(SAA) Min } \alpha \sum_{j \in N} \sum_{t=0}^{T-1} c_{jt} x_{jt} + \beta \, \frac{1}{|\Omega|} \sum_{w \in \Omega} \left(\sum_{t=0}^{T-1} \left[\delta z_t^w + \sum_{k \in K} \delta_{kt} z_{kt}^w \right] \right) \qquad (10)$$

subject to

$$(7), (8), (9)$$

$$\sum_{j \in N} \sum_{s=\max(t-\xi_j^w+1,0)}^{t} x_{js} \leq C_t + z_t^w,$$

$$\forall w \in \Omega, \quad \forall t \in [0, T-1] \qquad (11)$$

$$\sum_{j \in F^k} \sum_{s=\max(t-\xi_j^w+1,0)}^{t} x_{js} \leq C_{kt} + z_{kt}^w,$$

$$\forall w \in \Omega, \quad \forall k \in K, \quad \forall t \in [0, T-1] \qquad (12)$$

$$z_t^w \geq 0, z_{kt}^w \geq 0, \qquad \forall w \in \Omega, \quad \forall k \in K, \quad \forall t \in [0, T-1] \qquad (13)$$

The objective function (10) is the weighted sum of two components: the first component is the same as the first component of objective function (6), while the second component is the sample average of the penalties for the additional train-sets. We call the first component and second component of objective function (10), the *earliness tardiness cost* (ETC) and the *resource violation cost* (RVC), respectively. For every scenario w, constraints (11) and (12) describe the additional train-sets, represented by the slack variables z_t^w (z_{kt}^w), based on the limits C_t and C_{kt}. Constraint (13) is the non-negativity constraint.

4.1 Property of SAA Model

SAA has been extensively used in literature to solve stochastic optimisation problems (see for example, [7]). However, solving SAA model by itself can be computationally challenging as the number of scenarios increases. To demonstrate the complexity of SAA model, we perform a preliminary experiment with various number of scenarios. The result in term of computation time is reported in Table 1. In the same table, we also present the computation time required for solving SAA model if a sequence is given. Given a set of train-sets, a sequence is the order in which the train-set arrives at the maintenance centre. Let E be a set of arcs representing the precedence relations obtained from the sequence. The SAA model with sequence (SAA-sequence) includes the addition of the following constraint:

$$\sum_{t=0}^{T-1} t \, x_{it} \leq \sum_{t=0}^{T-1} t \, x_{jt}, \quad \forall (i, j) \in E \qquad (14)$$

Constraint (14) enforces the precedence relations among train-sets according to the given sequence.

Table 1. Computation time (in seconds) required for solving SAA model with and without sequence

Number of scenarios	SAA model	SAA-sequence model
5	288.348	9.375
10	–	21.146
20	–	44.947
30	–	62.048
40	–	81.929
50	–	129.034
100	–	421.410

From Table 1, we note that SAA model cannot be solved within the time limit of one hour when the number of scenarios is larger than or equal to 10 scenarios. On the other hand, solving SAA model with given sequence is relatively easier. Problem instance with 100 scenarios can be solved in less than 500 s.

5   The Solution Method

Motivated by the result in Sect. 4, a genetic search with SAA (GS-SAA) is proposed to solve the problem. The idea of GS-SAA is simple. It combines genetic algorithm for global search and solving SAA-sequence model for a solution. The genetic algorithm is inspired by [5] and has previously been studied in [2]. Detail of the proposed GS-SAA is given below.

Algorithm

1: Generate initial population
2: Decode the chromosomes into solutions by solving SAA-sequence model
3: Evaluate the solutions
4: **for** the number of generations $< IT_{max}$, and time $< T_{max}$ **do**
5:      Select parents
6:      Generate offspring (two-point crossover)
7:      Diversify offspring (mutation)
8:      Decode offspring into solutions by solving SAA-sequence model
9:      Evaluate the solutions
10: **end for**
11: **return** the best solution

The proposed GS-SAA starts with the generation of an initial population of size μ. Then, each chromosome in the initial population is decoded into a sequence. The sequence decoding procedure works as follow. An arrival plan, or a solution P is represented by a chromosome. Each gene in the chromosome corresponds to a train-set and the gene value is generated randomly from the uniform distribution $U(0,1)$. The gene values determine the priority of the train types. Respecting this priority list, the priority of train-sets (i.e. the sequence of train-sets) is obtained by sorting the train-sets in the same train family based on its earliest desired arrival date (see Fig. 1 for an example).

Train-set Index	1	2	3	4	5
Train type	V	K	T	T	T
Due window	[0, 28]	[19, 47]	[30, 58]	[34, 62]	[41, 69]
Chromosome	0.57	0.08	0.84	0.12	0.23
Sorted gene values	0.08	0.12	0.23	0.57	0.84
Priority of train types	K	T	T	V	T
Decoded sequence	2	3	4	1	5

Fig. 1. Example of sequence decoding by train types.

Once the sequence is known, an arrival plan, or a solution P is generated by solving the SAA-sequence model. Next, the solution P is evaluated according to the objective function (6). The evaluation is the same as that given by [2].

Given the current generation, the next generation is produced through elite selection, crossover, and mutation.

In the elite selection phase, some of the best individuals of the current generation will continue to exist in the next generation. The number of surviving individuals is equal to N_{elite}. The elite selection strategy is motivated by the "Survival of the fittest" [10]. However, the drawback is that it can lead to premature convergence (i.e. catch in a local optimum).

In the crossover phase, a child is produced based on the two-point crossover as in [2]. The two parents are randomly selected from a pool of parents. This pool consists of the top N_{elite} best individuals of the current generation and the remaining candidates are created from the Roulette Wheel Selection. Each proportion of the wheel is given to an individual of the current generation. The size of the proportion is equivalent to the probability of selection, which is defined as

$$\text{Probability of selection} = \frac{\text{fitness}}{\text{sum of fitness}},$$

where fitness is equal to the inverse of the value of the objective function (6), and sum of fitness is equal to the total fitness of all individuals.

In the mutation phase, some gene values of the child chromosome will be replaced by a random number sampled from the uniform distribution $U(0,1)$. The replacement is initiated when the random number is less than P_m. Detail of the mutation process is discussed in [2].

6   Computational Results

The GS-SAA method was coded in Python 2 on an Intel Core i5-6500 CPU @3.2 GHz with 8 GB of RAM. The SAA-sequence problem was solved using the Branch-and-Cut method in IBM ILOG CPLEX 12.6.1 on the same computer.

We adopt the test problems in [2] where there are 3 train families with 35 train-sets in total. The planning horizon is set to one year. The parameters associated with the train families are shown in Table 2. Furthermore, we set C_t = 5, $\delta_t = 1$, $\lambda_1 = \lambda_2 = 1$, $\alpha = 1$, and $\beta = 1000$. For the setting of our genetic algorithm, we have $\mu = 20$, $N_{elite} = 2$, $P_m = 0.05$, $IT_{max} = 40$, and $T_{max} = 3600$ (s). The information on the probability distributions of the cycle time by train family is given in Table 3.

Table 2. Parameters for the train families

Train family	$|F^k|$	p_k	Norm		Special	
			C_{kt}	δ_{kt}	C_{kt}	δ_{kt}
1	25	4	3	1	1	10
2	5	5	2	1	1	10
3	5	5	1	1	1	10

Norm refers to normal day and Special refers to special days.

Table 3. Probability distribution information for cycle time by train family.

Train family	Minimum	Most likely	Maximum	Distribution
1	20	25	40	beta-PERT
2	27	30	46	beta-PERT
3	29	30	52	beta-PERT

In using SAA, one question to ask is how many scenarios are required in order to obtain solution with good quality within acceptable computation time. To answer this question, we extend the preliminary experiment in Sect. 4.1 for the SAA-sequence model such that both the solution quality and computation time are presented. The SAA-sequence model is used to solve the test problems with 5, 10, 20, 30, 40, 50, and 100 scenarios. The scenarios are randomly generated. The average objective value and average computation time obtained after 5 runs are reported in Table 4.

From Table 4, it can be seen that the computation time increases significantly as the number of scenarios increases. As a good trade-off between solution quality and computation time, SAA-sequence with up to 10 scenarios is sufficient. Therefore, we set $|\Omega| = 1, 5, 10$.

Table 5 shows the performance of our GS-SAA method with 1, 5, and 10 scenarios, in comparison with the method proposed in [2] (denoted as MIM) and solving SAA model by itself with 5 scenarios (denoted as SAA-5). For GS-SAA and MIM solution approaches, the best solution of the initial population is reported under the column titled "In.", while the best solution of the final population is reported under the column titled "obj.". For SAA-5 solution approach,

Table 4. Average results for 5 runs of SAA-sequence for different number of scenarios

Number of scenarios	Computation time (s)	Objective value	ETC	RVC
5	9.375	206,760	22,447	184.318
10	21.146	205,086	22,613	182.473
20	44.947	199,781	22,743	177.038
30	62.048	199,046	22,743	176.030
40	81.929	199,046	22,743	176.030
50	129.034	199,046	22,743	176.030
100	421.410	199,046	22,743	176.030

the optimal solution obtained by CPLEX is reported under the column titled "CPLEX obj.". For each solution approach, 5 runs of experiment are performed, the average, maximum and minimum of the objective values are also shown in Table 5.

On average, GS_SAA with 10 scenarios produces better solution for the initial population. This result is not surprising since the solution quality of the approximation of the objective function (6) by its sample average increases with increasing number of scenarios as demonstrated in Table 4.

To test the impact of the number of scenarios on the performance of GS_SAA, we consider the relative improvement in the objective function value over the best solution of the initial population and it can be calculated by $(In. - obj.)/In. \times 100\%$. On average, the relative improvement in the objective function value over the best solution of the initial population is approximately 27.5%, 28.8%, 19.9%, respectively, for 1, 5, 10 scenarios. The small relative improvement of GS_SAA with 10 scenarios is due to the amount of time required for solving SAA-sequence model; as a result, only a few generations are explored and the search capability of genetic algorithm is not fully employed. This observation suggests that GS_SAA with 10 scenarios can potentially give better solution if it is allowed to run for a longer period of time.

Furthermore, as shown in Table 5, the performance of GS_SAA is consistently better than MIM irrespective of the number of scenarios used. The method used in MIM to transform a sequence into a feasible arrival plan is just a simple greedy heuristic and it suffers from poor performance as the idle time inserted between train-sets is not optimal. On the other hand, given the sequence, GS_SAA generates an optimal arrival plan by solving the corresponding SAA-sequence model using CPLEX.

Compare GS_SAA with SAA-5, it can be noted that the latter outperforms GS_SAA in all runs of experiment. This reveals that a good sequence is crucial to obtaining a good solution. However, as demonstrated in Table 1, solving SAA model by itself is computationally challenging if the number of scenarios increases or the problem size becomes large (CPLEX cannot obtain optimal solution in 1 h given the problem size in this paper). This shows the limitation of SAA model in large applications.

Table 5. Comparison of the performance of the different solution approaches (solution quality).

Run	MIM		GS_SAA						SAA-5
			1 scenario		5 scenarios		10 scenarios		
	In.	obj.	In.	obj.	In.	obj.	In.	obj.	CPLEX obj.
1	607,466	475,601	343,337	248,974	290,096	247,636	331,497	230,787	218,949
2	533,746	475,601	340,894	227,894	344,851	238,421	330,219	246,854	226,197
3	624,286	475,601	379,608	242,879	370,441	238,200	314,281	280,038	213,755
4	629,515	475,601	344,171	331,267	338,932	234,148	331,583	282,617	212,428
5	593,304	475,601	320,719	300,493	345,213	243,890	348,886	286,488	216,267
Average	597,663	475,601	345,745	270,301	337,906	240,495	331,293	265,357	217,519
Max	629,515	475,601	379,607	331,267	370,441	247,636	348,886	286,488	226,197
Min	533,746	475,601	320,719	227,894	290,096	234,148	314,281	230,787	212,428

In. is the best solution of the initial population.
obj. is the objective value. CPLEX obj. is the optimal solution obtained by CPLEX

Table 6 shows the computation times in seconds obtained by MIM, GS_SAA with 1, 5 and 10 scenarios, and SAA-5. On average, SAA-5 can be solved to optimality in 288.35 s, whereas MIM takes more than 1288.55 s, and GS_SAA takes 1972.81, 3844.88, 4137.58 s, respectively, for 1, 5, 10 scenarios. Both MIM and GS_SAA require significant computation time because both methods could only terminate after the predefined maximum number of generations and time limit are reached.

Table 6. Comparison of the performance of the different solution approaches (Computation time).

Run	MIM	GS_SAA			SAA-5
		1 scenario	5 scenarios	10 scenarios	
1	1263.52	2053.31	3796.17	4132.10	137.55
2	1255.65	2036.93	3921.30	4104.96	202.65
3	1439.97	1903.90	3803.61	4051.59	135.22
4	888.78	1881.32	3829.91	4097.07	254.10
5	1594.85	1988.57	3873.41	4300.58	712.23
Average	1288.55	1972.81	3844.88	4137.58	288.35
Max	1594.85	2053.31	3921.30	4300.58	712.23
Min	888.78	1881.32	3796.18	4051.59	135.22

7   Conclusions

This paper examines the overhaul maintenance scheduling problem with stochastic cycle time. We present the mathematical formulation of SOMSP and show how it can be formulated as a mixed integer programming model using the sample average approximation approach. A combined genetic search with SAA is developed to solve the problem. The result shows that solving the SAA model by itself is computationally challenging as the number of scenarios become large. However, if a sequence of train-sets is given, the computation time decreases significantly. Computational results reveal that the proposed GS-SAA can produce good solution within acceptable time for test problem consisting of 35 train-sets and a planning horizon of one year.

In conclusion, SAA can be used for small instances since CPLEX can obtain optimal solution within acceptable computation time. For larger application, GS_SAA is the preferred choice. It is noted that the performance of GS_SAA depends on having both a good sequence and good inserted idle time. Future work can investigate methods to further reduce the computation time of SAA-sequence model to enable the search capability of the genetic algorithm.

References

1. Doganay, K., Bohlin, M.: Maintenance plan optimization for a train fleet. WIT Trans. Built Environ. **114**, 349–358 (2010)
2. Gu, H., Joyce, M., Lam, H.C., Woods, M., Zinder, Y.: A genetic algorithm for assigning train arrival dates at a maintenance centre. In: 9th IFAC Conference on Manufacturing Modelling, Management and Control (MIM), Berlin, Germany, 28–30 August 2019 (2019)
3. Keller, B., Bayraksan, G.: Scheduling jobs sharing multiples resources under uncertainty: a stochastic programming approach. IIE Trans. **42**, 16–30 (2010)
4. Lai, Y.C., Fan, D.C., Huang, K.L.: Optimizing rolling stock assignment and maintenance plan for passenger railway operations. Comput. Ind. Eng. **85**, 284–295 (2015)
5. Li, H., Demeulemeester, E.: A genetic algorithm for the robust resource leveling problem. J. Sched. **19**(1), 43–60 (2016)
6. Lin, B., Wu, J., Lin, R., Wang, J., Wang, H., Zhang, X.: Optimization of high-level preventive maintenance scheduling for high-speed trains. Reliab. Eng. Syst. Saf. **183**, 261–275 (2019)
7. Mancilla, C., Storer, R.: A sample average approximation approach to stochastic appointment sequencing and scheduling. IIE Trans. **44**(8), 655–670 (2012)
8. Sriskandarajah, C., Jardine, A.K.S., Chan, C.K.: Maintenance scheduling of rolling stock using a genetic algorithm. J. Oper. Res. Soc. **49**(11), 1130–1145 (1998)
9. Tsuji, Y., Kuroda, M., Kitagawa, Y., Imoto, Y.: Ant colony optimization approach for solving rolling stock planning for passenger trains. In: 2012 IEEE/SICE International Symposium on System Integration (SII), Fukuoka, Japan, 16–18 December 2012 (2012)

10. Valente, J.M.S., Moreira, M.R.A., Singh, A., Alves, R.A.F.S.: Genetic algorithms for single machine scheduling with quadratic earliness and tardiness costs. Int. J. Adv. Manufact. Technol. **54**(1), 251–265 (2011)
11. Wan, L., Yuan, J.: Single-machine scheduling to minimize the total earliness and tardiness is strongly NP-hard. Oper. Res. Lett. **41**(4), 363–365 (2013)

Data Mining and Data Processing

eDTWBI: Effective Imputation Method for Univariate Time Series

Thi-Thu-Hong Phan[1]([envelope]) [ORCID], Émilie Poisson Caillault[2], and André Bigand[2] [ORCID]

[1] Vietnam National University of Agriculture, Trau Quy, Gia Lam, Hanoi, Vietnam
ptthong@vnua.edu.vn
[2] Univ. Littoral Côte d'Opale, EA 4491-LISIC, 62228 Calais, France
{emilie.caillault,bigand}@univ-littoral.fr

Abstract. Missing data frequently occur in many applied domains and pose serious problems such as loss of efficiency and unreliable results for various approaches. Many real applications require complete data, thus, the filling procedure is a mandatory and precursory pre-processing step. DTWBI is a previously proposed method to estimate missing data in univariate time series with recurrent data. This paper introduces an extension of DTWBI, namely eDTWBI. Firstly, we simultaneously find the two most similar windows to the sub-sequences before and after a gap using DTWBI. Secondly, we impute the gap by average values of the following and previous sub-sequence of the most similar values. Experimental results on three datasets show that our approach outperforms than seven related methods in case of time series having effective information.

Keywords: Imputation · Missing data · Univariate time series · Dynamic time warping · Similarity

1 Introduction

Lots of useful information can be exploited from collected time series and they are used in different domains such as economics [24], finance area [3], health-care [7], meteorology [4,19] and traffic engineering [16]. But the collected data are usually incomplete for various reasons as sensor errors, transmission problems, incorrect measurements, bad weather conditions (outdoor sensors) to manual maintain, etc. Missing data can generate inaccurate data interpretation, biased and unreliable results [8]. Moreover, most of proposed models for time series analysis suffer from one major drawback, which is their inability to process incomplete datasets, despite their powerful techniques. An easy way is to delete or ignore missing data. But this solution comes at high price because of losing valuable information especially for time series where considered values depend on the past ones. So, replacing missing data is a mandatory and precursory pre-processing task. The imputation technique is a conventional method to handle the this problem [11].

© Springer Nature Switzerland AG 2020
H. A. Le Thi et al. (Eds.): ICCSAMA 2019, AISC 1121, pp. 121–132, 2020.
https://doi.org/10.1007/978-3-030-38364-0_11

Imputation methods can be categorized into 2 types: (1) multivariate imputation techniques and (2) univariate imputation approaches. For the first type,
these techniques take advantages of relations between variables to estimate missing data [6,7,21,22]. These methods handle incomplete data by filling missing
features based on observable ones. They usually train separate models, such as
missForest [21], ELM (extreme learning machines) [20], MLP (multi-layer per-
ceptron) [10], etc., for estimating the unobserved attributes.

However, when dealing with missing data in univariate time series, we can
only exploit available observations of this variable to predict incompleteness
data. Moritz et al. pointed out this task is a particularly challenging [13]. And,
they performed a review of various methods for univariate time series and showed
limitations of some other approaches in [13]. Fewer studies investigates to fill
missing data in univariate time series. Simple methods are often used as mean [1],
median [5], locf (last observation carried forward), linear interpolation or spline
interpolation. For the interpolation methods, missing data are estimated from
preceding and succeeding values of the univariate time series. These techniques
are effective when the missing data type is isolated (one missing point) or small
gap. But when the gap is large, i.e., many consecutive missing values, they do
not give good results. For example, if a gap has a sine wave shape, the linear
interpolation would complete the gap by a straight line. In addition, we also
use statistical methods (e.g. ARMA or ARIMA) to complete missing data in
univariate time series but these models require linear data after differencing [2].

Therefore, it is necessary to propose effective imputation methods for univariate time series and consider the characteristics of data, especially for complex
distribution data.

In our previous study [15], we proposed DTWBI approach which enables
to impute large consecutive missing values in univariate time series. DTWBI
is based on the combination of the shape-feature extraction algorithm [14] and
Dynamic Time Warping method [17]. In this study, we define a large gap when
number of consecutive missing points is larger than the known-process change, so
it depends on each application. In order to improve imputation ability, we introduce a novel and effective method for univariate time series, namely eDTWBI
which is an extension of DTWBI. Besides, we compare the proposed method
with heuristic approach (called Random method) and study the performance of
conserving frequency information of all considered methods after the imputation.

This paper is organized as follows. Section 2 focuses on the proposed method.
Next, Sect. 3 introduces our experiments, results and discussion. Finally, conclusions are drawn and future work is presented.

2   The Proposed Method: eDTWBI

In the DTWBI algorithm we only envisaged one query either before or after the
considered gap. In this study, we modify DTWBI by taking into account two
queries, one query before and one query after this gap. Moreover, data before
and data after the gap will be treated as two referenced univariate time series.

This would, on the one hand, enrich the learning base and, consequently, increase the prediction ability of the method. On the other hand, this allows to consider dynamics (important key) of data before and after the considered gap to estimate imputation values and to relax temporal constraints between two queries.

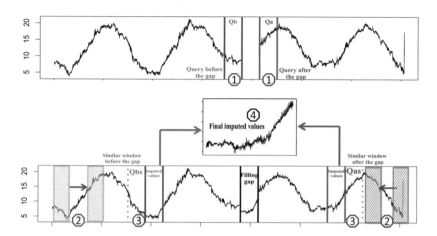

Fig. 1. Scheme of eDTWBI for the imputation task: 1-Queries building, 2-Sliding windows comparison, 3-Most similar windows selection, 4-Gap filling

The eDTWBI algorithm is implemented in order to always ensure accurate results. First, if the position of a gap (with size T) is in the first $2 \times T$ of database, only DTWBI is applied on the remaining data after the gap. If the gap position is in the last $2 \times T$ of series, only DTWBI is performed on data before the gap. In the case of missing data locates in the middle of the series, i.e between $2 \times T$ and $N - 2 \times T$ (where N is the length of the time series), eDTWBI is applied to impute missing data. Figure 1 illustrates the mechanism of eDTWBI to fill large missing data in univariate time series and the detailed algorithm is described in Algorithm 1. This approach consists of three main phases described as follows:

The first phase - Queries building (cf. 1 in Fig. 1): For each T-gap, two referenced time series are extracted from the original signal and two queries are created. The data before the gap (namely Db) and the data after this gap (noted Da) are treated as two separated time series. We noted Qb is the sub-sequence before the gap and Qa is the sub-sequence after the gap, respectively. Qa and Qb queries have the same size T as the gap.

The second phase - Retrieving the most similar windows (cf. 2 & 3 in Fig. 1): For the Da database, sliding reference windows (denoted R) of size T are built. From these R windows, we use DTWBI method [15] to find the most similar window (Qas) to Qa. The same process is carried out to retrieve the most similar window Qbs in Db data.

A key-point of the eDTWBI approach is to envisage the dynamics and shape of data before and after a gap. This means that two queries before and after the

---

Algorithm 1. eDTWBI algorithm

Input: $X = \{x_1, x_2, \ldots, x_N\}$: incomplete time series
  t: index of a gap (position of the first missing of the gap)
  T: size of the gap
  θ_cos: cosine threshold (≤ 1)
  $step\_threshold$: increment for finding a threshold
  $step\_sim\_win$: increment for finding a similar window
Output: Y - completed (imputed) time series

 1: For each gap ContainsMissing(X) do:
 2: Step 1: Divide X into two separated time series Da, Db: $Da = X[t+T:N], Db = X[1:t-1]$
 3: Step 2: Construct queries Qa, Qb - temporal window after and before the missing data $Qa = Da[1:T]; Qb = Db[t-T+1:t-1]$
 4: For Db data do
 5: Step 3: Find the threshold on the Db data
 6:   $i \leftarrow 1; DTW\_costs \leftarrow NULL$
 7: **while** $i <= length(Db)$ **do**
 8:     $k \leftarrow i + T - 1$
 9:     Create a reference window: $R(i) = Db[i:k]$
10:     Calculate global feature of Qb and $R(i)$: $gfQb, gfR$
11:     Compute cosine coefficient: $cos = cosine(gfQb, gfR)$
12:     **if** $cos \geq \theta\_cos$ **then**
13:         Calculate DTW cost: $cost = DTW\_cost(Qb, R(i))$
14:         Save the cost to DTW_costs
15:     $i \leftarrow i + step\_threshold$
16: $threshold = min\{DTW\_costs\}$
17: Step 4: Find similar windows on the Db data
18: $i \leftarrow 1; Lopb \leftarrow NULL$
19: **while** $i <= length(Db)$ **do**
20:     $k \leftarrow i + T - 1$
21:     Create a reference window: $R(i) = Db[i:k]$
22:     Calculate global feature of Qb and $R(i)$: $gfQb, gfR$
23:     Compute cosine coefficient: $cos = cosine(gfQb, gfR)$
24:     **if** $cos \geq \theta\_cos$ **then**
25:         Calculate DTW cost: $cost = DTW\_cost(Qb, R(i))$
26:         **if** $cost < threshold$ **then**
27:             Save position of $R(i)$ to $Lopb$
28:     $i \leftarrow i + step\_sim\_win$
29: **return** Qbs - the most similar window to Qb having the minimum DTW cost in the $Lopb$ list.
30: For Da data do
31: Perform step 3 and 4 with Da data
32: **return** Qas - the most similar window to Qa
33: Step 5: Replace the missing values at the position t by average vector of the window after the Qbs and the one previous the Qas

---

studied gap we considered. This allows to detect windows that have the most similar dynamics and shape to the queries.

The third phase - Completing the gap (cf. 4 in Fig. 1): When the two most similar windows are found, we impute the gap by averaging values of the previous window of Qas and the following window of Qbs. In the eDTWBI approach, the average values are used because Schomaker and Heumann indicated that model averaging makes the final results more stable and unbiased [18].

3 Experiments

To illustrate performance of the proposed method, we evaluate it and compare with other imputation methods including DTWBI [15], Kalman [12], na.interp [9], na.locf, na.aggregate and na.spline [25] and heuristic method. To perform the last comparison, we randomly chose 10 windows having the same size of the gap, then compute the average values to fill in the gap.

3.1 Data Description

Four time series are utilized to perform experiments including monthly mean C02 concentrations [23], daily mean air temperature at the Cua Ong meteorological station, monthly mean air temperature and humidity at the Phu Lien meteorological station, in Vietnam. In order to obtain useful information from the datasets and to make the datasets easily exploitable, we analyzed these series. Table 1 summarizes their characteristics. These datasets have a seasonality component (i.e. an annual cycle) without any linear trend. The seasonality component that would be respected after the imputation but they don't have regular amplitude.

Table 1. Characteristics of time series

No	Dataset name	Period	#Samples	Seasonality (Y/N)	Trend (Y/N)	Frequencey
1	CO2 concentrations	1974–1987	160	Y	N	Monthly
2	Phu Lien humidity	1961–2015	692	Y	N	Monthly
3	Phu Lien air temperature	1961–2014	684	Y	N	Monthly
4	Cua Ong air temperature	1973–1999	9859	Y	N	Daily

1. CO2 concentrations - This dataset contains monthly mean CO2 concentrations at the Mauna Loa Observatory from 1974 to 1987 ([23]).
2. Phu Lien humidity - This dataset, containing monthly mean air humidity at the Phu Lien meteorological station in Vietnam, was collected from 1/1961 to 8/2015.
3. Phu Lien air temperature - This dataset is composed of monthly mean air temperature at the Phu Lien meteorological station in Vietnam from 1/1961 to 12/2014.

4. Cua Ong temperature - daily mean air temperature at the Cua Ong meteorological station in Vietnam from 1/1/1973 to 31/12/1999.

3.2 Experiment Process

Actually, assessing the performance of imputation methods can not be done
because the real values are missing. Thus, we must generate artificial missing
data on complete time series in order to compare the ability of imputation methods. A technique of three steps is used to conduct experiments described in detail
as follows:

- *The first step*: Simulated missing data are produced by deleting data segments
  from each time series with different size of consecutive values.
- *The second step*: All imputation algorithms are applied to estimate the missing values
- *The third step*: The true values and imputed data (generated from different
  approaches above-mentioned) are compared.

Here, 5 missing data levels are considered on 4 datasets. For CO2 and Phu
Lien series, the imputation size ranges from 6%, 7.5%, 10%, 12.5% and 15% of
their size respectively. For Ong Ong series, this is a quite big dataset, so gaps are
created with size of 3%, 3.75%, 5%, 6.25% and 7.5% dataset length (the largest
gap of this time series is 739 missing points i.e. equivalent to more than 2 years
of missing data).

For each missing rate in a dataset, 10 missing positions are randomly chosen
and all the algorithms are conducted.

3.3 Imputation Performance Indicator

After completing missing data, experiment results are discussed in two parts
viz., quantitative performance and visual ability. Specially, the quantitative performance is analyzed in amplitude, shape and frequency criteria. To compare the
amplitude between imputation values and actual ones, we use Similarity (Sim),
an adapted Normalized Mean Absolute Error (NMAE), Root Mean Square Error
(RMSE). Fractional Bias (FB) is applied to compare the shape between prediction data and real data. To assess the ability of frequency conservation of
imputation methods, we perform a comparison between the seasonality components of the full series and the imputed signal using NMAE (denoted NMAE(s)).
These indicators are computed as following:

1. Similarity - defines the similar percentage between the imputed value (Y) and
   the respective true values (X). It is calculated by:

$$Sim(Y, X) = \frac{1}{T} \sum_{i=1}^{T} \frac{1}{1 + \frac{|y_i - x_i|}{max(X) - min(X)}} \tag{1}$$

Where T is the number of missing values. A higher similarity ($\in [0, 1]$) high-
lights a better ability to complete missing values.

2. NMAE, the Normalized Mean Absolute Error between the imputed value Y and the respective true value time series X is computed as:

$$NMAE(Y, X) = \frac{1}{T} \sum_{i=1}^{T} \frac{|y_i - x_i|}{V_{max} - V_{min}} \qquad (2)$$

where V_{max}, V_{min} are the maximum and the minimum value of original time series. A lower NMAE means better performance method for the imputation task.

3. RMSE: The Root Mean Square Error is defined as the average squared difference between the imputed value Y and the respective true value time series X. This indicator is very useful for measuring overall precision or accuracy. In general, the more effective method would have a lower RMSE.

$$RMSE(Y, X) = \sqrt{\frac{1}{T} \sum_{i=1}^{T} (y_i - x_i)^2} \qquad (3)$$

4. FB (Fractional Bias) is defined by:

$$FB(Y, X) = 2 * \left| \frac{mean(Y) - mean(X)}{mean(Y) + mean(X)} \right| \qquad (4)$$

A model is considered perfect when its FB tends to 0.

3.4   Experiments Results

This part presents experiment results obtained from the proposed approach and compares its ability with the seven published methods.

Tables 2 and 3 show the averaged performance of different imputation methods on 4 datasets for the 5 indicators previously mentioned. These results confirm that eDTWBI is more effective than compared methods in most of the cases, especially in relative high missing rate scenario.

On the CO2 and Phu Lien temperature series, eDTWBI provides the highest Similarity, the lowest NMAE, RMSE and NMAE(s) at nearly every missing ratio (excluding NMAE(s) at 7.5% on the CO2 series). The results demonstrate that the imputation values using eDTWBI method are close to the real values, especially for large missing size (at 12.5% and 15% on Phu Lien temperature and CO2 series). In addition, our method is capable of preserving frequency, which is disclosed on the NMAE(s) index. FB is a quantitative index that allows a shape comparison between predicted and true values. When looking at FB index on Tables 2 and 3, FB values of eDTWBI are the smallest in the majority of missing rates and ranked the second at some levels like 10%, 12.5% on CO2 series, at 15% on Phu Lien temperature, and ranked the 3rd at 7.5%, 10% and 15% on Phu Lien humidity signal. Again, the results show that the improved ability to estimate missing values of eDTWBI method in terms of shape.

Table 2. Average imputation performance indices of various imputation algorithms on CO2 and Phu Lien temperature datasets

Method	Gap size	CO2					Phu Lien temperature				
		Sim	NMAE	RMSE	FB	NMAE (s)	Sim	NMAE	RMSE	FB	NMAE (s)
Random	6%	0.625	0.196	5.22	0.013	0.03	0.779	0.237	4.82	0.018	0.021
Kalman		0.754	0.097	2.768	0.006	0.019	0.58	0.733	14.924	0.48	0.02
DTWBI		0.832	0.055	1.509	0.004	0.009	0.878	0.114	2.576	0.023	0.009
eDTWBI		**0.919**	**0.024**	**0.693**	**0.001**	**0.006**	**0.916**	**0.075**	**1.7**	**0.01**	**0.005**
na.interp		0.731	0.106	2.973	0.006	0.022	0.778	0.244	5.28	0.062	0.02
na.locf		0.721	0.114	3.22	0.006	0.024	0.775	0.257	5.718	0.15	0.019
Aggregate		0.636	0.18	4.802	0.012	0.028	0.791	0.216	4.379	0.016	0.019
na.spline		0.764	0.092	2.66	0.006	0.019	0.599	0.694	14.379	0.433	0.02
Random	7.5%	0.671	0.153	4.042	0.009	0.022	0.798	0.227	4.607	0.014	0.026
Kalman		0.726	0.126	3.607	0.008	0.024	0.534	0.993	19.84	1.273	0.027
DTWBI		0.798	0.068	1.731	0.004	**0.008**	0.883	0.119	2.631	0.022	0.012
eDTWBI		**0.889**	**0.034**	**0.924**	**0.001**	0.01	**0.913**	**0.086**	**1.983**	**0.011**	**0.008**
na.interp		0.737	0.105	3.026	0.005	0.026	0.772	0.281	6.071	0.144	0.026
na.locf		0.725	0.115	3.359	0.008	0.024	0.776	0.273	5.8	0.152	0.025
Aggregate		0.681	0.14	3.846	0.009	0.024	0.797	0.228	4.605	0.013	0.026
na.spline		0.741	0.117	3.414	0.008	0.023	0.547	0.957	19.432	1.241	0.027
Random	10%	0.644	0.196	5.145	0.013	0.041	0.797	0.236	4.802	0.013	0.035
Kalman		0.71	0.15	4.572	0.009	0.033	0.484	1.3	26.479	1.58	0.035
DTWBI		0.735	0.122	3.595	0.009	0.035	0.885	0.12	2.691	0.021	0.016
eDTWBI		**0.804**	**0.082**	**2.271**	0.004	**0.025**	**0.912**	**0.089**	**2.065**	**0.009**	**0.011**
na.interp		0.777	0.09	2.54	**0.002**	0.031	0.787	0.255	5.395	0.029	0.035
na.locf		0.74	0.114	3.202	0.006	0.03	0.775	0.293	6.49	0.189	0.034
Aggregate		0.629	0.204	5.344	0.014	0.038	0.799	0.23	4.644	0.014	0.034
na.spline		0.658	0.591	19.906	0.046	0.043	0.475	1.322	27.19	3.809	0.035
Random	12.5%	0.634	0.199	5.262	0.014	0.047	0.797	0.234	4.756	0.009	0.043
Kalman		0.761	0.107	3.238	**0.003**	0.041	0.622	0.722	15.389	0.372	0.042
DTWBI		0.731	0.122	3.508	0.008	0.036	0.879	0.128	2.835	0.013	0.021
eDTWBI		**0.804**	**0.083**	**2.43**	0.005	**0.031**	**0.901**	**0.101**	**2.304**	**0.008**	**0.016**
na.interp		0.767	0.1	2.886	0.006	0.044	0.78	0.282	6.263	0.171	0.042
na.locf		0.744	0.117	3.338	0.007	0.048	0.763	0.315	6.95	0.229	0.043
Aggregate		0.63	0.206	5.446	0.014	0.043	0.803	0.225	4.547	0.009	0.042
na.spline		0.756	0.113	3.533	0.005	0.041	0.537	1.112	23.34	1.149	0.042
Random	15%	0.674	0.199	5.308	0.014	0.047	0.798	0.234	4.731	**0.007**	0.052
Kalman		0.651	0.233	6.319	0.015	0.053	0.45	1.45	29.078	9.551	0.052
DTWBI		0.747	0.124	3.313	0.008	0.033	0.886	0.12	2.684	0.012	0.023
eDTWBI		**0.831**	**0.082**	**2.297**	0.005	**0.031**	**0.897**	**0.107**	**2.388**	0.008	**0.021**
na.interp		0.771	0.115	3.311	0.007	0.05	0.782	0.277	6.192	0.149	0.051
na.locf		0.744	0.135	3.794	0.008	0.05	0.777	0.288	6.404	0.191	0.05
Aggregate		0.699	0.176	4.778	0.011	0.05	0.801	0.227	4.585	0.008	0.05
na.spline		0.662	0.223	6.14	0.015	0.051	0.527	1.097	22.954	1.278	0.051

Cua Ong time series is long so we pay special attention to the shape and dynamics of the imputation values. This is very important when we fill in large missing data. Therefore, we take into account another index, FA2. It represents the fraction of data points that satisfied smoothing amplitude cover. This indicator is calculated as $FA2(Y,X) = \frac{length(0.5 \leq \frac{Y}{X} \leq 2)}{length(X)}$. For the imputation task, if FA2 is closer to 1, the imputation values are closer to the real values. When

Table 3. Average imputation performance indices of various imputation algorithms on Phu Lien humidity and Cua Ong series

Method	Gap size	Phu Lien humidity					Cua Ong temperature				
		Sim	NMAE	RMSE	FB		Sim	NMAE	RMSE	FB	FA2
Random	6%	0.858	0.135	6.8	0.021	0.021	0.83	0.18	54.7	0.042	0.98
Kalman		0.845	0.153	7.4	0.041	0.019	0.83	0.19	58.6	0.152	0.97
DTWBI		0.861	0.132	6.2	0.023	**0.011**	0.901	0.10	33.3	**0.022**	0.993
eDTWBI		**0.877**	**0.114**	**5.6**	0.018	**0.011**	**0.906**	**0.09**	**31.3**	0.028	**1.00**
na.interp		0.828	0.176	8.3	0.054	0.02	0.83	0.19	58.6	0.152	0.97
na.locf		0.786	0.236	10.4	0.096	0.021	0.80	0.23	72.2	0.18	0.95
Aggregate		0.865	0.126	6.3	0.019	0.019	0.82	0.19	56.3	0.042	0.98
na.spline		0.534	0.908	37.4	0.4	0.02	0.39	2.43	727.6	2.077	0.21
Random	7.5%	0.84	0.125	5.9	0.02	0.019	0.84	0.18	53.2	**0.014**	0.98
Kalman		0.834	0.132	6.0	0.026	0.02	0.81	0.22	68.5	0.145	0.95
DTWBI		0.851	0.118	5.7	0.016	0.011	0.89	0.10	34.7	**0.016**	0.99
eDTWBI		**0.859**	**0.11**	**5.3**	0.022	**0.01**	**0.912**	**0.09**	**30.2**	0.02	**0.994**
na.interp		0.844	0.125	5.9	0.022	0.02	0.81	0.22	68.5	0.145	0.95
na.locf		0.821	0.149	6.9	0.047	0.019	0.81	0.21	67.4	0.154	0.96
Aggregate		0.84	0.124	5.8	0.02	0.019	0.83	0.18	54.1	**0.014**	0.98
na.spline		0.459	1.193	51.9	0.437	0.02	0.31	3.91	1153.9	2.256	0.15
Random	10%	0.865	0.124	6.1	0.008	0.031	0.83	0.19	57.8	0.052	0.98
Kalman		0.85	0.143	6.9	0.037	0.031	0.83	0.20	62.1	0.056	0.97
DTWBI		0.859	0.134	6.8	0.015	0.025	0.903	0.10	32.6	0.025	0.99
eDTWBI		0.864	0.127	6.3	0.012	**0.023**	**0.912**	**0.09**	**28.7**	0.022	**0.999**
na.interp		0.84	0.155	7.3	0.047	0.031	0.83	0.20	62.1	0.056	0.97
na.locf		0.834	0.163	7.6	0.05	0.031	0.83	0.20	64.1	0.134	0.97
aggregate		**0.872**	**0.116**	**5.8**	**0.008**	0.03	0.83	0.18	54.9	0.051	0.98
na.spline		0.423	1.817	74.8	0.69	0.032	0.34	3.42	1005.6	3.035	0.18
Random	12.5%	0.866	0.122	6.0	0.012	0.036	0.82	0.21	61.7	0.031	0.97
Kalman		0.85	0.143	6.8	0.026	0.037	0.82	0.21	66.8	0.153	0.97
DTWBI		0.867	0.122	6.0	0.013	**0.019**	0.9	0.11	35.9	0.023	0.991
eDTWBI		**0.874**	**0.115**	**5.8**	0.009	**0.019**	**0.91**	**0.09**	**31.7**	**0.02**	**0.998**
na.interp		0.821	0.179	8.3	0.048	0.037	0.82	0.21	66.8	0.153	0.97
na.locf		0.786	0.221	9.5	0.086	0.035	0.82	0.22	67.9	0.159	0.96
Aggregate		0.873	0.116	5.8	**0.009**	0.036	0.84	0.18	54.3	0.027	0.98
na.spline		0.39	2.103	87.0	1.833	0.04	0.28	5.38	1625.1	2.045	0.13
Random	15%	0.859	0.135	6.2	0.019	0.042	0.84	0.18	53.9	0.013	0.98
Kalman		**0.871**	0.125	6.0	**0.017**	0.039	0.82	0.22	68.2	0.157	0.96
DTWBI		0.863	0.133	6.4	0.026	**0.023**	0.91	0.10	33.1	0.018	0.99
eDTWBI		0.87	**0.123**	**5.9**	0.02	**0.023**	**0.913**	**0.09**	**30.4**	0.018	**0.999**
na.interp		0.865	0.133	6.3	0.02	0.039	0.82	0.22	68.2	0.157	0.96
na.locf		0.854	0.148	7.0	0.041	0.039	0.82	0.21	66.4	0.148	0.97
Aggregate		0.867	0.125	**5.9**	0.018	0.039	0.84	0.18	53.9	**0.01**	0.98
na.spline		0.314	2.674	111.0	1.816	0.04	0.23	6.55	1901.0	6.482	0.08

looking at the results in Table 3, eDTWBI method proves its superior ability compared to other methods for the task of completing missing data on large datasets. It provides the highest Similarity and FA2, the lowest NMAE and RMSE at every missing levels.

Besides, the visualization performance of imputed values generated from different methods is studied. Figure 2 presents the shape of imputation data using 7 different methods on CO2 series. DTWBI well respects the shape of real values.

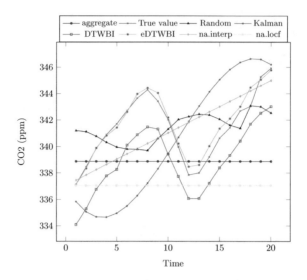

Fig. 2. Visual comparison of imputed values of different univariate methods with true values on CO2 series with the gap size of 20 (12.5%) at postion 89.

But when comparing it with eDTWBI, this approach proves again its ability to deal with missing data. The dynamics of prediction values yielded by eDTWBI is almost identical to the form of true values.

Although the FB value of Kalman method is the 1st on the CO2 and Phu Lien humidity series at 12.5% missing ratio (Tables 2 and 3), but when looking at Fig. 2, it clearly shows that the amplitude and shape of imputation values produced by Kalman differ greatly from the acutal values.

Figure 2 also shows that three methods, including na.aggregate, na.locf and na.interp, always provide a straight line even though quantitative indicators are quite good (Table 2). This means that they do not respect the shape of true values. Random is heuristic method but it provides better results than Kalman or spline in most cases on the Phu Lien series (Table 2).

4   Conclusions and Future Work

This paper proposes a new method, namely eDTWBI, for imputing missing data in univariate time series. The eDTWBI method is an extension of DTWBI by finding the similar values in both databases before and after each gap. It is evaluated and compared with seven other methods on 4 datasets using different criteria (amplitude, frequency and shape constraints). The obtained results clearly demonstrate that our method provides improved performance. However, eDTWBI is based on an assumption of recurring data. In the future, we intend to combine eDTWBI with other algorithms such as interpolation methods to effectually complete missing data in every type of univariate time series.

References

1. Allison, P.D.: Missing Data, Quantitative Applications in the Social Sciences, vol. 136. Sage Publication, Thousand Oaks (2001)
2. Ansley, C.F., Kohn, R.: On the Estimation of ARIMA Models with Missing Values. Springer, New York (1984)
3. Bauer, S., Schlkopf, B., Peters, J.: The arrow of time in multivariate time series, p. 9 (2016)
4. Billinton, R., Chen, H., Ghajar, R.: Time-series models for reliability evaluation of power systems including wind energy. Microelectron. Reliab. **36**(9), 1253–1261 (1996). https://doi.org/10.1016/0026-2714(95)00154-9
5. Bishop, C.M.: Pattern Recognition and Machine Learning. Information Science and Statistics. Springer, New York (2006)
6. Buuren, S., Groothuis-Oudshoorn, K.: mice: Multivariate imputation by chained equations in R. J. Stat. Softw. **45**(3), 1–67 (2011)
7. Che, Z., Purushotham, S., Cho, K., Sontag, D., Liu, Y.: Recurrent neural networks for multivariate time series with missing values. Sci. Rep. **8**(1) (2018). https://doi.org/10.1038/s41598-018-24271-9, http://www.nature.com/articles/s41598-018-24271-9
8. Hawthorne, G., Hawthorne, G., Elliott, P.: Imputing cross-sectional missing data: comparison of common techniques. Aust. N. Z. J. Psychiatry **39**(7), 583–590 (2005)
9. Hyndman, R., Khandakar, Y.: Automatic time series forecasting: the forecast package for R, used package in 2016. J. Stat. Softw. 1–22 (2008). http://www.jstatsoft.org/article/view/v027i03
10. Jerez, J.M., Molina, I., Garca-Laencina, P.J., Alba, E., Ribelles, N., Martn, M., Franco, L.: Missing data imputation using statistical and machine learning methods in a real breast cancer problem. Artif. Intell. Med. **50**(2), 105–115 (2010). https://doi.org/10.1016/j.artmed.2010.05.002, http://www.sciencedirect.com/science/article/pii/S0933365710000679
11. Junninen, H., Niska, H., Tuppurainen, K., Ruuskanen, J., Kolehmainen, M.: Methods for imputation of missing values in air quality data sets. Atmos. Environ. **38**(18), 2895–2907 (2004). https://doi.org/10.1016/j.atmosenv.2004.02.026
12. Moritz, S., Bartz-Beielstein, T.: imputeTS: Time series missing value imputation in R. R J. **9**(1), 207–218 (2017). https://journal.r-project.org/archive/2017/RJ-2017-009/index.html
13. Moritz, S., Sardá, A., Bartz-Beielstein, T., Zaefferer, M., Stork, J.: Comparison of different methods for univariate time series imputation in R. arXiv preprint arXiv:1510.03924 (2015)
14. Phan, T.T.H., Caillault, E.P., Bigand, A.: Comparative study on supervised learning methods for identifying phytoplankton species. In: 2016 IEEE Sixth International Conference on Communications and Electronics (ICCE), pp. 283–288. IEEE, July 2016. https://doi.org/10.1109/CCE.2016.7562650
15. Phan, T.T.H., Caillault, E.P., Lefebvre, A., Bigand, A.: Dynamic time warping-based imputation for univariate time series data. Pattern Recognit. Lett. (2017). https://doi.org/10.1016/j.patrec.2017.08.019
16. Ran, B., Tan, H., Feng, J., Liu, Y., Wang, W.: Traffic speed data imputation method based on tensor completion. Comput. Intell. Neurosci. **2015**, 1–9 (2015). https://doi.org/10.1155/2015/364089
17. Sakoe, H., Chiba, S.: Dynamic programming algorithm optimization for spoken word recognition. IEEE Trans. Acoust. Speech Signal Process. **16**, 43–49 (1978)

18. Schomaker, M., Heumann, C.: Model selection and model averaging after multiple imputation. Comput. Stat. Data Anal. **71**, 758–770 (2014)
19. Shi, X., Chen, Z., Wang, H., Yeung, D.Y., Wong, W.K., Woo, W.C.: Convolutional LSTM network: a machine learning approach for precipitation nowcasting, p. 9 (2015)
20. Sovilj, D., Eirola, E., Miche, Y., Bjrk, K.M., Nian, R., Akusok, A., Lendasse, A.: Extreme learning machine for missing data using multiple imputations. Neurocomputing **174**, 220–231 (2016). https://doi.org/10.1016/j.neucom.2015.03.108, http://www.sciencedirect.com/science/article/pii/S0925231215011182
21. Stekhoven, D.J., Bühlmann, P.: MissForest—non-parametric missing value imputation for mixed-type data. Bioinformatics **28**(1), 112–118 (2012). https://doi.org/10.1093/bioinformatics/btr597
22. Su, Y.S., Gelman, A., Hill, J., Yajima, M., et al.: Multiple imputation with diagnostics (mi) in R: opening windows into the black box. J. Stat. Softw. **45**(2), 1–31 (2011)
23. Thoning, K.W., Tans, P.P., Komhyr, W.D.: Atmospheric carbon dioxide at Mauna Loa observatory. II - analysis of the NOAA GMCC data 1974–1985. J. Geophys. Res.: Atmos. **94**, 8549–8565 (1989)
24. Yang, Y.: Modelling nonlinear vector economic time series, p. 212 (2012)
25. Zeileis, A., Grothendieck, G.: zoo: S3 infrastructure for regular and irregular time series, used package in 2016 (2005). https://doi.org/10.18637/jss.v014.i06, https://www.jstatsoft.org/v014/i06

Reweighted ℓ_1 Algorithm for Robust Principal Component Analysis

Hoai Minh Le[1,2(✉)] and Vo Xuanthanh[3]

[1] Computer Science and Application Department, LGIPM,
University of Lorraine, Metz, France
[2] Institute for Research and Applications of Optimization (VinOptima),
VinTech, VinGroup, Hanoi, Vietnam
v.minhlh5@vinoptima.org
[3] Phuoc Quang, Tuy Phuoc, Binh Dinh, Vietnam

Abstract. In this work, we consider the Robust Principal Components Analysis, a popular method of dimensionality reduction. The corresponding optimization involves the minimization of l_0-norm which is known to be NP-hard. To deal with this problem, we replace the l_0-norm by a non-convex approximation, namely capped l_1-norm. The resulting optimization problem is non-convex for which we develop a reweighted l_1 based algorithm. Numerical experiments on several synthetic datasets illustrate the efficiency of our algorithm and its superiority comparing to several state-of-the-art algorithms.

Keywords: Robust principal component analysis · Sparse optimization · Non-convex optimization · Reweighted-l_1

1 Introduction

Principal Component Analysis (PCA) is certainly one of the most popular methods of dimensionality reduction with a very wide range of applications such as data visualization, image compression, bio-informatics, etc. PCA tries to interpret the data by assuming that it lies on a space of lower dimension. Formally, let $M \in \mathbb{R}^{m \times n}$ be the data matrix whose each column is a data point. One assumes that M can be approximatde by the sum of two matrices $L + S$ where L is a rank-r matrix with $r << \min(m, n)$ and S represents a small noisy perturbation of each data point of L. Thus, the PCA which consists on seeking the best rank-r matrix L can be reformulated as follows

$$\begin{cases} \min_{L, S \in \mathbb{R}^{m \times n}} \|M - L\|_F \\ s.t \quad M = L + S, \\ \quad \text{rank}(L) \leq r, \end{cases} \tag{1}$$

where $\|.\|_F$ is the Frobenius norm. The above optimization problem can be efficiently solved using the singular value decomposition (SVD). Moreover, under

© Springer Nature Switzerland AG 2020
H. A. Le Thi et al. (Eds.): ICCSAMA 2019, AISC 1121, pp. 133–142, 2020.
https://doi.org/10.1007/978-3-030-38364-0_12

the assumption that the data matrix M is corrupted by a small Gaussian noise S, PCA enjoys several optimality properties [10]. However, it is well known that the biggest drawback of PCA is that it is highly sensitive to outliers, e.g., even a single grossly corrupted entry in the data may break down PCA [12]. Unfortunately, in many machine learning applications, it happens often that data contains gross errors. Numerous methods have been developed such as multivariate trimming, alternating minimization, random sampling techniques, etc in order to improve the robustness of PCA [1]. Robust PCA (RPCA) is arguably one of most efficient approaches for this purpose. Unlike PCA, in PCA, one considers that the noise matrix S is sparse and can contain highly corrupted measurements. RPCA aims to recover the data matrix M with the lowest-rank matrix L and the sparsest error matrix S. Thus, RPCA can be formulated as:

$$\begin{cases} \min\limits_{L,S\in\mathbb{R}^{m\times n}} \operatorname{rank}(L) + \lambda\|S\|_0 \\ s.t \quad M = L + S, \end{cases} \tag{2}$$

where $\lambda > 0$ is the trade-off parameters between two terms. Moreover, let $p = \min(m,n)$ and $\sigma(L) = (\sigma_1(L),\ldots,\sigma_p(L))$ be singular values of L in descending order. It is obvious that $\operatorname{rank}(L) = \|\sigma(L)\|_0$. Therefore, the problem (2) can be equivalently expressed in the form

$$\begin{cases} \min\limits_{L,S\in\mathbb{R}^{m\times n}} \|\sigma(L)\|_0 + \lambda\|S\|_0 \\ s.t \quad M = L + S \end{cases} \tag{3}$$

Hence, the RPCA problem involves the minimization of $\|.\|_0$ which is known to be NP-hard. Sparse optimization plays a very important role, especially in machine learning. Numerous methods have been developed for sparse optimization. The readers are referred to the paper [4] for an extensive survey on existing methods for sparse optimization. Among the methods for RPCA, most of the recent researches focus on solving an approximate problem of RPCA by replacing the non-convex functions $\operatorname{rank}(L)/\|\sigma(L)\|$ and $\|S\|_0$ by convex functions. In [1], Candes et al. proposed the so called Principal Component Pursuit (PCP) defined as

$$\begin{cases} \min\limits_{L,S\in\mathbb{R}^{m\times n}} \|L\|_* + \lambda\|S\|_1 \\ s.t \quad M = L + S, \end{cases} \tag{4}$$

where $\operatorname{rank}(L)$ is approximated by the nuclear norm $\|L\|_* = \|\sigma(L)\|_1 = \sum\limits_{i=1}^{p} \sigma_i(L)$ and $\|S\|_0$ is replaced by its convex approximation $\|S\|_1$. Thus, PCP problem is a convex problem for which tractable convex optimization techniques can be investigated. Candes et al. [1] developed an Alternating Direction Multiplier Method (ADMM) for solving (4). Furthermore, the authors proved that, under some weak assumptions, solving (4) can exactly recovers the low-rank L and the sparse error matrix S. Different methods such as Inexact Augmented Lagrangian method (Inexact ALM) [6], Alternating Direction Method [6], etc. were also developed for solving (3). In [10], Wright et al. considered a variant of PCP, namely Stable RPCA, where the constraint $M = L + S$ is replaced by $\|M - L - S\|_F^2 \leq \epsilon$. Hence (4) becomes

$$\min_{L,S \in \mathbb{R}^{m \times n}} \|L\|_* + \lambda \|S\|_1 + \frac{\mu}{2} \|M - L - S\|_F^2 \leq \epsilon. \tag{5}$$

Wright et al. [10] proved that as $\mu \searrow 0$, the solutions of (5) approach the solution set of (4).

Although PCP, Stable RPCA and their variants improve the robustness of PCA in case of grossly corrupted error matrix S, they still show limitations. The use of convex approximations to replace non-convex terms rank(L) and $\|S\|_0$ in the RPCA makes the resulting problem convex, thus easier to solve. However, it is not enough to simply use l_1-norm or l_2-norm to model the noise, since unfortunately the real noise has often more complex structures than simple Gaussian or Laplacian error [11]. In [9], for the first time, Sun et al. used a non-convex function, namely the capped l_1-norm, to replace the l_0-norm. The capped l_1 function is then decomposed as a DC (Difference of two Convex) function and Sun et al. developed a DCA (DC Algorithm) to solve the resulting problem. DCA is well-known as an efficient approach in the nonconvex programming framework thanks to its versatility, flexibility, robustness, inexpensiveness and their adaptation to the specific structure of considered problems [3,7,8]. DCA solves a sequence of convex problems instead of solving the considered non-convex problem. In [9], the proposed DCA requires to solve convex sub-problems for which Augmented Lagrange Method of Multipliers (ADMM) was developed. Unfortunately, ADMM for solving the convex sub-problem is quite slow.

Convinced by the necessity and efficiency of non-convex approximations of l_0-norm, in this work, we will also consider non-convex approximation. More precisely, we deal with the following formulation of RPCA:

$$\min_{L,S \in \mathbb{R}^{m \times n}} \|\sigma(L)\|_0 + \lambda \|S\|_0 + \frac{\mu}{2} \|M - L - S\|_F^2. \tag{6}$$

The remainder of the paper is organized as follows. In Sect. 2, we reformulate the RPCA formulation (6) using the capped ℓ_1-norm then develop a reweighted ℓ_1 based algorithm to solve the resulting problem. Computational experiments are reported in Sect. 3 and Sect. 4 concludes the paper.

2   Reweighted-l_1 Based Algorithm for RPCA

Conventionally, to deal with the minimization of ℓ_0-norm, we replace the hard discontinuous ℓ_0-norm with an appropriate continuous approximation. Then the approximation problem would be more amenable to optimization methods, and in many cases, we can obtain an equivalent reformulation [4]. Following successes of the non-convex approximations in previous works (see [4] and references therein), we will approximate the ℓ_0-norm in (6) by the capped-ℓ_1 function which is defined as follows

$$\varphi_\theta(x) = \theta \min\left\{|x|, \frac{1}{\theta}\right\}, \qquad \forall x \in \mathbb{R},$$

where $\theta > 0$ is the approximation parameter. Specifically, for $\theta > 0$ and $x \in \mathbb{R}^n$, we have

$$\|x\|_0 \approx \sum_{i=1}^{n} \varphi_\theta(x_i).$$

Hence for $\theta_1 > 0$ and $\theta_2 > 0$, we consider the following approximation of problem (6), which will be called the capped principle component analysis (CaPCA) problem,

$$\min_{L,S} \quad \sum_{i=1}^{p} \varphi_{\theta_1}(\sigma_i(L)) + \lambda \sum_{i=1}^{m} \sum_{j=1}^{n} \varphi_{\theta_2}(S_{ij}) + \frac{\mu}{2}\|M - L - S\|_F^2. \qquad (7)$$

In the next sections, we will present the reweighted-l_1 for spare optimization and then develop it to solve the CaPCA problem.

2.1   Reweighted-l_1 for Spare Optimization

In this section, we outline a reweighted-l_1 procedure for solving sparse optimization problems. More details on the subject can be found in [4] (Sect. 6.2 and line r_{cap} of Table 3).

Consider a sparse optimization problem of the form

$$\min_{x \in \mathbb{R}^n} \sum_{i=1}^{n} \varphi_\theta(x_i) + f(x),$$

where $f : \mathbb{R}^n \to \mathbb{R}$ is a finite convex function.

The idea of reweighted-l_1 for iteratively solving the above problem is described as follows. At each iteration k, we replace the nonconvex term $\sum_{i=1}^{n} \varphi_\theta(x_i)$ with the weighted l_1 term $\sum_{i=1}^{n} z_i^k |x_i|$, where x^k is the current iterate and weights z_i^k's are determined by $z_i^k = \theta$ if $|x_i^k| < 1/\theta$ and 0 otherwise. Then the next iterate x^{k+1} will be a solution of the subproblem

$$\min_{x \in \mathbb{R}^n} \sum_{i=1}^{n} z_i^k |x_i| + f(x).$$

The above two steps are repeated until convergence.

2.2   Reweighted-$l1$ for Solving the CaPCA Problem

For solving problem (7), we will adapt the reweighted-ℓ_1 procedure described earlier by regarding $\sigma_i(L)$'s as variables. At iteration k, given the current iterate (L^k, S^k) we need to solve the following reweighted-ℓ_1 subproblem to get the next iterate (L^{k+1}, S^{k+1})

$$\min_{L,S} \quad \sum_{i=1}^{p} \alpha_i^k \sigma_i(L) + \lambda \sum_{i=1}^{m} \sum_{j=1}^{n} \beta_{ij}^k |S_{ij}| + \frac{\mu}{2} \|M - L - S\|_F^2, \qquad (8)$$

where

$$\alpha_i^k = \begin{cases} \theta_1 & \text{if } \sigma_i(L^k) < \frac{1}{\theta_1} \\ 0 & \text{otherwise,} \end{cases} \quad \text{and} \quad \beta_{ij}^k = \begin{cases} \theta_2 & \text{if } |S_{ij}^k| < \frac{1}{\theta_2} \\ 0 & \text{otherwise.} \end{cases} \qquad (9)$$

Problem (8) is generally nonconvex. Fortunately, if either L or S is fixed, the other can be computed explicitly as described below. Thus, we can solve problem (8) by iteratively alternating between L and S. Before proceeding, let $S_\tau : \mathbb{R} \to \mathbb{R}$ (with $\tau > 0$) denote the soft-thresholding operator

$$S_\tau(x) = \text{sgn}(x) \max(|x| - \tau, 0) = \max(x - \tau, 0) + \min(x + \tau, 0). \qquad (10)$$

Computing S with fixed L. Let $A = M - L$, then computing S amounts to solving the problem

$$\min_{S \in \mathbb{R}^{m \times n}} \quad \sum_{i=1}^{m} \sum_{j=1}^{n} \left[\frac{\lambda \beta_{ij}^k}{\mu} |S_{ij}| + \frac{1}{2}(S_{ij} - A_{ij})^2 \right]$$

which has closed-form solution given by

$$\bar{S}_{ij} = S_{\frac{\lambda \beta_{ij}^k}{\mu}}(A_{ij}), \qquad \forall i, j. \qquad (11)$$

Computing L with fixed S. With fixed S, problem (8) reduces to

$$\min_{L \in \mathbb{R}^{m \times n}} \quad \sum_{i=1}^{p} \frac{\alpha_i^k}{\mu} \sigma_i(L) + \frac{1}{2} \|L - B\|_F^2, \qquad (12)$$

where $B = M - S$.

Assume that $B = U \Sigma_B V^T$ is any singular value decomposition (SVD) and $(\sigma_1(B), \ldots, \sigma_p(B)) = \text{diag}(\Sigma_B)$ are nonincreasingly ordered. Due to [2, Corollary 7.4.1.3], we have

$$\sum_{i=1}^{p} \frac{\alpha_i^k}{\mu} \sigma_i(L) + \frac{1}{2} \|L - B\|_F^2 \geq \sum_{i=1}^{p} \frac{\alpha_i^k}{\mu} \sigma_i(L) + \frac{1}{2} \sum_{i=1}^{p} (\sigma_i(L) - \sigma_i(B))^2, \qquad (13)$$

for any $L \in \mathbb{R}^{m \times n}$.

It is easy to show that minimizing the right-hand side of (13) in $\sigma_i = \sigma_i(L)$ with constraints $\sigma_i \geq 0$, $\forall i$, will give the solution $\bar{\sigma} = (\bar{\sigma}_1, \ldots, \bar{\sigma}_p)$, where

$$\bar{\sigma}_i = \max\left(\sigma_i(B) - \frac{\alpha_i^k}{\mu}, 0 \right) = S_{\frac{\alpha_i^k}{\mu}}(\sigma_i(B)), \qquad \forall i = 1, \ldots, p. \qquad (14)$$

By letting $\bar{L} = U \text{diag}(\bar{\sigma}) V^T$, we see that equality holds in (13), while the right-hand side of (13) is minimized. Thus, $\bar{L}$ is the optimal solution of problem (12).

We summarize the proposed algorithm for solving problem (7) below.

Algorithm 1. CaPCA

Require: $M \in \mathbb{R}^{m \times n}$, $\lambda > 0$, $\mu > 0$, $\theta_1 > 0$, $\theta_2 > 0$, $\varepsilon_1 > 0$, $\varepsilon_2 > 0$
$\quad L^0 = S^0 = 0$, $k = 0$, $p = \min(m, n)$
$\quad$ **repeat**
$\quad\quad$ Compute weights $\alpha^k \in \mathbb{R}^p$ and $\beta^k \in \mathbb{R}^{m \times n}$ using (9)
$\quad\quad$ Set $t = 0$, $L^{k,0} = L^k$, $S^{k,0} = S^k$
$\quad\quad$ **repeat**
$\quad\quad\quad$ Calculate $S_{ij}^{k,t+1} = \mathcal{S}_{\frac{\lambda \beta_{ij}^k}{\mu}} \left(M_{ij} - L_{ij}^{k,t} \right)$ for all i, j
$\quad\quad\quad$ Compute a SVD $M - S^{k,t+1} = U \Sigma V^T$ and $(\sigma_1, \ldots, \sigma_p) = \mathrm{diag}(\Sigma)$
$\quad\quad\quad$ Calculate $\bar{\sigma}_i = \max \left(\sigma_i - \frac{\alpha_i^k}{\mu}, 0 \right)$ for any $i = 1, \ldots, p$
$\quad\quad\quad$ Let $L^{k,t+1} = U \mathrm{diag}(\bar{\sigma}_1, \ldots, \bar{\sigma}_p) V^T$
$\quad\quad\quad$ Set $t = t + 1$
$\quad\quad$ **until** $\|\sigma(L^{k,t}) - \sigma(L^{k,t-1})\| / (1 + \|\sigma(L^{k,t})\|) < \varepsilon_1$
$\quad\quad$ Set $L^{k+1} = L^{k,t}$, $S^{k+1} = S^{k,t}$
$\quad\quad$ Set $k = k + 1$
$\quad$ **until** $\|\sigma(L^k) - \sigma(L^{k-1})\| / (1 + \|\sigma(L^k)\|) < \varepsilon_2$

The following result, whose proof is omitted due to paper's limited length, gives the validity of Algorithm 1.

Theorem 1. *Let $F(L, S)$ denote the objective function in (7). Suppose that (L^{k-1}, S^{k-1}) and (L^k, S^k) with $k \geq 1$ be two consecutive iterates of Algorithm 1. Then we have*

$$F(L^k, S^k) \leq F(L^{k-1}, S^{k-1}).$$

3   Numerical Experiments

Dataset. For each quadruple (m, n, r, c_p) where $r << \min(m, n)$ is the dimension of target space and $c_p \in (0, 1)$ represents the sparsity ratio of error matrix, the data matrix $M \in \mathbb{R}^{m \times n}$ is generated as follows. We first generate the rank–r matrix $L_0 = UV^T$ where $U \in \mathbb{R}^{m \times r}$ and $V \in \mathbb{R}^{n \times r}$ have entries independently and identically distributed (i.i.d.) according to the standard Gaussian distribution. We generate the sparse matrix $S_0 \in \mathbb{R}^{m \times r}$ such that $\|S_0\|_0 \approx c_p mn$ and its nonzero entries are i.i.d. from the Rademacher distribution. Finally, the data matrix M is computed as $M = L_0 + S_0$. In the result tables, we explicitly indicate the value of $\|S_0\|_0$.

Comparative Algorithms. We compare our algorithm with the 4 following algorithms:

- **DCA-ADMM** [9] is a DCA based algorithm for solving the following formulation of RPCA

$$\begin{cases} \min_{L,S\in\mathbb{R}^{m\times n}} \operatorname{rank}(L) + \lambda\|S\|_0 \\ s.t \quad \|M - L - S\|_2^F \leq \epsilon. \end{cases} \tag{15}$$

Recall that in [9], Sun et al. replaced the $\|.\|_0$ by capped l_1-norm. In the same work, the authors also developed **AL**, an alternating algorithm to solve (15).
- **EALM** and **IALM** are two version of augmented Lagrange multipliers (ALM) based algorithms, presented in [5], for solving the PCP formulation (4). **EALM** stands for Exact ALM while **IALM** is the inexact version of **EALM**.

Comparative Criteria. To evaluate the performance of algorithms, we consider the following criteria (($\hat{L}, \hat{S}$ is the obtained solution of comparative algorithms)

- the rank of matrix $\hat{L}$,
- $\|\hat{S}\|_0$ - the number of nonzero components of $\hat{S}$,
- the error relative between $\hat{L}$ and L_0 computed by $\frac{\|\hat{L}-L_0\|_F}{\|L_0\|_F}$,
- the error relative between $\hat{S}$ and S_0 defined as $\frac{\|\hat{S}-S_0\|_F}{\|S_0\|_F}$,
- computation time.

The experiments are performed on a Intel Core i5 3.60 GHz PC with 8 GB of RAM and the codes were written in MATLAB. The limited CPU time for the algorithms is set to 3 h.

In Table 1, we report the comparative results with $m = n = 500$, $r = 25$ and different values of sparsity ratio c_p. We observe that for in all 4 settings, our algorithm **CaPCA** recovers the true rank of matrix L_0 while only **IALM** successfully finds the true value of $\operatorname{rank}(L_0)$ for $c_p = 0.25$. When c_p increases, e.g., the error matrix S_0 is less sparse, the other comparative algorithms (**DCA-ADMM**, **AL**, **EALM** and **IALM**) perform badly. Overall, **CaPCA** gives better results than the other 4 algorithms in all comparative criteria. Especially in terms of CPU time, **CaPCA** is by far the fastest algorithm comparing to **DCA-ADMM**, **AL** and **EALM**.

In Tables 2 and 3, we present the comparative results with $(m, n, r) = (1000, 1000, 50)$ and $(m, n, r) = (2000, 2000, 100)$. Similarly to the results in the Table 1, we can see that our algorithm **CaPCA** outperforms the others 4 algorithms.

Table 1. $m = n = 500$, $r = \text{rank}(L_0) = 25$

$\|S_0\|_0$	Algorithm	rank($\hat{L}$)	$\|\hat{S}\|_0$	$\frac{\|\hat{L}-L_0\|_F}{\|L_0\|_F}$	$\frac{\|\hat{S}-S_0\|_F}{\|S_0\|_F}$	Time(s)
62646 ($c_p = 0.25$)	DCA-ADMM	46	62454	2.4e−6	1.8e−5	48.2
	AL	25	63234	3.2e−6	**1.2e−5**	20.3
	EALM	58	64234	1.2e−6	1.4e−5)	14.1
	IALM	**25**	**62646**	**1.0e−6**	1.4e−5	4.7
	CaPCA	textbf25	**62646**	6.4e−6	2.8e−5	**1.0**
75195 ($c_p = 0.30$)	DCA-ADMM	287	79512	2.6e−6	2.5e−5	209
	AL	55	89421	2.6e−6	2.5e−5	13
	EALM	113	82442	2.5e−6	**2.4e−5**	60.2
	IALM	299	144055	2.9e−3	2.6e−2	4.9
	CaPCA	**25**	**75195**	**1.8e−6**	**2.4e−5**	**3.9**
87605 ($c_p = 0.35$)	DCA-ADMM	134	119973	2.1e−4	2.4e−4	1273
	AL	98	104936	2.2e−4	2.8e−4	176
	EALM	227	129976	**1.8e−5**	1.5e-4	203.0
	IALM	297	150091	2.3e−2	1.9e−1	3.7
	CaPCA	**25**	**87605**	**1.8e−5**	**1.2e−4**	**1.7**
100247 ($c_p = 0.40$)	DCA-ADMM	388	121716	2.1e−2	4.2e−2	1593
	AL	124	134723	7.4e−2	3.7e−2	265
	EALM	311	164725	4.4e−2	3.5e−1	155
	IALM	289	160162	4.7e−2	3.7e−1	3.6
	CaPCA	**25**	**100252**	**1.1e−3**	**9.5e−3**	**2.9**

Table 2. $m = n = 1000$, $r = \text{rank}(L_0) = 50$

$\|S_0\|_0$	Algorithm	rank($\hat{L}$)	$\|\hat{S}\|_0$	$\frac{\|\hat{L}-L_0\|_F}{\|L_0\|_F}$	$\frac{\|\hat{S}-S_0\|_F}{\|S_0\|_F}$	Time(s)
249920 ($c_p = 0.25$)	DCA-ADMM	93	204382	6.5e−7	2.9e−5	510
	AL	**50**	256323	**4.5e−7**	2.6e−5	294
	EALM	116	254487	9.6e−7	1.7e−5	113.8
	IALM	**50**	**249920**	8.5e−7	**1.6e−5**	34.0
	CaPCA	**50**	**249920**	5.4e−6	2.6e−5	**10.2**
299047 ($c_p = 0.30$)	DCA-ADMM	172	3059222	3.6e−6	2.6e−5	1243
	AL	53	315234	2.9e−6	2.6e−5	682
	EALM	198	314445	**1.3e−6**	**2.0e−5**	327.1
	IALM	59	305231	1.6e−4	2.1e−3	105.1
	CaPCA	**50**	**299047**	2.6e−6	2.6e−5	**12.3**
349412 ($c_p = 0.35$)	DCA-ADMM	392	469608	5.4e−6	3.2e−4	5304
	AL	143	489328	**3.9e−6**	3.3e−4	1983
	EALM	520	568268	6.4e-6	**7.7e−5**	1045
	IALM	589	596800	1.5e−2	1.7e−1	38.8
	CaPCA	**50**	**349412**	7.9e-6	**7.7e−5**	**15.8**
399835 ($c_p = 0.40$)	DCA-ADMM	763	469835	4.3e−4	3.5e−2	10932
	AL	211	499336	4.3e−4	2.3e−2	2421
	EALM	570	597514	3.1e−2	3.5e−1	1376.4
	IALM	575	647112	3.3e−2	3.7e−1	37.4
	CaPCA	**50**	**399835**	**4.4e−5**	**4.2e−4**	**21.5**

Table 3. $m = n = 2000$, $r = \mathrm{rank}(L_0) = 100$

$\|S_0\|_0$	Algorithm	$\mathrm{rank}(\hat{L})$	$\|\hat{S}\|_0$	$\frac{\|\hat{L}-L_0\|_F}{\|L_0\|_F}$	$\frac{\|\hat{S}-S_0\|_F}{\|S_0\|_F}$	Time(s)
1001120 ($c_p = 0.25$)	DCA-ADMM	539	1051425	2.5e−06	2.5e−05	11800
	AL	**100**	**1001120**	6.1e−07	1.6e−05	2532
	EALM	232	1017204	2.8e−07	7e−06	845
	IALM	**100**	**1001120**	**2.1e−07**	1.9e−05	245
	CaPCA	**100**	**1001120**	7e-07	**1.5e−05**	**81**
1201715 ($c_p = 0.3$)	DCA-ADMM	314	1213012	4.3e−6	2.7e−5	11800
	AL	**100**	1213035	**1.2e−6**	2.5e−5	11800
	EALM	406	1262762	1.3e−6	2.7e−5	2337
	IALM	101	1203035	1.3e−6	2.7e−5	301
	CaPCA	**100**	**1201715**	4.7e−6	3.6e−5	**97**
1399371 ($c_p = 0.35$)	DCA-ADMM	1983	2438876	3.2e−4	3.6e−3	11800
	AL	541	1569374	7.2e−5	9.5e−5	11800
	EALM	1101	2639010	5.6e−6	9.5e−5	10174
	IALM	1175	2378633	0.01	0.17	303
	CaPCA	**100**	**1399371**	**5.4e−6**	**7.4e−5**	**11**
1601671 ($c_p = 0.4$)	DCA-ADMM	1987	2894311	0.09	1.21	11800
	AL	1784	2874512	0.09	0.98	11800
	EALM	1163	2434515	0.02	0.37	10438
	IALM	1151	2599415	0.02	0.37	321
	CaPCA	**100**	**1601671**	**1e−5**	**1e−4**	**154**

4 Conclusion

We have studied the Principal Component Analysis (PCA), one of the most popular methods of dimensionality reduction. In order to deal with highly corrupted noisy data, we have considered a variant of PCA, namely Robust PCA (RPCA). The RPCA can be formulated as an optimization problem which involves the minimization of ℓ_0-norm. The discontinuity of ℓ_0 makes the corresponding problem hard to solve. Thus, we approximate the l_0-norm by a continuous non-convex function, namely the capped ℓ_1 norm. The resulting problem is a non-convex for which we developed a reweighted-ℓ_1 based algorithm. The proposed algorithm consists in solving iteratively a non-convex sub-problem of two variables L and S. Fortunately, if either L or S is fixed, the other can be computed explicitly.

   We have carefully conducted numerical experiments on several synthetic datasets. The numerical results showed that our algorithm $CaPCA$ can recover exactly the rank of original matrix L_0 in all instances. Moreover, the number of nonzero components of S given by our algorithm is very closed to the one of the original matrix S_0. Overall, $CaPCA$ outperforms several state-of-the-art algorithm for RPCA, with respect to all comparative criteria.

References

1. Candès, E.J., Li, X., Ma, Y., Wright, J.: Robust principal component analysis? J. ACM **58**(3), 1–37 (2011)
2. Horn, R.A., Johnson, C.R.: Matrix Analysis, 2nd edn. Cambridge University Press, New York (2013)
3. Le Thi, H.A., Pham Dinh, T.: The DC (Difference of Convex Functions) programming and DCA revisited with DC models of real world nonconvex optimization problems. Ann. Oper. Res. **133**(1–4), 23–46 (2005)
4. Le Thi, H.A., Pham Dinh, T., Le, H.M., Vo, X.T.: DC approximation approaches for sparse optimization. Eur. J. Oper. Res. **244**(1), 26–46 (2015)
5. Lin, Z., Chen, M., Ma, Y.: The augmented lagrange multiplier method for exact recovery of corrupted low-rank matrices. arXiv:1009.5055 (2010)
6. Lin, Z., Ganesh, A., Wright, J., Wu, L., Chen, M., Ma, Y.: Fast convex optimization algorithms for exact recovery of a corrupted low-rank matrix. In: Intl. Workshop on Comp. Adv. in Multi-Sensor Adapt. Processing, Aruba, Dutch Antilles (2009)
7. Pham Dinh, T., Le Thi, H.A.: Convex analysis approach to dc programming: theory, algorithms and applications. Acta Math. Vietnamica **22**(1), 289–355 (1997)
8. Pham Dinh, T., Le Thi, H.A.: A DC optimization algorithm for solving the trust-region subproblem. SIAM J. Optim. **8**(2), 476–505 (1998)
9. Sun, Q., Xiang, S., Ye, J.: Robust principal component analysis via capped norms. In: Proceedings of the 19th ACM SIGKDD International Conference on Knowledge Discovery and Data Mining, KDD 2013, pp. 311–319 (2013)
10. Wright, J., Ganesh, A., Rao, S., Peng, Y., Ma, Y.: Robust principal component analysis: exact recovery of corrupted low-rank matrices via convex optimization. Adv. Neural Inf. Process. Syst. **22**, 2080–2088 (2009)
11. Zhao, Q., Meng, D., Xu, Z., Zuo, W., Zhang, L.: Robust principal component analysis with complex noise. In: Proceedings of the 31st International Conference on International Conference on Machine Learning - Volume 32, ICML 2014, pp. 55–63 (2014)
12. Zhou, Z., Li, X., Wright, J., Candes, E., Ma, Y.: Stable principal component pursuit. In: 2010 IEEE International Symposium on Information Theory, pp. 1518–1522 (2010)

A Probability-Based Close Domain Metric in Lifelong Learning for Multi-label Classification

Thi-Ngan Pham[1,2], Quang-Thuy Ha[1(✉)], Minh-Chau Nguyen[1],
and Tri-Thanh Nguyen[1(✉)]

[1] Vietnam National University, Hanoi (VNU), VNU-University of Engineering
and Technology (UET), No. 144, Xuan Thuy, Cau Giay, Hanoi, Vietnam
{nganpt.di12, thuyhq, 15021766, ntthanh}@vnu.edu.vn
[2] The Vietnamese People's Police Academy, Hanoi, Vietnam
ptngan2012@gmail.com

Abstract. Lifelong machine learning has recently become a hot topic attracting the researchers all over the world by its effectiveness in dealing with current problem by exploiting the past knowledge. The combination of topic modeling on previous domain knowledge (such as *topic modeling with Automatically generated Must-links and Cannot-links,* which exploits must-link and cannot-link of two terms), and *lifelong topic modeling* (which employs the modeling of previous tasks) is widely used to produce better topics. This paper proposes a close domain metric based on probability to choose valuable knowledge learnt from the past to produce more associated topics on the current domain. This knowledge is, then, used to enrich features for multi-label classifier. Several experiments performed on review dataset of hotel show that the proposed approach leads to an improvement in performance over the baseline.

Keywords: Close domain · Lifelong learning · Multi-label classification · Lifelong topic modeling

1 Introduction

Latent Dirichlet Allocation (LDA) [1, 2], and Probabilistic latent semantic analysis (pLSA) [12] are the two popular topic models for discovering the hidden topics in a text corpus following an unsupervised learning approach. In general, topic modeling assumes that, probabilistically, each document is a multinomial distribution over a fixed number of topics; and each topic has a multinomial distribution over all the observed words in the corpus. Therefore, the relationship between document-topic distribution and topic-word distribution are defined. These models typically need an enormous amount of data (thousands of documents) to describe meaningful statistics for extracting good quality topics.

Recently, *lifelong topic modeling* (LTM), a type of lifelong unsupervised learning, has been widely used to exploit the prior domain knowledge to build the model inference to produce more sensible and reasonable topics [4–8]. The existing work exploits the previous knowledge in two forms including *must-links* (i.e., two words that

© Springer Nature Switzerland AG 2020
H. A. Le Thi et al. (Eds.): ICCSAMA 2019, AISC 1121, pp. 143–149, 2020.
https://doi.org/10.1007/978-3-030-38364-0_13

often appear in the same topic), and *cannot-links* (i.e., two words that rarely appear in the same topic) in the previously generated topics from several domains (even a big number of domains called big data) to generate better topics in the current domain. This knowledge-based learning approach imitates the same way as human does, i.e., storing the learnt knowledge and using them to infer in the future.

In [3], instead of collecting all the data from previous domains, they assume that the data from close domains may more effectively exploit the distributions of document-topic and topic-word thanks to the topic overlapping between the domains. In addition, the close domains may share the same features at different levels which are selected to improve the current topic learning. The LTM method proposed in [3] is a learning bias approach on domain level by identifying the close domains based on their vocabularies, top words, and topics. By focusing on exploiting prior knowledge of the close domains for enriching features of classifier, this approach showed little improvement in comparison with traditional *topic modeling with Automatically generated Must-links and Cannot-links* (AMC) [5].

In many approaches, the distance is used to determine whether two objects are close or not. In this paper, we offer an approach to find close domains by exploiting the relationship among label and features, i.e., the probability for labeling documents based on the features on labeled domains. This is due to the fact that each label may be presented by specific features and certain features in many presentations may help to explore the already known concepts or similar ones.

This paper has two main contributions, i.e., (i) proposing the close domain measures based on features of probability; and (ii) performing an application of multi-label classification using proposed approaches.

The remaining of this paper consists of following sections: We deliver some definitions to identify close domains based on probability distribution in the next section. Section 3 describes a multi-label classification framework using proposed lifelong topic model. Our experiments and discussions on the results will be discussed in the Sect. 4. We also mention and present the differences in our proposed approach to some related work in Sect. 5. The sum-up and coming work will be shown in the last section.

2 Definitions

2.1 Problem Formulation

Let D_1, D_2, ..., D_N be N datasets of of T_N previously tasks. Let S be the *knowledge base*, which includes all the knowledge and information from N previous tasks. S is empty when $N = 0$. Let D_{N+1} be the dataset of *current task* T_{N+1}. The goal is to determine a set of previously datasets D_{close}, which includes previous datasets D_i closed with D_{N+1}, then using the part of knowledge of S, which related with D_{close} for solving the current task T_{N+1}.

Assume that there exists a general feature space F, in which the data from all domains can be represented. Assume that the sets of values of all features of F are discrete. Let $sim(x, y)$ be a similarity measure of a data element pair x, y, e.g., cosine measure. Let X be a subset of elements, $sim(x,X)$ be defined as the maximum similarity between x and all the elements of X, i.e., $sim(x, X) = \max_{y \in X} sim(x, y)$, this is also called the *complete link* similarity between two datasets.

2.2 The Closeness in Probability of Two Datasets

Assume that all the data of *the ith previous dataset D_i* are labelled; and L_i is the label set. Let $\theta_i = (\theta_{i1}, \theta_{i2}, \ldots, \theta_{i||L_i|})$ be the probability threshold vector of the dataset D_i, $i = 1, 2, \ldots N$. The probability threshold vector θ_i is a previous knowledge, which is determined based on the set of the posterior probability vector $\{(prob(l_{i1}|x), prob(l_{i2}|x), \ldots, prob(l_{i|L_i|}|x))| \ x \in D_i\}$.

Definition 1. Let x be an element belonging to the current dataset D_{N+1}, let θ_i be the probability threshold vector of *the ith previous dataset D_i*, $i = 1, 2, \ldots, N$. D_i is called close to x iff at least one posterior probability of x in D_i is greater than or equal to the predetermined probability threshold θ_{ij}, i.e. $\exists j \in \{1, 2, \ldots, |L_i|\}$ such that $prob(l_{ij}|x) \geq \theta_{ij}$.

Definition 2 (closeness between to datasets). The *i^{th} previous dataset D_i* is called close in the probability to dataset D_{N+1} iff

$$\frac{|x \in D_{N+1} : D_i \text{ is close} - prob \text{ with } x|}{|D_{N+1}|} \geq \theta_{prob} \tag{1}$$

where θ_{prob} is a predefined threshold for deciding whether two datasets are close to each other or not.

3 Proposed Model of Lifelong Topic Modeling Using Close Domain Knowledge for Multi-label Classification

All stages in the multi-label classification framework are illustrated in Fig. 1:

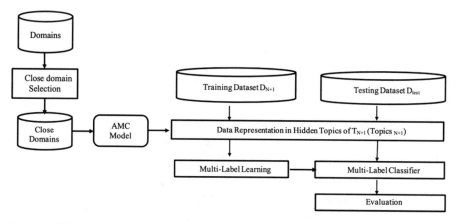

Fig. 1. A lifelong topic model using prior knowledge from close domains for multi-label classification framework

- Assume that, we already have a knowledge base of several previous domains, the current task including a training dataset and a testing dataset (in current domain).
- Firstly, we find close domains to the current domain using the closeness in probability approach.
- After that, the close domains will be used to train AMC model to exploit the prior domain knowledge and to adjust the distribution hidden topic on the current domains.
- Then, the knowledge of Hidden Topics derived from the close domains is transferred to new features for representing the texts. These features are considered better than those extracted from all previous domains.
- Finally, a multi-label classifier will be built for new documents in current domain D_{N+1}. We use different classifiers to take a careful look at the improvement in performance of system when applying the proposed approach.

4 Experiments and Discussion

4.1 The Datasets

In the experiments, we focus on the impact of different domains on the model especially when the current domain has few data (small training dataset). We used a multi-label dataset of about 1350 reviews on hotels with a associated set of 5 labels including *Place and cost*, *Services*, *Equipment*, *Room standard*, and *Food*. We divided the original dataset into 5 sub-datasets named D_1, D_2, D_3 (the three previous domains with 400 reviews of each dataset), D_4 (the current domain which is set up with different size of 20, 30, 50, 80, 100 and 150 reviews) and D_{test} (dataset for testing with 100 reviews).

4.2 Experimental Scenarios

Different configurations were set up in experiments below:

- The Term Frequency (TF) feature is used for data presentation.
- The number topics for the AMC model is set to 10, 15 and 25.
- For multi-label classifier, we use Binary Relevant method with core algorithms of Naïve Bayes, k-Nearest Neighbor (kNN), Gaussian Process, and Random Forest.
- The threshold θ_{prob} to decide the close domain is set to 0.1, 0.3 and 0.7. In oder to reduce the result, the evarage of the results of the experiments with three different values of θ_{prob} is calculated.

We took the effectiveness of system into considerations with the popular measures, i.e., precision, recall and F1. We perform two groups of experiments with different settings on each dataset of D_1, D_2, D_3, D_4 and D_{test} as follows:

- The baseline (denoted by the OF): run multi-classifiers on the original features of TF without using any previous domain knowledge.
- Our approach: run multi-classifiers on the original features with the knowledge from the close domains with different number of AMC topics.

4.3   Discussions on Results of Experiments

The proposed approach aims to find close domains to the current domain. Then, the found close domains are taken into AMC model to build topic modelling features for enriching features of the current domain. We show results of experiments based on the size of D_4 and number of topics for AMC model in Table 1. In comparison with the baseline, the results with better performance are highlighted with yellow and the best results are formatted in bold.

Table 1. Results of multi-label text classification using proposed approach in finding close domain enriching features for the classifiers

D4	Experiment	Gaussian Process			kNN			Naïve Bayes			Random Forest		
		Precision	Recall	F1	Precision	Recall	F1	Precision	Recall	F1	Precision	Recall	F1
20	OF	100.00%	4.55%	8.70%	67.65%	29.87%	41.44%	56.07%	62.99%	59.33%	86.65%	7.79%	14.30%
	10 topics	90.91%	6.49%	12.12%	55.88%	37.01%	44.53%	56.50%	64.94%	60.42%	56.72%	24.68%	34.39%
	15 topics	51.81%	27.92%	36.29%	55.06%	31.82%	40.33%	55.17%	62.34%	58.54%	50.00%	20.13%	28.70%
	25 topics	100.00%	2.60%	5.06%	65.22%	29.22%	40.36%	56.80%	62.34%	59.44%	63.64%	18.18%	28.28%
30	OF	96.97%	20.78%	34.22%	57.79%	57.79%	57.79%	56.74%	65.58%	60.84%	79.05%	11.69%	20.37%
	10 topics	72.13%	28.57%	40.93%	50.38%	42.86%	46.32%	56.04%	66.23%	60.71%	58.33%	13.64%	22.11%
	15 topics	80.60%	35.06%	48.87%	55.63%	54.55%	55.08%	54.64%	64.94%	59.35%	55.17%	20.78%	30.19%
	25 topics	91.38%	34.42%	50.00%	61.70%	56.49%	58.98%	56.59%	66.88%	61.31%	51.67%	20.13%	28.97%
50	OF	93.10%	35.06%	50.94%	66.25%	34.42%	45.30%	58.96%	66.23%	62.39%	82.93%	6.67%	12.35%
	10 topics	93.22%	35.71%	51.64%	68.00%	33.12%	44.54%	58.96%	66.23%	62.39%	83.33%	12.99%	22.47%
	15 topics	88.14%	33.77%	48.83%	68.18%	38.96%	49.59%	58.96%	66.23%	62.39%	71.43%	6.49%	11.90%
	25 topics	93.22%	35.71%	51.64%	71.65%	59.09%	64.77%	58.96%	66.23%	62.39%	50.00%	5.84%	10.47%
80	OF	90.24%	48.05%	62.71%	57.23%	59.09%	58.15%	56.83%	67.53%	61.72%	79.86%	3.13%	6.02%
	10 topics	85.71%	38.96%	53.57%	59.18%	56.49%	57.81%	56.83%	67.53%	61.72%	100.00%	1.95%	3.82%
	15 topics	85.88%	47.40%	61.09%	70.18%	51.95%	59.70%	56.83%	67.53%	61.72%	86.67%	8.44%	15.38%
	25 topics	90.70%	50.65%	65.00%	63.70%	55.84%	59.52%	56.83%	67.53%	61.72%	71.43%	6.49%	11.90%
100	OF	91.11%	53.25%	67.21%	53.61%	57.79%	55.63%	53.54%	68.83%	60.23%	57.41%	1.12%	2.20%
	10 topics	90.59%	50.00%	64.44%	48.95%	60.39%	54.07%	53.54%	68.83%	60.23%	92.31%	7.79%	14.37%
	15 topics	91.36%	48.05%	62.98%	61.49%	64.29%	62.86%	53.54%	68.83%	60.23%	81.25%	8.44%	15.29%
	25 topics	91.67%	50.00%	64.71%	56.17%	59.09%	57.59%	53.54%	68.83%	60.23%	50.00%	1.95%	3.75%
150	OF	92.47%	55.84%	69.64%	61.90%	59.09%	60.47%	53.61%	67.53%	59.77%	98.96%	3.36%	6.51%
	10 topics	92.50%	48.05%	63.25%	59.09%	59.09%	59.09%	53.61%	67.53%	59.77%	87.50%	4.55%	8.64%
	15 topics	92.86%	50.65%	65.55%	58.43%	62.99%	60.63%	53.61%	67.53%	59.77%	77.78%	9.09%	16.28%
	25 topics	92.05%	52.60%	66.94%	62.50%	61.69%	62.09%	53.61%	67.53%	59.77%	100.00%	8.44%	15.57%

The results show that the proposed model leads to a very promising performance in lots of experiments in comparison with the baseline. The classifications using kNN and Random Forest algorithms provide a higher performance in all groups of experiments (may be in different configuration of topic number). Especially when the current training dataset (D_4) is very small (20 and 30 reviews). In other words, the previous knowledge from close domains may take reasonable contribution to the text classification of supervised algorithms.

When the current training dataset (D_4) is increased (to 100 and 150 reviews), the classification results are not improved or even lower than baseline. This may due to the fact that the larger the dataset (D_4) is, the more useful features it gets and these useful features may conflict with the features from close domains that leads to the decrease performance of the model.

5   Related Works

In lifelong learning, the past knowledge will be used to enhance the performance of the current task. However, in most lifelong learning models, which data domains of the past knowledge should be used, are still an open problem. In many applications, a model which is effective in one domain may be transformed to another domain (i.e., cross domain learning problem). This approach leads to lots of advantages in saving time and cost consumption for manually labeling new data. It is clear that the quality of past knowledge or known domains will have an impact on the current model performed on new domain (dataset).

To our best knowledge, there has not been much research about selecting the close data domains to the current dataset for better enhancing the learning performance. In [3], the similarity between two domains is defined based on the measure of weighted word bags, and topics (derived from the probability distribution of hidden topics in domain) without mining the label set of training data. The close domains found in [3] were used to enrich the knowledge base of lifelong topic modeling (LTM) [4], which improves the hidden topic features of the current task. Our proposed approach exploits the probability features of both data and label set to form the closeness between two domains. And then the close domains will become input of the AMC model to extract high quality features of hidden topics for the current task.

Probabilistic viewpoint plays an important role in many areas of scientific research. In a probabilistic model, the relationship among the observables is formed to describe the fundamental possibility based on their behaviors even they are not assumed to hold exactly for each observation. Therefore, in statistical estimation problems, it is significant to find out and evaluate a close probability distribution. In [9, 11] the closeness between two probability distribution is defined based on an information measure and the criterion of maximum entropy. In [10], the measures between two probability distributions were used including Hellinger coefficient, Chernoff coefficient and Jeffreys distance, J-divergence, etc. In our research, we define the distance of two different domain datasets via the relationship between an element and a dataset which is determined by comparing the posterior probability distribution of the element in the dataset to a predefined threshold. This is an extended definition of the closeness between two probability distributions to present the relationship between two domain datasets.

6   Conclusions

This paper provides a multi-label classification framework using lifelong learning technique to use the previous knowledge of the close domains. This paper makes the contribution in term of the proposal of a close domain measure to determine the closeness of two different domain datasets based on their aspect of probabilities. This measure is used for selecting only the datasets that are deemed to close to the current dataset for enhancing the current task's learning. The results of our experiments demonstrate the reasonable improvement of the supervised classification using the proposed approach, especially in case of having a small labeled dataset.

This work will be upgraded in the future in several ways. Firstly, the threshold parameters should be chosen accounting for features of the datasets themselves instead of fixing them. Secondly, more experiments should be performed (especially in other datasets) to get more evaluation for the proposed approaches. Finally, the technique of mining close domain should be improved in AMC model to get more effectiveness.

References

1. Blei, D.M.: Probabilistic topic models. Commun. ACM **55**(4), 77–84 (2012)
2. Blei, D.M., Ng, A.Y., Jordan, M.I.: Latent Dirichlet allocation. J. Mach. Learn. Res. **3**, 993–1022 (2003)
3. Ha, Q.T., Pham, T.N., Nguyen, V.Q., Nguyen, T.C., Vuong, T.H., Tran, M.T., Nguyen, T.T.: A new lifelong topic modeling method and its application to vietnamese text multi-label classification. In: Asian Conference on Intelligent Information and Database Systems, pp. 200–210. Springer, Cham (2018)
4. Chen, Z., Liu, B.: Topic modeling using topics from many domains, lifelong learning and big data. In: ICML, pp. 703–711 (2014)
5. Chen, Z., Liu, B. Mining topics in documents: standing on the shoulders of big data. In: KDD, pp. 1116–1125 (2014)
6. Chen, Z., Mukherjee, A., Liu, B., Hsu, M., Castellanos, M., Ghosh, R.: Discovering coherent topics using general knowledge. In: CIKM, pp. 209–218 (2013)
7. Andrzejewski, D., Zhu, X., Craven, M.: Incorporating domain knowledge into topic modeling via Dirichlet forest priors. In: ICML, pp. 25–32 (2009)
8. Chen, Z., Mukherjee, A., Liu, B., Hsu, M,. Castellanos, M., Ghosh, R.: Exploiting Domain Knowledge in Aspect Extraction. In: EMNLP, pp. 1655–1667 (2013)
9. Higashi, M., Klir, G.J.: On the notion of distance representing information closeness: possibility and probability distributions. Int. J. Gen Syst **9**(2), 103–115 (1983)
10. Lewis II, P.M.: Approximating probability distributions to reduce storage requirements. Inf. Control **2**(3), 214–225 (1959)
11. Chow, C., Liu, C.: Approximating discrete probability distributions with dependence trees. IEEE Trans. Inf. Theory **14**(3), 462–467 (1968)
12. Hofmann T.: Probabilistic latent semantic indexing. In: Proceeding SIGIR 1999 Proceedings of the 22nd Annual International ACM SIGIR Conference on Research and Development in Information Retrieval, pp. 50–57 (1999)

Applying MASI Algorithm to Improve the Classification Performance of Imbalanced Data in Fraud Detection

Thi-Lich Nghiem[1]([✉]) [iD] and Thi-Toan Nghiem[2]

[1] Thuongmai University, 79 Ho Tung Mau, Hanoi, Vietnam
lichnt72@tmu.edu.vn
[2] LyNhanTong High School, Bacninh, Vietnam

Abstract. Imbalanced data is recognized as one of the most attractive matters to many researches. It is shown by numerous publications on this which is a growing interest. The hardest challenge is the failure of generalizing inductive rules by learning algorithms. such as difficulty in forming good classification on decision boundary over more features but fewer samples and risk of overfitting of the sampling. So many solutions have been applied to deal with these problems. In our article, we propose a novel method called MASI (Moving to Adaptive Samples in Imbalanced) in term of changing majority class samples' label into minor class samples based on data distribution. This proposed method rebalances the classes before training a model in order to improve the classification performance in imbalanced data. We tested on some unbalanced datasets from data of UCI. The empirical results showed that our method has a significant achievement in Sensitivity and G-mean values than other classification models, such as Random Over-sampling, Random Under-sampling, SMOTE, and Borderline SMOTE in using different machine learning approaches, including SVM, C5.0, and RF.

Keywords: Classification performance · Fraud detection · Imbalanced data

1 Introduction

In recent years, fraud detection is one of the most interesting topics. The imbalanced class problem being difficult challenge faced by machine learning has received considerable attention of a significant amount of research, especially in the fraud detection domain [1–6]. It also has been pointed out as one of the 10 most challenging problems in the domain of data mining research [7, 35]. Balanced data sets are said to be balanced if there are, approximately, as many positive examples of the concept as there are negative ones. Otherwise, data imbalance often occurs when there significantly differs in the number of samples between classes. It means that the number of examples representing a class is much larger than other classes [8]. The class that occupies the samples in the major class is called the negative class, whereas the class with few examples is called the positive class. Data imbalance being a popular issue in the classification appears in various realistic areas, such as Fraud Detection, Medical Diagnosis, Network Intrusion

© Springer Nature Switzerland AG 2020
H. A. Le Thi et al. (Eds.): ICCSAMA 2019, AISC 1121, pp. 150–162, 2020.
https://doi.org/10.1007/978-3-030-38364-0_14

Detection, Detection of oil spills from radar images of the ocean surface, etc. As an illustration, the distribution of unbalanced class distribution known as samples in the major class significantly be more than the number of samples in the minor class [9]. The primary issue in imbalanced class is that popular classification models are often biased to the major class It leads to increase accurately classification of the majority class samples, whereas many samples in the minority class are incorrectly classified.

In some real applications, the misclassified samples minority class leads to serious consequences. Detecting frauds seriously affects business organizations. Take telecommunication fraud in the United States as an example, it cost millions of dollars per year [10]. In order to detect fraudulent transactions early, people often analyze the information in the existing transaction database, thereby discovering unusual transactions early. However, the number of none-fraudulent transactions is significantly higher than that of fraudulent cases. False prediction of frauds can cause huge economic losses. In medicine, the clinical database contains a large amount of patients' information s and their pathology. The data mining algorithms used in these databases tend to explore relationships, patterns in clinical and pathological databases to predict the progress and characteristics of some diseases. From there, it is possible to predict a person who is ill or not [10]. However, the number of patients is usually much smaller than those who are not sick. If a patient is misdiagnosed not being ill, there will be no timely treatment. Moreover, the combination of various classifiers is considered a typical idea to improve performance. The biggest problem is that how to choose correctly the classifiers in the myriad of diverse classifiers [4].

Dealing with imbalanced dataset, we should consider classifiers with adjusting the output threshold instead of using standard machine learning algorithms. This article presents a new method called MASI being an integrated method is combined of ADASYN and SPY methods in term of changing the majority class samples' label into minor class samples' label based on data distribution. Basing on this approach, our method rebalances the classes before training a model. As a result, classifiers' performance in imbalanced data can be improved. This article is organized as following: Part 2 continues with some information on related work; Methodology is presented in part 3; Part 4 analyzes the experimental results and compares to other classification approaches, including Random Over-sampling (ROS), Random Under-sampling (RUS), SMOTE, Borderline SMOTE, SPY, and finally part 5 ends up with conclusions.

2   Related Work

The imbalanced of data is a common subject in data mining and machine learning. Some approaches such as decision tree, SVM, K-nearest, Naïve Bayes, etc. have been developed and successfully applied in many areas. However, these approaches face some difficulties with some imbalanced datasets.

2.1   The Imbalanced Data in the Fraud Detection

In fraud detection, there are myriad of datasets used. Although these datasets differ in their sizes, the types of fraud and the numbers of fraudulent cases, they also have a common feature, like the number of fraudulent cases accounting for a very small ratio compared to non-fraudulent cases. In other words, the datasets often lose their balance. In the field of fraud detection, the data imbalance means there is a big gap of quantity between classes. For example, in binary classification, the given two-class datasets have much less representatives of one class (minor class) than of another (major class).

For a bi-class issue, the grade of imbalance class distribution can be represented by the proportion of the size of minor class samples for that of the major class. To give practical application as a clear example, 1:100, 1:1000, or even larger can be showed as drastic ratio [10]. The class that occupies the major examples is called the negative class, whereas the class with few examples is called the positive class. In classification algorithm, the major class usually has been achieved high accuracy. On the contrary, the accuracy in minor class is tend to be low.

In financial fraud detection, according to [11–14], these studies reviewed that financial frauds can be divided into three main areas: internal, insurance and credit However, internal frauds can break into two sub-categories, namely financial statement fraud and transaction fraud. Financial statement fraud has been known as management fraud while transaction fraud captures the process of snatching organizational assets. In addition, [9] presented the data imbalance in biomedical data. They showed that some methods have been achieved to rebalance the data imbalance, but this issue has still not been solved completely, such as reducing the classification performance. Some imbalanced datasets can be seen in Table 1.

It can clearly be seen in Fig. 1, the data imbalance exists in both financial domain and biomedical domain. The first set of data has a highest rate of imbalance ~1:20 with 23 transitional fraudulent cases in total 500 transactions, etc.

2.2   The Approaches for Imbalanced Data Classification

To deal with the imbalanced dataset, the researches have used many techniques such as supervised learning, unsupervised learning, etc. The techniques have been devided into two groups, including the data and algorithmic levels [3].

Data - Level Approach
At the data-level, the imbalanced strategies are that the techniques are used to adjust the data distribution by rebalancing data rate, or removing the noise between the minor class and major class before applying any algorithm. To take an example, this method can

Table 1. Some imbalanced datasets.

Data	Total records	Cheat	Legal
UCSD-FICO	500	23	477
Carclaim	15,420	923	14,497
Yeast	1484	163	1321
Haberman	306	81	225
German credit data	1,000	300	700
Australian credit approval	690	307	383

increase the quantity of minor class, reduce the number of majority class, or combine many techniques. One of the advantages of this approach is flexibility. It means the transformed data can be used to train with different classifiers [5, 10, 15]. Data-level method can be grouped into two categories, including over-sampling methods and under-sampling methods [16].

Over-Sampling Methods

Over-sampling methods create a dataset which is larger than initial dataset [8]. Over-sampling consists upsizing the minority class and decreasing the rate of the unbalanced data by randomized samples, selected samples or added synthetic samples. This method may increase the overfitting and biasing problems because of replicating the minor class until creating an equal frequency rate between two classes [5, 17, 18].

It is simple to understand both methods, including ROS, and RUS. ROS idea randomly selects the set S from the samples in minority class, then replicate the samples selected and add them to samples in minority class. Otherwise, the set S instances is randomly chosen in major class to remove these S instances from the prototype dataset [19] (Fig. 2).

Efficacious approaches, such as SMOTE, has a positive impact on diverse applications. The main idea of this algorithm is that SMOTE oversamples the minority class by generating synthetic samples based on the feature space, rather than data space [20]. These artificial examples created along the line segments joining a portion or all of the k-nearest neighborhood defined as the k-elements of observed minority class and based on: (a) Euclidian distance between (b) Generating a new sample, one of the K-nearest neighbors randomly is selected before finding the ratio between the number of the selected sample and its nearest neighbor. (c) Multiply this ratio by a number generated uniformly from 0 to 1. (d) Create a vector based on the selected samples relabeled into the minority class. SMOTE will alter the original class distribution in order to enhance the number of minor samples correctly classified.

After using SMOTE, although there are no false negatives in the dataset, it still exists a larger number of false positives. In imbalanced classification, almost classifiers expected that there are many samples to correctly classify all minor samples by reducing the number of false negatives. It is easy to understand because missing positive samples'

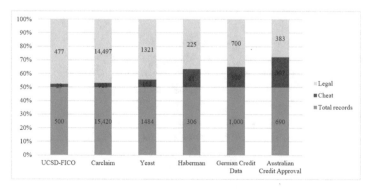

Fig. 1. The imbalance of some datasets

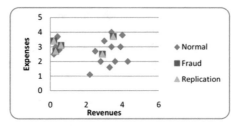

Fig. 2. The basic technique of over-sampling a minority class by replication its own samples *Synthetic Minority Over-Sampling Technique (SMOTE)*

cost (a false negative) outweighs that of negative sample (a false positive). The idea of SMOTE can be illustrated in Fig. 3.

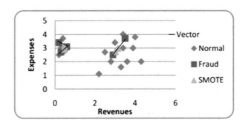

Fig. 3. An example of SMOTE technique

The figure shows the artificial examples are generated along the line segment between green triangle examples. These examples can rebalance the original class distribution and gradually enhances learning. But there is biggest problem in SMOTE method is that it still has over generalization of minority class space.

Under-Sampling Methods

Under-sampling method generates a subset of the original dataset by reducing the number of majority classes [8]. It is the simplest way to remove the samples randomly. In the fraud detection problem, this method selects randomly the samples representing the non-fraudulent cases and removes them from the original dataset as following (Fig. 4).

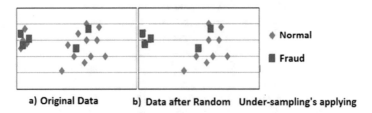

a) Original Data   b) Data after Random Under-sampling's applying

Fig. 4. Demonstrations of the under-sampling's applying

Although, this method can reduce the data imbalance, randomly removing samples could loss the important information which is useful for building model [8].

Algorithmic - Level Approaches
Another approach is algorithm-level approaches. At this level, the main purpose of classifiers is tuned to enhance learning the features of the minority classes by adjusting the error cost and misclassified samples of minorities (fraud cases classified as non-fraudulent) which mean it has higher costs than the misclassification of majority class (usual cases are classified as fraud).

For instance, decision tree classifier such as C5.0 use Information Gain as splitting criteria. The modification would be implemented by changing predicted probability at leaf nodes or development new pruning methods. With the support vector machine (SVM) method, adjustment is done by adding different finite constants to different classes, or setting class borderlines based on the idea of a kernel link. In tier-learning approach, the model is constructed only with samples of the target class. This method does not attempt to find the boundary that distinguishes the majority and minority classes but try to find the boundary that surrounds the target class. For this target class, elements that need to classified will be measured its similarity with the target class' elements using threshold boundary between two classes. If this threshold is too small then the large number of majority samples will be filtered, otherwise it will be kept. Therefore, setting an effective threshold is very important for tier-learning approach.

3   Methodology

3.1   SPY and ADASYN Methods

ADASYN
Haibo et all proposed a novel adaptive learning algorithm from the extension of SMOTE called ADASYN [19]. ADASYN tend to mitigate imbalanced class issue in a dataset by generating the number of minority instances based on the amount of its majority nearest neighbor [20]. The number of synthetic data samples is determined by a parameter called β used to rebalance data level after synthetic process. In ADASYN, density distribution is used as a criterion to estimate the number of artificial sample hass while each minority sample has equally likely chance to be selected for artificial process in SMOTE. However, ADASYN does not notice noise samples. It is possible to generate a large number of synthetic data around those examples, which may create an unrealistic minority space for the learner.

SPY
The samples tending to be more mistakenly classified than the others far from the borderline is located on the borderline and nearby ones. Based on this matter, Dang et all [9] introduced a new approach called SPY to adjust the balance data ratio by changing the majority class' labels in the nearest k-neighbor into the minority class labels [9]. By doing so the number of the major instances decreases and the number of the minor class samples increases. When using SPY method, this majority class samples is called "SPY".

The advantage of SPY method is that it determines the vital importance of borderline samples and focuses on the strength of minority boundary samples to increase classification performance. Moreover, SPY method adjusts data distribution without changing data size. On the one hand, in some cases, the number of samples selected for relabeling does not correspond to the distribution needs of each specific data. Therefore, SPY does not improve classification efficiency, even reduces accuracy in these cases.

3.2 MASI Algorithm

Idea

The proposed algorithm is an improvement of ADASYN [19] and SPY [9]. Based on data density distribution, MASI will relabel the majority class into minority class samples. The number of rechanged labels is affected by two factors. Firstly, the ratio between the number of majority class and minority class influences on the minority class relabeled. In addition, for each minority class, the nearest neighbor samples are chosen to change labels differently. This depends on the proportion of the number of nearest neighbor between two classes: minor and major class. In other words, the number of the nearest neighbor of its majority class changed label will be proportional to this rate (Fig. 5).

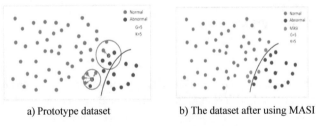

a) Prototype dataset   b) The dataset after using MASI

Fig. 5. Data representation used MASI (before and after)

Algorithm MASI (T, β, k).

Input: Training dataset T based on original set $S = \{(x_i, y_i), y_i \in \{-1,1\}\}$ is the sample's label, $x_i \in \Re^n$ where $i \in [1, n]$, k is the number of nearest neighbors, and β is a parameter to estimate the desired equilibrium ratio.

Output: New training T'

1. Select randomly 90% instances to train, the rest for validation set as follows: T=90%xS, V=S-T

    n_{maj} = majority(T), n_{min} = minority(T)

2. Calculate the synthetic data samples in majority need to rechange their label to minority class samples as follows: $G = (n_{maj} - n_{min}) * \beta$ (1)

3. For i:=1 to n_{min} do

    In each data sample of $p_{i:}$
    o Find its nearest neighbors in the entire training dataset
    o Compute the proportion of the nearest neighbor in the majority class (x_i) to k nearest neighbor. This rate called $r_{i:}$ $r_i = \frac{x_i}{k}$ where $r_i \in [0,1]$ (2)

4. Calculate the density distribution of data, called $r_{i'}$:
    $$r_i' = \frac{r_i}{\sum_{i=1}^{n_{min}} r_i}$$ (3)
    Where $\sum_i r_i' = 1$

5. For i:=1 to n_{min} do
    In each data sample of $p_{i:}$
    o Compute the number of nearest neighbors g_i to be labeled around each minority by the following formula: $g_i = r_i' * G$ (4)

    Where G is taken in equation (1).

Evaluation Measures

A confusion matrix (contingency table) is a typical evaluation measure to represent classification performance [21]. As an illustration, the minor class and major class called positive and negative, respectively. Table 2 shows this matrix of bi-class.

Table 2. A confusion matrix

	Forecasted as positive	Forecasted as negative
Genuine positive	TP	FN
Genuine negative	FP	TN

In this matrix, the rows and the columns are genuine class labels and forecasted class labels, respectively. TN (True Negative) is the number of class samples that are mostly accurately classified; FN (False Negative): is the number of minor class samples mistakenly classified as major class samples; TP (True Positive): is the number of minor class samples to be correctly classified; FP (False Positive): the number of major class samples that are mistakenly classified as minor class samples.

For balanced datasets, the efficiency of a classification model can be evaluated by accuracy (or error rate) which is the ratio of the number correctly classified samples. It is defined as:

$$G - mean = \sqrt{SE * SP} \qquad (5)$$

According to [9, 17, 22], there are two values being the SE (sensitivity) and SP (specificity) to measure the performance of classification. The correct forecasted positive samples felt to minority class is called SE value while the ratio of forecasted correctly negative samples is called SP. In imbalanced data classification, the using k-fold Cross-Validation (k = 10) is to validate the classifiers' performance. In our paper, G-mean is used as a measurement to evaluate the balance between SE and SP. Moreover, we averaged every G-mean out at 20 G-mean values taken from every k-fold Cross-validation.

4   Experimental Results

4.1   Datasets

We tested four imbalanced datasets which are divided into two types in UCI Machine learning respiratory, including Financial datasets and biomedical datasets. To give a clear example, Table 3 shows the rate of imbalanced data.

Table 3.   The detailed imbalanced datasets from UCI

Dataset	Samples	Attributes	Ratio minor/major
German credit card	1,000	20	1:2.33
UCSD-FICO	500	19	1:20.74
Haberman	306	3	1:2.28
Yeast	1,484	8	1:28.1

4.2   Results

We experimented on four imbalanced datasets with MASI, and compared to five alternative adjustment algorithms, including ROS, RUS, SMOTE, Borderline SMOTE1 (BSO1), SPY, and proposed MASI. The classification models using artificial samples are SVM, C5.0, and RF. These algorithms are determined as suitable resampling methods for selected classifiers SVM, C5.0 and RF. In addition, SE, SP, and G-mean are used as criteria to assess classification performance between these approaches. This study also used k-fold Cross-validation method in testing to reduce the limitation of small dataset, k was chosen at 10. The Table 4 showed experiment of all classifiers.

Table 4. Classifiers' performance

Data		SVM			C5.0			RF		
		SE	SP	G-mean	SE	SP	G-mean	SE	SP	G-mean
German credit data	Original	39.75	**91.53**	60.3	47.62	**83.7**	63.1	41.9	**91.76**	62
	ROS	67.72	74.56	71.05	52.62	76.87	63.58	50.82	87.38	66.63
	RUS	**73.48**	68.92	71.16	**65.57**	65.74	65.63	**73.37**	69.38	71.34
	SMOTE	69.83	71.49	70.65	56.05	73.14	64	56.65	82.24	68.25
	BSO1	65.55	75.64	70.41	50.32	80.65	63.68	48.38	87.8	65.17
	SPY	70.23	72.26	71.22	63.13	69.53	66.24	71.02	70.61	70.81
	MASI	70.08	73.67	**71.85**	60.35	72.81	**66.27**	69.82	73.11	**71.44**
UCSD -FICO	Original	0	**100**	0	0.65	**99.61**	3.12	23.7	**99.62**	48.37
	ROS	56.3	91.57	71.73	46.74	96.21	66.9	35.43	99.34	59.27
	RUS	71.3	62.53	66.65	68.04	59.95	63.72	**80**	68.5	73.96
	SMOTE	49.78	95.07	68.73	31.3	95.34	54.45	32.17	98.85	56.29
	BSO1	48.26	**95.21**	67.6	37.39	94.62	59.02	32.83	98.95	56.86
	SPY	**73.26**	73.44	73.31	73.04	75.31	74.07	68.7	79.77	73.97
	MASI	**73.26**	76.35	**74.76**	**73.26**	79.01	**76.04**	79.78	70.46	**74.95**
Haberman	Original	18.70	**92.93**	41.62	17.28	**91.36**	39.14	23.46	**89.91**	45.85
	ROS	48.70	72.04	59.19	55.86	67.18	61.14	38.52	78.49	54.95
	RUS	54.51	70.27	61.84	55.12	72.04	62.97	58.27	68.69	63.22
	SMOTE	**67.22**	57.24	62.01	54.63	77.42	65.00	44.07	77.00	58.22
	BSO1	63.40	59.89	61.61	**61.67**	68.78	65.09	38.27	81.93	55.93
	SPY	55.86	76.67	**65.40**	60.09	66.00	63.30	58.89	74.13	**66.02**
	MASI	60.99	70.07	65.34	57.28	74.42	**65.26**	**60.99**	70.07	65.34
Yeast	Original	3.73	**99.98**	19.55	26.76	**99.14**	51.33	14.12	**99.73**	37.39
	ROS	62.45	90.69	75.22	41.76	96.65	63.45	31.08	98.93	55.41
	RUS	62.06	94.69	76.63	65.98	93.42	78.48	51.47	96.11	70.29
	SMOTE	58.73	93.6	74.1	62.06	92.51	75.73	54.8	95.99	72.51
	BSO1	42.45	97.6	64.34	30.39	99.04	54.78	24.41	99.06	49.1
	SPY	70.59	92.92	80.98	70.49	92.19	80.6	68.73	92.11	79.54
	MASI	**85.39**	81.7	**83.52**	**90.39**	73.76	**81.63**	**84.02**	83.05	**83.52**

It can be seen clearly that almost all datasets used MASI method have significant achievement in G-mean comparing to other methods. Especially, the proposal method had an outstanding performance in Yeast dataset regarding G-mean rate, it is much better than well-known SPY method (83.52 versus 80.98; 66.27 versus 63.1; 81.63 versus 80.60, and 83.52 versus 79.54 respectively). In UCSD dataset, testing classifiers failed to identify fraud in original dataset, while its results improved after resampling. Moreover, MASI still presented the highest G-mean in all classifiers. This can be explained that both datasets (UCSD and Yeast) have high ratio in imbalanced of 1:20.74 and 1:28.1 respectively (see in Table 2). In three classifiers, C5.0 seems to have the best results compared to SVM, and RF.

Overall, MASI method illustrates the highest G-mean in almost imbalanced dataset in all classifiers. Especially, G-mean is highest value in three datasets (German, UCSD, and Yeast) in three classification algorithms and G-mean has approximately result in Haberman when using MASI.

5 Conclusion

The data imbalance issue is becoming common in many practical applications and has attracted great interest from researchers. Numerous different methods try to solve this problem to enhance the classification performance. In addition, minority boundary samples are misclassified more than other samples. Therefore, we propose MASI method to increase minority borderline samples based on data distribution by changing their label. We also implemented and experimented to compare this MASI with other methods. Our results showed that the proposed method outperform other methods like ROS, RUS, SMOTE, Borderline SMOTE, SPY, and MASI using classifiers SVM, C5.0, and RF. Although our method has significant results, there are still some shortcomings in our method such as algorithm complexity, imbalanced regression, etc. Interesting idea is that a combination between MASI and the previous methods, for example, SPY, or SMOTE, is new idea to address these challenges in the future.

References

1. Matías, M.D., Federico, D., Juan, M.: Improving electric fraud detection using class imbalance strategies. In: International Conference on Pattern Recognition Applications and Methods (ICPRAM), pp. 135–141 (2012)
2. Japkowicz, N., Stephen, S.: The class imbalance problem: a systematic study. Intell. Data Anal. 6, 429–449 (2002)
3. Nistesh, C.V.: Data mining for imbalanced datasets: an overview. In: Data Mining and Knowledge Discovery Handbook, pp. 853–868. Springer, Boston (2005)
4. Rafiq, M.A., Kok, W.W., Mohd, S.F., Xuequn, W.: Improving fraud prediction with incremental data balancing technique for massive data streams. CoRR, pp. 1–8 (2019)
5. Fei, W., Xiao-Yuan, J., Shiguang, S., Wangmeng, Z., Jing-Yu, Y.: Multiset feature learning for highly imbalanced data classification. In: Proceedings of the Thirty-First AAAI Conference on Artificial Intelligence (AAAI), pp. 1583–1589 (2017)
6. Chao, C., Andy, L., Leo, B.: Using random forest to learn imbalanced data, pp. 1–12. University of California, Berkeley (2004)

7. Qiang, Y., Xindong, W.: 10 challenging problems in data mining research. Int. J. Inf. Technol. Decis. Mak. **5**(4), 597–604 (2006)
8. Enislay, R., Yailé, C., Rafael, B.: SMOTE-RSB∗: a hybrid preprocessing approach based on oversampling and undersampling for high imbalanced data_sets using SMOTE and rough sets theory. Knowl. Inf. Syst. **33**(2), 245–265 (2011)
9. Dang, T.X., Tran, H.D., Osamu, H., Kenji, S.: SPY: a novel resampling method for improving classification performance in imbalanced data. In: Seventh International Conference on Knowledge and Systems Engineering (KSE), pp. 280–285 (2015)
10. Yanmin, S., Andrew, W.K., Mohamed, K.S.: Classification of imbalanced data: a review. Int. J. Pattern Recognit. Artif. Intell. **23**(4), 687–719 (2009)
11. Lich, N.T., Thuy, N.T., Toan, N.T.: MASI: moving to adaptive samples in imbalanced credit card dataset for classification. In: International Conference on Innovative Research and Development (ICIRD), pp. 133–137 (2018)
12. Alireza, P., Majid, K., Alireza, N.: Fraud detection in E-banking by using the hybrid feature selection and evolutionary algorithms. IJCSNS Int. J. Comput. Sci. Netw. Secur. **17**(8), 271–279 (2017)
13. Aastha, B., Rajan, G.: Financial frauds: data mining based detection – a comprehensive survey. Int. J. Comput. Appl. **156**(10), 20–28 (2016)
14. Anuj, S., Prabin, P.K.: A review of financial accounting fraud detection based on data mining techniques. Int. J. Comput. Appl. **39**(1), 37–47 (2012)
15. Kaizhu, H., Haiqin, Y., Irwin, K., Michael, L.: Machine Learning: Modeling Data Locally and Globally. Springer (2008)
16. Federica, M., Marco, B., Gianfranco, B., Francesca, C.: Peculiar genes selection: a new features selection method to improve classification performances in imbalanced data sets. PLoS ONE **12**, 1–18 (2017)
17. Alberto, F., Salvador, G., Francisco, H.: SMOTE for learning from imbalanced data: progress and challenges, marking the 15-year anniversary. J. Artif. Intell. Res. **61**, 863–905 (2018)
18. Arnis, K., Sergei, P., Henrihs, G.: Entropy-based classifier enhancement to handle imbalanced class problem. Procedia Comput. Sci. **104**, 586–591 (2017)
19. Haibo, H., Yang, B., Edwardo, G.A., Shutao, L.: ADASYN: adaptive synthetic sampling approach for imbalanced learning. In: IEEE International Joint Conference on Neural Networks, pp. 1322–1328 (2008)
20. Nitesh, C.V., Kevin, B.W., Lawrence, H.O., Philip, K.W.: SMOTE: synthetic minority oversampling technique. J. Artif. Intell. Res. **16**, 321–357 (2002)
21. Ivan, T.: Two modifications of CNN. Trans. Syst. Man Commun. **6**(11), 769–772 (1976)
22. Masoumeh, Z., Pourya, S.: Application of credit card fraud detection: based on bagging ensemble classifier. In: International Conference on Computer, Communication and Convergence (ICCC), vol. 48, pp. 679–685 (2015)
23. Sheng, G., Min, C., Hsin, Y.H., Shu, C.C., Mei, S.L., Chengde, Z.: Deep learning with MCA-based instance selection and bootstrapping for imbalanced data classification. In: IEEE Conference on Collaboration and Internet Computing (CIC), pp. 288–295 (2015)
24. Reshma, D.K., Banait, S.: Imbalanced time series data classification using oversampling technique. Int. J. Electron. Commun. Soft Comput. Sci. Eng. 75–80 (2015). ISSN 2277-947
25. Mousa, A.: Detecting financial fraud using data mining techniques: a decade review from 2004 to 2015. J. Data Sci. **14**, 553–570 (2016)
26. Adrian, B.: Detecting and preventing fraud with data analytics. Procedia Econ. Financ. **32**, 1827–1836 (2015)
27. Yiyang, B., Min, C., Chen, Y., Yuan, Y., Qing, L., Leon, Z., Liang, L.: Financial fraud detection: a new ensemble learning approach for imbalanced data. In: Pacifc Asia Conference on Information Systems (PACIS), pp. 315–326 (2016)

28. Rafiq, M.A., Kok, W.W., Mohd, S.F., Xuequn, W.: Scalable machine learning techniques for highly imbalanced credit card fraud detection: a comparative study. In: Pacific Rim International Conference on Artificial Intelligence (PRICAI), pp. 237–246 (2018)
29. Mario, A., Firas, M., Elli, A., Stefan, S., Andreas, M.: The random forest classifier in weka: discussion and new developments for imbalanced data. Comput. Vis. Pattern Recognit. 1–6 (2019)
30. Ludmila, K.I., Álvar, A.-G., José-Francisco, D.-P., Iain, G.A.D.: Instance selection improves geometric mean accuracy: a study on imbalanced data classification. Prog. Artif. Intell. **8**, 215–228 (2018)
31. Ila, D., Shantanu, D., Bijan, R.: Detecting financial restatements using data mining techniques. Expert Syst. Appl. **93**, 374–393 (2017)
32. Leila, G., Mohammad, T.J.: Survey of detecting fraud in automobile insurance using data mining techniques. Int. J. Comput. Inf. Technol. (IJOCIT) **4**(4), 111–125 (2016)
33. Maciej, M.A., Piotr, H.A., Jacek, Z.M., Joseph, L.Y., Jay, B.A.: Training neural network classifiers for medical decision making: the effects of imbalanced datasets on classification performance. Neural Netw. **21**, 427–436 (2008)
34. Masoumeh, Z., Pourya, S., Deepak, J.K., Haoxiang, W.: Kernelized support vector machine with deep learning: an efficient approach for extreme multiclass dataset. Pattern Recognit. Lett. **115**, 4–13 (2018)
35. Wei-Chao, L., Shih-Wen, K., Chih-Fong, T.: Top 10 data mining techniques in business applications: a brief survey. Kybernetes **46**(7), 1158–1170 (2017)

Learning Rough Set Based Classifiers
Using Boolean Kernels

Hung Son Nguyen[1]([✉]) [iD] and Sinh Hoa Nguyen[2] [iD]

[1] University of Warsaw, Banacha 2, 02-097 Warsaw, Poland
son@mimuw.edu.pl
[2] Polish-Japanese Academy of Information Technology,
Koszykowa 86, 02-008 Warsaw, Poland
hoa@mimuw.edu.pl

Abstract. In this paper we present a hybridization of Rough Set (RS) theory and Support Vector Machine (SVM). Both approaches to data analysis employ the area between positively and negatively labeled examples, i.e. the "boundary region" in RS and the "margin" in SVM, but they offer different ways to use this concept in the classification problem. We will show that despite differences, many Rough Set methods can be also implemented by SVM. In particular we will show that the rough set methodology to discretization problem can be also solved by SVM with a special Boolean kernel. At the end we propose a compound classification method that aggregates the feature selection method in RS and object selection method in SVM.

Keywords: Classification problems · Rough sets · Support vector machine · Boolean kernels · Hybrid systems

1 Introduction

In machine learning, classification is the problem of construction of the algorithm that can categorize the new unseen data objects into some predefined classes. Such algorithms are called *the classification algorithms* or shortly *the classifiers*. It is a typical supervised learning task, because the classifiers are constructed (learned) from labeled training data sets. There are at least three ways to define the partition of the instance space, i.e. (1) using a Logical expression, (2) Using the Geometry of the instance space and (3) using Probability. Thus the existing classification methods can be categorized into Logical, Geometrical and Probabilistic approaches to Machine Learning [6].

In this paper we present some comparison analysis between the two techniques for classifier construction offered by rough set theory (also called Rough Sets or briefly RS) and Support Vector Machine (SVM). The rough classifiers are rather Logical approach, while the second are typical Geometrical approach to Machine Learning. The choice of those classification techniques was dictated

© Springer Nature Switzerland AG 2020
H. A. Le Thi et al. (Eds.): ICCSAMA 2019, AISC 1121, pp. 163–173, 2020.
https://doi.org/10.1007/978-3-030-38364-0_15

by the fact that both Rough sets and Support Vector Machine pay a deep attention on the border area between decision classes. However this concept (called "boundary region" in rough set theory and "margin" in SVM) is used in different ways in order to form the final classifiers.

Rough set methodology to classification is looking for minimal subsets of features that maintain as much as possible discriminant power as the set of all features. Usually this idea leads to creating the methods that minimize the boundary region between the decision classes. Many existing methods in Rough Sets perform the Minimal Description Length (MDL) principle and hierarchical concept approximation idea. Boolean Reasoning methodology is also a popular technique in rough set theory [14]. The problem of learning rough classifier from training can be decomposed into simpler tasks like searching for minimal reducts, for minimal decision reducts or for a set of irreducible decision rules.

Rough Sets and Support Vector Machine are realized by different computation techniques. Support Vector Machine(SVM) follows the maximal margin principle. SVM approach to classification is founded on two mathematical tricks. The first trick applies the Lagrange multiplier strategy to find linear classifiers with the maximal margin. This method transforms the original problem into a corresponding Quadratic Optimization. The second trick, also call the Kernel trick, implements an embedding of the input space in a very large feature space without increasing the time and memory complexity. Moreover, searching for reducts in Rough Sets can be interpreted as a feature selection method, while searching for support vectors in SVM is a method for object reduction.

There are few ideas to combine RS with SVM. In [3,5,9] the authors proposed to use attribute reduction method from RS as a pre-processing step for SVM. The other researchers proposed to apply two SVM models to present the lower and upper approximations of a concept [1,10–12,19]. In [15], we presented another approach to combine SVM and RS. We recalled a the concept of Boolean kernels that was designed for learning the monomial Boolean formulas and show that the fundamental rough set techniques like searching for minimal reducts, extraction of decision rules or construction of rule based classification algorithms can be also implemented by SVM with different Boolean kernels. However, the proposed method was designed for decision tables with symbolic values only. This paper extends the previous results by showing that the methods proposed in [15] can be modified so that they become applicable for data sets with real value attributes. We present a special kernel function that realizes the discretization method based on Boolean reasoning and Rough set theory. We also propose a hybrid method that combines the feature selection technique in RS and object selection technique in SVM.

The paper is organized as follows. Section 2 contains some basic notions of RS, SVM, Boolean kernels and revises the main results in [15]. In Sects. 3 and 4 we present the SVM with Boolean kernels that implements the discretization problem in Rough set theory as well as a hybrid method that combines SVM with Boolean kernel and Rough Set method for classification. The paper ends with conclusions and future plan in Sect. 5.

2 Preliminaries

Rough Set theory has been proposed by Z. Pawlak in [16] as a data driven method for the concept approximation problem under vague, imprecise, inconsistent, incomplete information and knowledge. In Rough Set Theory the uncertainty and imprecision is expressed by mean of the boundary region of a set. Formally, in the simplest case, when we have a finite set of objects, and each object is described by a finite set of attributes, the *information system* is a tuple $\mathcal{S} = (U, A)$, where U is a set of objects, A is a set of attributes, i.e. the set of functions of form $a : U \rightarrow V_a$, where V_a is the domain of attribute $a \in A$. For any subset of attributes $B \subset A$ we define the *B-indiscernibility relation* and denote it by $IND_{\mathcal{S}}(B))$ as follows

$$IND_{\mathcal{S}}(B) = \{(x_i, x_j) \in U \times U : a(x_i) = a(x_j) \text{ for all } a \in B\}.$$

A pair of different objects satisfies the indiscernibility relation if we are not able to discern those objects using only available information about them.

The training data sets for classification problem are the special case of information system called the *decision table* because it contains some special attributes that defines the partition of objects into decision classes. Formally, decision table is a tuple $\mathcal{S} = (U, A \cup \{dec\})$, where $dec : U \rightarrow V_{dec}$ is called decision attribute. In cases of binary classification problem, $V_{dec} = \{-1, 1\}$.

Feature selection refers to the methods that look for subsets of features (attributes) that are most relevant for the classification task, or that eliminate the features with little or no predictive information. Feature selection can significantly improve the quality of the classification models and often build classifiers that generalizes better to the new unseen objects. The feature selection problem in Rough Set Theory is defined in terms of reducts [17].

For decision tables, we are interested in the classification power of the subsets of condition attributes. A set of attributes $B \subseteq A$ is called a *decision oriented reduct* (or a *relative reduct*) of decision table $\mathcal{S} = (U, A \cup \{dec\})$ if and only if B is a minimal subset (with respect to the inclusion) of A satisfying the property that for any pair (x_i, x_j) of objects, if $dec(x_i) \neq dec(x_j)$ and $(x_i, x_j) \notin IND_{\mathcal{S}}(A)$ then $(x_i, x_j) \notin IND_{\mathcal{S}}(B)$.

Rough set methodology to classification problems make use of the rule-based classifiers see [2]. Therefore, searching for a set of high quality rules is the one of the fundamental problems in Rough Sets. In fact, decision rules are implications in description logic which present the relationship between conditional and decision attributes. Lets consider the simplest form of decision rules:

$$\mathbf{r} \equiv (a_{j_1} = v_1) \wedge \ldots \wedge (a_{j_m} = v_m) \Rightarrow (dec = d) \tag{1}$$

Thus the premises of decision rules are the conjunctions of descriptors of form $a = v$ for $a \in A$ and $v \in V_a$.

Each decision rule $\mathbf{r}$ is characterized by its length(denoted by $length(\mathbf{r})$), which is the number of descriptors in the conjunction, by $strength(\mathbf{r})$ – the number of objects satisfying the premise of $\mathbf{r}$ and $confidence(\mathbf{r})$ - the ratio of the number of objects satisfying $\mathbf{r}$ to the strength of $\mathbf{r}$. We say that $\mathbf{r}$ is a *consistent rule* with the decision table $\mathcal{S}$ if and only if $confidence(\mathbf{r}) = 1$.

2.1   Support Vector Machine

Any decision table $\mathcal{S} = (U, A \cup \{dec\})$ with real value attributes and $U = \{u_1, \ldots, u_n\}$ can be interpreted as a training set containing object-decision pairs

$$\mathbb{D} = \{(\mathbf{x}_1, d_1), \ldots, (\mathbf{x}_n, d_n)\}$$

where $\mathbf{x}_i \in R^m$ is the information vector of object $u_i \in U$ and $d_i = dec(u_i) \in \{-1, 1\}$ for $1 \leq i \leq n$. The support vector machine (SVM) problem is defined in [18] as the problem of searching for a linear classifier $L : R^m \rightarrow \{-1, 1\}$, of the form $L(\mathbf{x}) = \mathbf{w} \cdot \Phi(\mathbf{x}) + b$, where Φ is a function that embeds the objects from R^m into a higher (even infinite) dimensional space, such that L properly classifies the objects from $\mathbb{D}$ and the margin of L is maximal. These conditions can be described by the following optimization problem

$$\min_{\mathbf{w}, \xi, b} \frac{\|\mathbf{w}\|^2}{2} + C \sum \xi_i$$
$$\text{subject to } d_i(\mathbf{w} \cdot \Phi(\mathbf{x}_i) + b) \geq 1 - \xi_i; \text{ for } i = 1, \ldots, n \qquad (2)$$

If Φ is a special embedding function, for which there exists a function $K : R^m \times R^m \rightarrow R$ (called the kernel function) such that $K(\mathbf{x}_i, \mathbf{x}_j) = \langle \Psi(\mathbf{x}_i), \Psi(\mathbf{x}_i) \rangle$ for any $\mathbf{x}_i, \mathbf{x}_j \in R^m$ then, applying the Lagrange multiplier strategy, this problem can be transformed to the following quadratic optimization problem:

$$\max_{\alpha} \sum_{i=1}^{n} \alpha_i - \frac{1}{2} \sum_{ij=1}^{n} \alpha_i \alpha_j d_i d_j K(\mathbf{x}_i, \mathbf{x}_j)$$
$$\text{subject to } C \geq \alpha_i \geq 0, i = 1, \ldots, n \quad and \quad \sum_{i=1}^{n} \alpha_i d_i = 0 \qquad (3)$$

This later formulation, called the *dual representation SVM problem*, changes the searching space, because it is looking for the optimal significance (non-negative) weight α_i for each object $u_i \in U$. Moreover, if te vector $(\alpha_1^o, \ldots \alpha_n^o)$ is the optimal solution of the problem (3), then the objects u_i corresponding to $\alpha_i^o = 0$ are said to be *redundant* and should be removed from the training decision table. The remaining objects (corresponding to $\alpha_i^o > 0$) are called *the support vectors* and they can be used to define the SVM classifier as follows:

$$d_{SVM}(\mathbf{x}) = \text{sgn} \left(\sum_{\text{sup. vectors}} d_i \alpha_i^0 K(\mathbf{x}_i, \mathbf{x}) + b_0 \right) \qquad (4)$$

where sgn is the sign function and $\mathbf{x} \in R^m$ is the information vector of an arbitrary new, unseen object.

2.2 SVM with Boolean Kernels in Learning Rule-Based Classifiers for Decision Tables with Symbolic Values

In rough set theory, classification problem can be solved by the same techniques used for the concept approximation problem. The idea is to construct some "rough classifiers" from the training data. Rough classifiers should contains the description of the lower and upper approximations of all decision classes. Instead of hard prediction for new unseen object, rough classifier can return a vector of positive real values that describe to which degrees we believe that the considered object belongs to decision classes.

For a given decision table $\mathcal{S} = (U, A \cup \{dec\})$, the typical process of building rough classifier starts with the *rule induction step* in which a set of short, strong (with high support) and high confidence decision rules $\mathcal{RULE}(\mathcal{S})$ is generated from decision table $\mathcal{S}$. Let $\mathcal{RULE}(\mathcal{S}) = \{R_1, \ldots, R_p\}$ and let w_i be the strength of rule R_i and $dec(R_i)$ the decision class of the decision rule R_i for $i = 1, \ldots, p$.

In order to classify a new unseen object $\mathbf{x} \notin U$ we should implement the function $Match(\mathbf{x}, R_i)$, which returns the degree in which the object $\mathbf{x}$ matches to the decision rule R_i. Let $MRULE(\mathcal{S}, \mathbf{x}) = \{R_i : Match(\mathbf{x}, R_i) > \delta\}$ (for some $\delta > 0$) denotes the set rules from $\mathcal{RULE}(\mathcal{S})$ which are satisfied by $\mathbf{x}$.

The final decision returned by the rough classifier based of the decision rule set $\mathcal{RULE}(\mathcal{S})$ can be calculated by a voting step, which can be formulated by the following formula:

$$Dec_{\mathcal{RULE}}(\mathbf{x}) = S \left(\sum_{R_i \in MRULE(\mathcal{S}, \mathbf{x})} dec(R_i) \cdot w_i \cdot Match(\mathbf{x}, R_i) \right) \qquad (5)$$

where S is a function that, similarly to the activation function in artificial neural networks, converts the total voting power into one of decision classes [14].

We can see that two Eqs. (4) and (5) are quite similar. This observation was the main motivation in our previous paper [15]. We had prove that for a symbolic value decision table $\mathcal{S} = (U, A \cup \{d\})$ with two decision classes, the SVM with the following kernel function can implement a rule-based classifier for $\mathcal{S}$.

$$K_2^\varepsilon(u_i, u_j) = (1 + \varepsilon)^{|N_{i,j}|} - 1 \qquad (6)$$

where u_i, u_j are the arbitrary objects of U, and $|N_{i,j}| = |\{a \in A : a(u_i) = a(u_j)\}|$ is the number of attributes, for which u_i and u_j are indiscernible.

Moreother, if ε is a very small positive value, for example $\varepsilon < 0.05$, one can approximate the value $(1 + \varepsilon)^{\frac{1}{\varepsilon}}$ by e, therefore

$$K_2^\varepsilon(\mathbf{x}_i, \mathbf{x}_j) \approx e^{\varepsilon \cdot |N_{i,j}|} - 1$$

3 Discretization Problem in Rough Set Theory

Many classification methods including the rule-based classification in Rough Sets are not applicable for data sets with numeric data. Moreover, the experimental results are showing that the accuracy of many classification methods can be significantly improved if the data set has been early discretized. In general, discretization is a partition the real axis into disjoint intervals. Such partition can be defined by a relevant set of cuts (i.e., boundary points of intervals).

Let $\mathcal{S} = (U, A \cup \{dec\})$ be a decision table where all attributes are numeric, i.e. $V_a \subseteq R$. For any real value $c \in V_a$, the pair (a, c) is called a *cut* on attribute a, because it defines a partition of real axis into two intervals. We will say that two objects $u_i, u_j \in U$ are *discernible* by a cut $(a; c)$ if they are lying on different sides of the cut (a, c). The discernibility can be verified easy by checking if $(a(u_i) - c)(a(u_j) - c) < 0$.

A set of cuts $\mathbf{C}$ is called $\mathcal{S}$−consistent if for any pair of objects $u_i, u_j \in U$ such that $dec(u_i) \neq dec(u_j)$ if $(u_i, u_j) \notin IND_{\mathcal{S}}(A)$ then u_i and u_j are discernible by at least one cut from $\mathbf{C}$. By *optimal discretiztion problem* we denote the problem of looking for a $\mathcal{S}$ − consistent set of cuts that is minimal with respect to inclusion.

One of the algorithms for optimal discretization has been proposed in [13, 14]. The idea was based on Boolean reasoning methodology. In this method, the optimal discretization problem for a given decision table $\mathcal{S} = (U, A \cup \{dec\})$ was transformed to a corresponding boolean function $\phi_{\mathcal{S}}$. This transformation has the following property: the set of cuts is $\mathcal{S}$-consistent $\Longleftrightarrow$ the corresponding set of Boolean variables is a prime implicant of $\phi_{\mathcal{S}}$.

In fact, the boolean function $\phi_{\mathcal{S}}$ that can encode the optimal discretization problem was defined by

$$\phi_{\mathcal{S}} = \prod_{dec(u_i) \neq dec(u_j)} \sum_{u_i, u_j \text{ are discernible by } (a,c)} p_{(a,c)} \qquad (7)$$

where $p_{(a,c)}$ is a boolean variable corresponding to a (a, c) from a given set of candidate cuts $\mathbf{C}$.

The presented above encoding indicates that all heuristics for prime implicant problem can be also applied to solve the optimal discretization problem. In particular, the mentioned above SVM with Boolean kernels can be also modified to learn the optimal discretization.

The main idea of our proposition is to modify the Boolean kernel K_2^ε described in the previous Section. Let us define the new Boolean kernel as follows:

$$K_{Disc}^\varepsilon(u_i, u_j) = (1 + \varepsilon)^{(2 \cdot |\mathbf{C}| - m_{i,j})} - 1 \qquad (8)$$

where $\varepsilon \in R^+$ is a real positive parameter, $\mathbf{C}$ is the set of candidate cuts, $m_{i,j}$ is the number of cuts from $\mathbf{C}$ that discern the objects u_i and u_j of the given decision table. In case of decision table with real value attributes, instead of the standard decision rules in Eq. 1, we should use the interval decision rules, i.e. the decision rules of the form:

$$\mathbf{r} \equiv (a_{i_1} \in (l_1, r_1)) \wedge \ldots \wedge (a_{i_m} = (l_m, r_m)) \Rightarrow (dec = d) \qquad (9)$$

We can see that interval decision rules can be used build rough classifiers instead of the standard decision rules.

We have the following:

Table 1. The "weather" daecision table (left) and its boolean representation (right)

ID	outlook	temp	humidity	windy	play	sunny	overcast	rainy	hot	mild	cool	high	normal	FALSE	TRUE	dec	alpha
1	sunny	hot	high	FALSE	no	1	0	0	1	0	0	1	0	1	0	-1	0,22
2	sunny	hot	high	TRUE	no	1	0	0	1	0	0	1	0	0	1	-1	0,00
3	overcast	hot	high	FALSE	yes	0	1	0	1	0	0	1	0	1	0	1	0,00
4	rainy	mild	high	FALSE	yes	0	0	1	0	1	0	1	0	1	0	1	6,20
5	rainy	cool	normal	FALSE	yes	0	0	1	0	0	1	0	1	1	0	1	0,00
6	rainy	cool	normal	TRUE	no	0	0	1	0	0	1	0	1	0	1	-1	2,60
7	overcast	cool	normal	TRUE	yes	0	1	0	0	0	1	0	1	0	1	1	0,00
8	sunny	mild	high	FALSE	no	1	0	0	0	1	0	1	0	1	0	-1	6,12
9	sunny	cool	normal	FALSE	yes	1	0	0	0	0	1	0	1	1	0	1	1,69
10	rainy	mild	normal	FALSE	yes	0	0	1	0	1	0	0	1	1	0	1	0,00
11	sunny	mild	normal	TRUE	yes	1	0	0	0	1	0	0	1	0	1	1	2,86
12	overcast	mild	high	TRUE	yes	0	1	0	0	1	0	1	0	0	1	1	2,79
13	overcast	hot	normal	FALSE	yes	0	1	0	1	0	0	0	1	1	0	1	0,00
14	rainy	mild	high	TRUE	no	0	0	1	0	1	0	1	0	0	1	-1	4,60

Theorem 1. *For a given real value decision table $\mathcal{S} = (U, A \cup \{dec\})$ with two decision classes. The SVM using the Boolean kernel shown in Eq. (8) can simulate a classifier using the interval decision rules for $\mathcal{S}$.*

One can prove this theorem by transformation of the original decision table $\mathcal{S} = (U, A \cup \{dec\})$ with real value attributes in to a new binary decision table $\mathcal{S}_{bin} = (U, A_{\mathbf{C}} \cup \{dec\})$ containing the same set of objects but instead of real value attributes from A, we create a new set of attributes $A_{\mathbf{C}}$ on the base of the set of $\mathbf{C}$. For each cut $(a, c) \in \mathbf{C}$ we define two boolean attributes namely $t_{a<c}$ and $t_{a\geq c}$, where for any object $u \in U$ we define $t_{a<c}(u) = 1$ iff $a(u) < c$ and $t_{a\geq c}(u) = 1$ iff $a(u) \geq c$. One can see that applying the Boolean kernel presented in Eq. (6), we receive the kernel function shown in Eq. (8).

The only problem that we should take under attention in implementation is the computational complexity. Let n be the number of objects and m be the number of attributes. Then, the boolean function shown in Eq. (7) consists of $O(n^2)$ clauses, each clause can consists $O(|\mathbf{C}|)$ variables. In the worse case, the time complexity for construction of the kernel matrix for SVM can be $O(n^2 \cdot |\mathbf{C}|)$.

4   Rough Sets and SVM Hybridization

As it has been mentioned in Sect. 2, most of rough classifiers operate a set of decision rules, and the main difference between methods of their constructions is based on the rule induction and the rule selection steps. Among the existing rule selection methods, one can recall for example the covering algorithm LEM2 [7], or sampling technique called "dynamic reducts" [2]. In this paper we propose another technique based on SVM with Boolean kernel. We will show that, the SVM with Boolean kernel can be applied as a pre-processing step. The rule induction method from Rough Sets can be restricted to the support vectors returned by SVM.

Consider decision table "weather" shown on the left hand side of Table 1. This table contains 14 objects and 4 attributes. Table 2 presents the set of all possible decision rules for this table using RSES system. For the new, unseen object $x = \langle$ sunny, mild, high, TRUE $\rangle$, we see that only rule nr 3 and rule nr 13 are exactly satisfied by x. The rule number 3 has higher strength (satisfied by 3 objects from decision table) and has a negative decision (we denote this fact by -3) while the rule nr 13 is satisfied by only one training object. The voting process should returns the decision "no". However, if we accept also a partial matching, the voting process should take under consideration 10 rules. However, in both cases the object x is classified into the class "no".

Table 2. The classification process of object $x = \langle$ sunny, mild, high, TRUE$\rangle$ using the set of decision rules generated by RSES system

Nr	Condition	$\Rightarrow$ Dec	Strength	Match
1	(outlook=overcast)	$\Rightarrow$ yes	4	0
2	(humidity=normal) $\wedge$ (windy=FALSE)	$\Rightarrow$ yes	4	0
3	(outlook=sunny) $\wedge$ (humidity=high)	$\Rightarrow$ no	-3	1
4	(outlook=rainy) $\wedge$ (windy=FALSE)	$\Rightarrow$ yes	3	0
5	(outlook=sunny) $\wedge$ (temp.=hot)	$\Rightarrow$ no	-2	1/2
6	(outlook=rainy) $\wedge$ (windy=TRUE)	$\Rightarrow$ no	-2	1/2
7	(outlook=sunny) $\wedge$ (humidity=normal)	$\Rightarrow$ yes	2	1/2
8	(temp.=cool) $\wedge$ (windy=FALSE)	$\Rightarrow$ yes	2	0
9	(temp.=mild) $\wedge$ (humidity=normal)	$\Rightarrow$ yes	2	1/2
10	(temp.=hot) $\wedge$ (windy=TRUE)	$\Rightarrow$ no	-1	1/2
11	(outlook=sunny) $\wedge$ (temp.=mild) $\wedge$ (windy=FALSE)	$\Rightarrow$ no	-1	2/3
12	(outlook=sunny) $\wedge$ (temp.=cool)	$\Rightarrow$ yes	1	1/2
13	(outlook=sunny) $\wedge$ (temp.=mild) $\wedge$ (windy=TRUE)	$\Rightarrow$ yes	1	1
14	(temp.=hot) $\wedge$ (humidity=normal)	$\Rightarrow$ yes	1	0

Table 1 also illustrates the SVM using boolean kernel $K_2^\varepsilon(.,.)$, with $\varepsilon = 0.2$. The original decision table consists of 10 descriptors and we can see in Table 1 the boolean representation of the original decision table. The last column represents the optimal α values of each objects. Let us remind that the α values can be interpreted as the importance of the objects in the classification process, and the objects related to positive α value are called the support vectors. Therefore, the decision table has 8 support vectors (four objects for each decision class).

If we remove from weather decision table 6 objects: $2, 3, 5, 7, 10, 13$ corresponding to zero value of the parameter α, and use the RSES system for this restricted table, we can obtain 12 among 14 decision rules from Table 2. We lost only two rules with quite low strength (rules nr 10 and 14). This observation some how supports our hypothesis that SVM can be applied as a pre-processing step for rough set methods to reduce the redundant objects.

This observation motivates the authors to propose the following hybrid classification method:

SVM-based rough classifier:

1. Run SVM algorithm with the kernel function presented in Eq. (8) to find the support vectors
2. Build the rough classifier the decision table restricted to the support vectors only.

This classification method combines both reduction techniques, i.e. feature reduction from rough set theory as well as the object selection from SVM. In case of larger data sets the SVM algorithm can be replaced by SVM light [8].

We performed some experiments on small benchmarking data sets to check the accuracy of Boolean kernels and compare their accuracy with other rule based classifications methods. For each data set, the accuracy was calculated by averaging using 5-fold cross validation.

The accuracy of existing methods are given in the columns marked by: CN2, RSlib and Other. This columns were taken from paper [2], where the author presented the accuracy of CN2 algorithm [4], the best possible rough set based classifier (using discretization, dynamic reduct, ...) and the best accuracy on this data achieved by other techniques.

The accuracy of three methods proposed by the authors: (1) SVM using Boolean kernel K_1 (described in [15]), (2) SVM using Boolean kernel K_2 and (3) hybrid method using kernel K_{Disc} are reported in last 3 columns of Table 3. During the experimental work on each data set, we checked the classification accuracy for different values of the parameter ε. We noticed a very interesting fact that the best classification accuracy were achieved when ε was nearby $\frac{1}{n}$, where n is the number of objects in the decision table.

In Table 3 present the accuracy of proposed algorithms for $\varepsilon = \frac{1}{n}$. It is interesting to notice that in all cases, the number of support vectors did not exceed 50% of the total number of objects.

Table 3. Accuracy of the SVM method using different Boolean kernels.

Dataset	Dimensions	CN2	RSlib	Other	K_1	K_2	K_{Disc}+RSlib
Iris	150×4	0.96	0.97	0.97	0.97	0.95	0.97
Diabetes	768×9	0.711	0.745	0.777	0.759	0.761	0.775
Australian	690×14	0.796	0.854	0.87	0.823	0.817	0.864

One can see that the rough classifiers build by the support vectors have at least the same or better accuracy comparing to the rough classifiers that were constructed using whole data sets.

5   Conclusions

We shown that many basic methods in rough set theory can be also simulated and approximately calculated by SVM with Boolean kernels. The main idea is based on the fact that most of rough set methods can be solved by Boolean Reasoning approach, i.e. the optimal problem in Rough Sets can be encoded by a boolean function. We also proposed a hybrid classification method that combine a object selection feature of SVM with feature selection and rule-based classification method in Rough Sets. The experimental results on some small data sets are showing that this combination is quite promising. We plane perform more experiments with different heuristics for SVM, including SVMlight [8] to make rough set methods more scalable.

References

1. Asharaf, S., Shevade, S.K., Murty, N.M.: Rough support vector clustering. Pattern Recogn. **38**(10), 1779–1783 (2005)
2. Bazan, J.: A comparison of dynamic and non-dynamic rough set methods for extracting laws from decision tables. In: Polkowski, L., Skowron, A. (eds.) Rough Sets in Knowledge Discovery 1: Methodology and Applications, Studies in Fuzziness and Soft Computing, vol. 18, pp. 321–365. Springer, Heidelberg (1998)
3. Chen, H.L., Yang, B., Liu, J., Liu, D.Y.: A support vector machine classifier with rough set-based feature selection for breast cancer diagnosis. Expert Syst. Appl. **38**(7), 9014–9022 (2011)
4. Clark, P., Niblett, T.: The CN2 induction algorithm. Mach. Learn. **3**(4), 261–283 (1989)
5. Feng, H., Liu, B., Cheng, Y., Li, P., Yang, B., Chen, Y.: Using rough set to reduce SVM classifier complexity and its use in SARS data set. In: KES (3), pp. 575–580 (2005)
6. Flach, P.: Machine Learning: The Art and Science of Algorithms that Make Sense of Data. Cambridge University Press, Cambridge (2012)
7. Grzymala-Busse, J.W.: LERS—A system for learning from examples based on rough sets. In: Słowinski, R. (ed.) Intelligent Decision Support. Handbook of Applications and Advances of the Rough Sets Theory, pp. 3–18. Kluwer Academic Publishers, Dordrecht (1992)
8. Joachims, T.: Making large-scale SVM learning practical. In: Scholkopf, B., Burges, C., Smola, A. (ed.) Advances in Kernel Methods - Support Vector Learning. MIT-Press (1999)
9. Li, Y., Cai, Y., Li, Y., Xu, X.: Rough sets method for SVM data preprocessing. In: Proceedings of the 2004 IEEE Conference on Cybernetics and Intelligent Systems, pp. 1039–1042 (2004)
10. Lingras, P., Butz, C.J.: Interval set classifiers using support vector machines. In: Proceedings of the North American Fuzzy Inform Processing Society Conference, pp. 707–710 (2004)

11. Lingras, P., Butz, C.J.: Rough set based 1-v-1 and 1-v-r approaches to support vector machine multi-classification. Inf. Sci. **177**, 3782–3798 (2007)
12. Lingras, P., Butz, C.J.: Rough support vector regression. Eur. J. Oper. Res. **206**, 445–455 (2010)
13. Nguyen, H.S., Nguyen, S.H.: Discretization methods for data mining. In: Polkowski, L., Skowron, A. (eds.) Rough Sets in Knowledge Discovery, pp. 451–482. Springer, Heidelberg (1998)
14. Nguyen, H.S.: Approximate boolean reasoning: foundations and applications in data mining. In: Peters, J.F., Skowron, A. (eds.) Transactions on Rough Sets V, pp. 334–506. Springer-Verlag, Heidelberg (2006)
15. Nguyen, S.H. Nguyen, H.S.: Applications of boolean kernels in rough sets. In: Proceedings of Rough Sets and Intelligent Systems Paradigms - Second International Conference, RSEISP 2014, Held as Part of JRS 2014, Granada and Madrid, Spain, 9–13 July 2014, pp. 65–76 (2014)
16. Pawlak, Z.: Rough sets. Int. J. Comput. Inf. Sci. **11**, 341–356 (1982)
17. Pawlak, Z.: Rough sets: theoretical aspects of reasoning about data. In: System Theory, Knowledge Engineering and Problem Solving, vol. 9. Kluwer Academic Publishers, Dordrecht (1991)
18. Vapnik, V.: The Nature of Statistical Learning Theory. Springer-Verlag, NY (1995)
19. Zhang, J., Wang, Y.: A rough margin based support vector machine. Inf. Sci. **178**(9), 2204–2214 (2008)

Using Support Vector Machine to Monitor Behavior of an Object Based WSN System

Nga Ly-Tu[1(✉)], Qui Vo-Phu[1], and Thuong Le-Tien[2]

[1] School of Computer Science and Engineering, International University-VNUHCM,
Ho Chi Minh City, Vietnam
ltnga@hcmiu.edu.vn, vophuqui2211@gmail.com
[2] Department of Electrical and Electronics Engineering, University of Technology,
VNU, Vietnam-VNUHCM, Ho Chi Minh City, Vietnam
thuongle@hcmut.edu.vn

Abstract. In this paper, we propose a new method to classify the bound error values of Kullback-Leibler Distance (KLD)-particle filter (PF) based Support Vector Machine (SVM) to reduce the mean number of particles used (sampling) as well as improve the performance of runtime in reality for monitoring an object. In wireless sensor network (WSN) system, the object location is calculated via the collected received signal strength (RSS) variations which are effected by furniture, walls or reflections. Therefore, we propose an architecture diagram to track an object and build the dataset model. By transforming the system state model from the 1D to 2D, the bound error value of KLD resampling can enhance estimation accuracy and convergence rate of declining number of particle used by generating a sample set near the high-likelihood region for ameliorating the effect of the RSS variations. Our proposal considers how to classify and find the bound error values of KLD PF for each iteration. The first iteration, using the observation information via KLD resampling optimal bound error to conduct a resampling on the basis of the initial bound error. From the second to the end iteration, we propose the SVM technique to search the predicted bound error value that fulfills the minimum of mean number of particle used between at the current and the next iteration. Our experiments confirm this technique to apply in reality system.

Keywords: KLD-resampling · WSN · KNN · LDA · SVM

1 Introduction

Using weighted particle set with assigned primary weights serves as the basic idea of a particle filter (PF) is one of these methods to improve the location of object in space, called a recursive Bayesian filter in [1]. Monte Carlo method is also used a set of weighted particle to realize the recursive Bayesian filter for an effective nonlinear non-Gaussian system, called suboptimal prediction method, applied to monitor behavior of object in [2].

© Springer Nature Switzerland AG 2020
H. A. Le Thi et al. (Eds.): ICCSAMA 2019, AISC 1121, pp. 174–185, 2020.
https://doi.org/10.1007/978-3-030-38364-0_16

The estimation accuracy of object is effected by the degenerate set of particles during the resampling step of PF. The authors in [3] introduced initially employed to combat sample impoverishment (or degeneracy) by supplanting high-weighted to low-weighted particles in [4]. It is useful to decrease the probability of losing object. Some authors investigated the effects of choosing metric and weight functional approach on PFs in [5–9]. Firstly, the PF based on Kullback-Leibler Distance (KLD)-sampling is used to determine the minimum number of particles needed to maintain the approximation quality in the sampling process. By fixed probability and given the upper bound error in sample set size of KLD-sampling in [9], the operation time of object is improved. Secondly, the predictive certainty state is used in the KLD-sampling in [7] to estimate the underlying posterior for improving the micro-ability and adaptability of particle set. The noise variance of the new information estimation system is determined based on reflect relationship between the accuracy of the target prediction and the uncertainty of the system. It uses to determine the sampling of the proposed distribution. Thirdly, the authors in [10] enhanced the ability to predict the particle set via the new information of observation to control the number of particles double sampling. Finally, the authors in [11] applied the trained network for KLD sampling to generate the new bin size though space division by KD Trees that helps balance between approximation error and runtime. The authors in [5, 8, 14] proposed KLD-resampling algorithms to find the number of particles based on the distribution of particles before and after process reaches a pre-specified bound error. Our researches in [8, 14] proposed an enhanced KLD-resampling PF by finding optimal bound error. But the optimal bound error in here is maintained during the online training data, it is still open problem.

Related Works

To overcome this problem, first we propose architecture diagram to collect and build dataset of the bound error values for online training phase an object in WSN in [8, 14, 18]. Next, our dataset model considers to transform the features (iterations of predicted bound error) to many hyperplanes in case of increasing noise variance during two adjacent iterations. Final, under supervised machine learning via SVM technique to separate the overlap bound error values. It helps to classify the training bound error value of KLD-PF based on the minimum weight vector by applied Karush-Kuhn-Tucker. As a result its mean number of particles used reduces significantly. Our experiments show that applied supervised machine learning in [13] to predict bound error for KLD resampling based RSS measurements in WSN system in [8, 9, 12, 8–9] improves the location and runtime (or the number of particles used). Our methods are also the another latest technique in [11] to apply the trained data for KLD-resampling by generating the predicted bound error based supervised machine learning that helps balance between approximation error (or Root Mean Square Error-RMSE) and runtime.

The paper is organized as follows. Introduction to system is given in Sect. 2. Our proposal is introduced in Sect. 3. All results based on PYTHON for monitoring are shown in Sect. 4. Finally, we recommend the future work in Sect. 5.

2 System Model

In this paper, we consider the robot carrying the sensor node and some static nodes. The robot moves along the determined path and a velocity in the long and thin region. This node can send the data to static nodes and receive the data from static nodes. Using PF algorithm to find its position. All sensor nodes that have own equal physical parameters, and its movement velocity remains the same at time. The random velocity follows Gaussian distribution. This system contains three models: 1. Mobility, 2. RSS statistical, and 3. System state models.

2.1 Mobility Model

The time is split into equal time section as the moving robot along the path at the steady value. The robot's velocity has some random noise that matches normal distribution. Figure 1 indicates and determines the random velocity. Let us denote v' (a random variable which conforms to normal distribution) and v'' to be the random velocity and the determined velocity, respectively. As result, the robot's velocity is presented by the dash line.

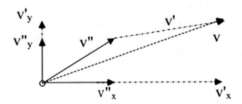

Fig. 1. The velocity of the robot in [12]

2.2 RSS Statistical Model

The statistical model for RSS shows the relationship between RSS and node distance. It is possible to express the popular mathematical model of WSN as follows

$$P(d) = P(d_0) - 10n \, log\left(\frac{d}{d_0}\right) + v_\sigma, \tag{1}$$

where $P(d)$ is the value of RSS at position d, and $P(d_0)$ is the RSS at reference (typically) position $d_0 = 1$ m. The path loss parameter related to the specific application environment is presented by n; and v_σ is a Gauss stochastic variable.

2.3 System State Model

The system state model for the mobile wireless sensor in [12] is defined as follows

$$x_k = x_{k-1} + V_k \Delta t + w_k, \tag{2}$$

$$z_k = Pref + Klog(x_k) + Rv_k, \tag{3}$$

where x_k is the position of a mobile node, Δt is the time segment, and z_k is the RSS measurement. V_k is the current velocity to determine velocity and random velocity in Eq. (1). The system state and measurement noises obey Gauss distributions. The reference value of RSS is *Pref* and the factor in path loss is K.

The system state model in Eqs. (2) and (3) can be rewritten Eqs. (3) and (4) for finding the required number of particles as

$$x_k = \begin{bmatrix} 1 & 1 & 0 & 0 \\ 0 & 1 & 0 & 0 \\ 0 & 0 & 1 & 1 \\ 0 & 0 & 0 & 1 \end{bmatrix} (x_{k-1} + V_k \Delta t) + \begin{bmatrix} 0.5 & 0 \\ 1 & 0 \\ 0 & 0.5 \\ 0 & 1 \end{bmatrix} Q \begin{bmatrix} w_{1,k} \\ w_{2,k} \end{bmatrix}. \tag{4}$$

$$z_k = Pref + Klog\left(arctan\left(x_{1,k}, x_{3,k}\right)\right) + Rv_k. \tag{5}$$

3 Proposal Techniques

3.1 Architecture Diagram

We propose the architecture diagram model to train the bound error of KLD PF in Eq. (7) via Machine learning in WSN systems, shown as in Fig. 2.

First, our diagram collects the observation information (such as RSSI from Zig-Bee, LoRa systems) based the system state model Eqs. (2)–(3) in [12]. Let us define $\varepsilon^l = \left[\varepsilon^l_0, \dots, \varepsilon^l_S\right]$ is the bound error that used to find the data of mean number of particles used in Eq. (7), where S is the length of ε^l. The data found as $N^f = \begin{bmatrix} N^f_{11} & \cdots & N^f_{1M} \\ \vdots & \ddots & \vdots \\ N^f_{S1} & \cdots & N^f_{SM} \end{bmatrix}$,

where f is the feature train.

The authors in [1] introduced weighted particle set with assigned primary weights serves as the basic idea of a particle filter (PF) to estimate the object location. This is a suboptimal prediction method applied in the field of object tracking in [2]. By sorting non-domination of these individuals in the population, the author in [7] proposed the fast KLD-sampling technique in the sampling process. At each iteration of the PF, it determines the number of samples with probability $1 - \delta$ of the error between the true posterior and the sample-based approximation is less than given threshold value (ε).

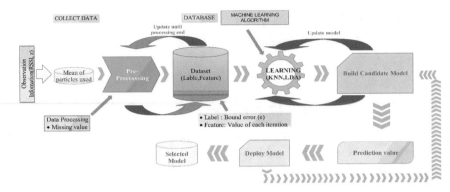

Fig. 2. Architecture diagram to track a target based supervised machine learning-KLD in WSN

Here, KLD is used to determine the number of samples (or the distance) between the sample-based maximum likelihood estimate and the true posterior does not exceed a pre-specified threshold ε. The KLD between the proposal q and p distributions can be defined in discrete form as follows

$$d_{KL}(p\|q) \triangleq \sum_{x} p(x) log\left(\frac{p(x)}{q(x)}\right). \tag{6}$$

The required number N_r of samples can be determined as follows $N_r = \frac{\chi^2_{k-1,1-\delta}}{2\varepsilon}$, where k is the number of bins with support, the quantizes of Chi-square distribution can be computed as follows $P\left(\chi^2_{k-1} \leq \chi^2_{k-1,1-\delta}\right) = 1 - \delta$.

The mean particle used criterion is collected as follows

$$N_r = \frac{k-1}{2\varepsilon}\left(1 - \frac{2}{9(k-1)} + \sqrt{\frac{2}{9(k-1)}}z_{1-\delta}\right)^3, \tag{7}$$

where $z_{1-\delta}$ is the upper quartile of the standard normal distribution.

3.2 Dataset Model

Our current work, Ly-Tu et al. in [8, 14, 18], introduced that the collected bound error range of KLD–resampling in [8] (see in Algorithm 6) from 0.7 to 0.975, the value of variances R and Q in Eqs. (2) and (3) from 0.1 to 0.9, in all cases of number particles N (N = 100, 200, 600) to track the target in WSN system. Based on this model, the mean number of particle used, bound error, and runtime are stored in one file excel, shown in Fig. 3.

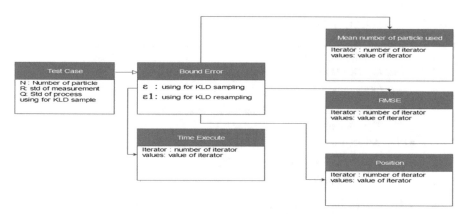

Fig. 3. Class diagram of our architecture

	Epsilon	Iter1	Iter2	Iter3	...	Iter37	Iter38	Iter39	Iter40
0	0.700	20.95	10.00	6.00	...	6.90	6.85	6.95	6.95
1	0.700	20.30	7.95	5.35	...	6.65	6.80	6.75	6.80
2	0.700	21.10	8.20	5.35	...	6.85	6.90	6.95	7.00
3	0.700	20.45	8.60	6.15	...	6.85	6.90	6.85	6.75
4	0.700	20.65	8.15	5.50	...	6.90	6.95	6.95	6.85
5	0.700	20.40	8.30	5.65	...	7.00	6.95	6.95	7.00
6	0.700	20.20	8.25	4.90	...	7.00	7.00	7.00	7.00
7	0.700	20.95	8.55	5.45	...	6.90	7.00	6.95	6.90
8	0.700	20.65	8.95	6.00	...	7.00	7.00	6.95	7.00
9	0.700	19.90	7.85	5.50	...	6.95	6.90	6.95	6.95
..	...	...	...	...	...	...	...	...	...
453	0.975	14.70	4.65	3.35	...	2.00	2.00	2.00	2.00
454	0.975	15.00	5.40	3.45	...	2.00	2.00	2.00	2.00
455	0.975	14.55	5.50	3.55	...	2.00	2.00	2.00	2.00
456	0.975	14.75	5.30	3.75	...	2.00	2.00	2.00	2.00
457	0.975	14.30	5.10	3.75	...	2.05	2.05	2.05	2.05
458	0.975	14.45	5.70	3.75	...	2.05	2.05	2.05	2.05

[459 rows x 41 columns]

Fig. 4. Dataset of 9 classes of Epsilon in case of N = 100

An example of values R = Q = 0.5, N = 100 for 9 classes of Label (called bound error in Fig. 2) namely *Epsilon* in Fig. 4. Each class has 51 rows and 41 columns. The first column is described the label of class (**Epsilon**) and the 40 next columns are assigned as 40 features. In order to overcome the process of these missing values (Pre-processing in Fig. 2), we follow the first method of four ones in [13] to remove these missing data.

Our Pseudo-code of finding predicted bound error using SVM method is mentioned in Table 1. Let us define T as a training set $T = \{f, \varepsilon^l\}$, which f is the feature train (line 7, Table 1) and f is a subset of N^f, size of f is $S \times [m : m + 1]$. Let us denote f^{test} is the feature test (line 8, Table 1). Hence, f & f^{test} will be updated in each iteration.

Table 1. Finding the predicted bound error based on SVM

1: **procedure** $(N_{max}, R, Q, m, N_m^{pred})$

2: Check N_{max}, R, Q to get patch

3: Load ε^l from patch ▷ Label train

4: Load N^f from patch ▷ Prepare data for feature train

5: $\Delta RMSE_{Pro}^{SIR} = \Delta RMSE_{Pro}^{KLDE} = 0$

6: **while** $(S!=\varnothing)$ **do**

7:      $f_{S(m:m+1)} = \left[N_{Sm}^f, N_{S(m+1)}^f\right]$ ▷ Feature train

8:      $f^{test} = \left[N_m^{pred}, \min(N_{S(m+1)}^f)\right]$ ▷Feature train

9:      idx=$index\left(\min\left(N_{S(m+1)}^f\right)\right)$ ▷ obtain index and remove missing data in [13]

10:     neigh = SVMClassifier(gamma =10) ▷ Run SVM model

11:     neigh.fit $\left(f, \varepsilon^l\right)$ ▷ traning set model

12:     $\varepsilon\_temp \leftarrow$ neigh.predict $\left(f^{test}\right)$ ▷ predicted bound error

13:     $RMSE_{\varepsilon\_temp}^{Pro}$ ▷ RMSE of proposal with the bound error at the mth iteration (ε_m^{pred})

14:     $RMSE^{SIR}$ ▷ RMSE of SIR

15:     $RMSE^{KLD}$ ▷RMSE KLD-resampling ($\varepsilon = 0.65$)

16:     $RMSE^{KLDE}$ ▷ RMSE KLDE ($\varepsilon_{opt} = 0.9$ in [8])

17:     **if** ($\Delta RMSE_{Pro}^{SIR} \geq 0$ & & $\Delta RMSE_{Pro}^{KLDE} \geq 0$) **then**

18:         $\varepsilon_m^{pred} = \varepsilon\_temp$ ▷ find ε_m^{pred}

19:         break;

20:     **else**   ▷ find new ε

21:         $\varepsilon^l = \varepsilon^l.remove.\varepsilon_{idx}^l$ ▷ remove and update label train

22:         $S = len(\varepsilon^l)$ ▷ Update **S**

23:         N^f = row.delete.N^f(idx) ▷remove and update mean number of particles used for new feature train

24:     **end if**

25: **end while**

26: **return** ε_m^{pred} ▷ find ε_m^{pred}

27: **end procude**

The objective of our proposal is to find the bound error (**Epsilon**) to reach the minimal of mean number of particle used based KLD-resampling adjusted bound error in [8] (see in Algorithm 6), therefore our works introduce the bound error algorithm with the initial bound error in [8] applied the first iteration and the predicted bound errors based KNN or LDA or SVM (line 10, Table 1) as shown in Fig. 5.

To more analysis this dataset in Fig. 5, three classes are selected as the first class (0.7), the middle (0.875) and the last ones (0.975) for evaluating the overlap them during the first four iterations vs. the effect of the variance noise Q in Eq. (2) or the fluctuation of RSS measurements in Eq. (3) as shown in Fig. 6 (see the next page), we can select the candidate model to deploy in reality. Here, if the predicted bound error is not satisfied

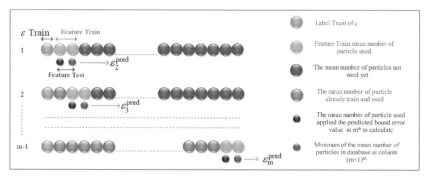

Fig. 5. Predict the bound error for our dataset model

conditions the performance of RMSE criterion in [8], it is removed out of the selected list. The output of our architecture diagram in Fig. 2 is the predicted bound error which fulfills both the mean number of particle used and RMSE.

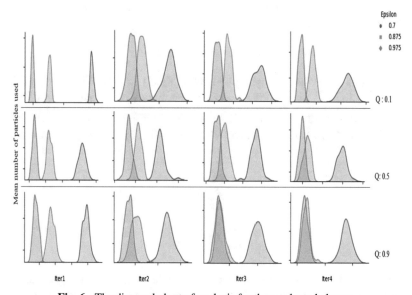

Fig. 6. The diagonal chart of analysis for three selected classes.

3.3 Support Vector Machine Technique to Classify the Bound Error Classes

There are many different hyperplanes which classify correctly all data points but the best choice will be the hyperplane that leaves the maximum of margin from both two classes. Following the example in Fig. 6, let us define Z_1 and Z_2 are the margin between two classes of *Epsilon* 0.85 (blue color) and 0.975 (green color), respectively, during two adjacent iterations shown in Fig. 7.

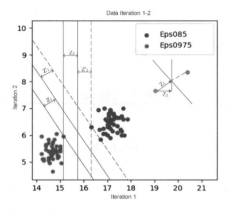

Fig. 7. The margin hyperplane of two classes

Figure 7 shows that the red and blue hyperplanes (separated by the red line and the blue line) with $Z_1 (Z_1 = Z'_1)$ and $Z_2 (Z_2 = Z'_2)$ are the margin of two classes, respectively. Here, the red hyperplane is chosen because of $Z_1 > Z_2$. The function of its hyperplane is determined as follows in [17].

$$g(\vec{x}) = \vec{w}^T \vec{x} + w_0 \tag{8}$$

where $g(\vec{x})$ is the label of hyperplane with $g(\vec{x}) \geq 1$, $\forall \vec{x}_{blue}$ or $g(\vec{x}) \leq -1$, $\forall \vec{x}_{green}$ and $\vec{w}^T$, w_0 are the weighted parameters of its hyperplane. The total margin is $\frac{1}{\|\vec{w}\|} + \frac{1}{\|\vec{w}\|} = \frac{2}{\|\vec{w}\|}$. To find the minimum of weight $\|\vec{w}\|$ for nonlinear optimization, Karush-Kuhn-Tucker (KKT) is applied by Lagrange multipliers λ_i as follows

$$\vec{w} = \sum_{i=0}^{N} \lambda_i y_i \vec{x}_i, \tag{9}$$

$$\sum_{i=0}^{N} \lambda_i y_i = 0, \tag{10}$$

Figure 8 (see the next page) shows an example to find the weight vector with two coordinates A(15, 6) and B(16, 6.5). The weight vector is calculated as $\vec{w} = (16, 6.5) - (15, 6) = (a, 0.5a)$.

$$g(A) = -1 \Leftrightarrow 15a + 6a + w_0 = -1 \tag{11}$$

$$g(B) = 1 \Leftrightarrow 16a + 3.25a + w_0 = 1 \tag{12}$$

From Equation (11)–(12), the value of a is 1.6 and $w_0 = -29.8$. Therefore the weight vector of SVM $\vec{w} = (1.6; 0.8)$ and $g(\vec{x}) = 1.6x_1 + 0.8x_2 - 29.8$. As a result, the blue hyperplane was contructed, and the predicted value will belong the class which represents for the bound error value of each iteration shown in Fig. 9 (see the next page).

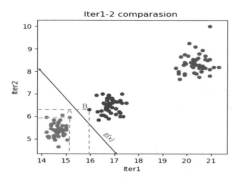

Fig. 8. An example to find weight vector

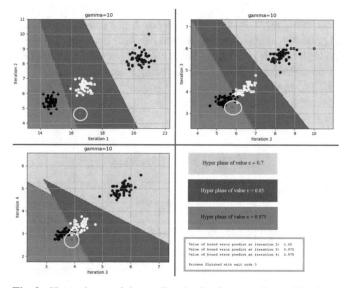

Fig. 9. Hyperplane and the predicted value for a small set of database

4 Simulation Results

Setting up to track an object of our systems in [8, 14, 18] as follows $R = 0.5$, $N_{max} = N$; $V_{max} = 5$; $V_{min} = 1$; $V_{init} = 5$; $P_{ref} = -23$; $K = -45$; a number of samples $N = 100$, a range of variances Q (0.1,0.3,0.5,0.7,0.9), length time is 40 for sample size variation in 20 trials as shown in Tables 2 and 3.

Table 2 verifies that SVM-KLD of our proposal dominates about the performance of the number of particles used when compared to others. This is because it can separate the overlap bound error value when increasing Q from 0.1 to 0.7 (shown in Fig. 6) by its margin hyperplane (shown in Fig. 7). This approach can classify the training bound error value in Eq. (7) based on the minimum weight vector (Fig. 8) in case of many hyperplanes (shown in Fig. 9) leading the mean number of particles used of our method reduces significantly.

Table 2. Mean number of particles used

Q	Mean number of particles used				
	KLD	KLDE	KNN-KLD	LDA-KLD	SVM-KLD
0.1	15.78	6.64	6.39	6.02	**5.54**
0.3	10.18	3.29	3.53	3.23	**3.11**
0.5	8.39	2.73	2.65	2.88	**2.54**
0.7	6.53	2.34	2.33	2.36	**2.31**
0.9	4.61	2.21	2.26	**2.20**	2.25

The performance of RMSE value and runtime for all methods are shown in Table 3. It shows that some values of RMSE of SVM-KLD, our technique, in case of $Q = 0.3$ and 0.9, is not good for others under supervised machine learning (KNN-KLD in [15] or LDA-KLD in [16]); but it is better than others without machine learning (KLD or KLDE in [14, 18]). This is thanks to the optimal choosing these values of the bound error that trained from KLDE in [14, 18]. However, the runtime of our method is very good due to the reduced number of particles used for resampling (see more detail in Table 2). It is useful in many fast reality.

Table 3. RMSE and runtime

Q	RMSE: runtime (ms)				
	KLD	KLDE	KNN-KLD	LDA-KLD	SVM-KLD
0.1	15.78:3.67	6.64:0.75	6.39:0.68	6.02:0.63	**5.54:0.63**
0.3	17.18:2.19	14.77:0.75	8.40:0.80	**7.13**:0.34	10.55:**0.30**
0.5	18.49:1.17	25.08:0.39	12.50:0.57	12.34:0.39	**12.30:0.39**
0.7	6.53:1.08	2.34:0.47	2.33:0.31	2.36:0.62	**2.31:0.24**
0.9	23.66:1.69	42.18:0.78	19.71:**0.52**	**15.89**:0.63	16.96:0.65

5 Conclusion

This paper, we propose the SVM method to find the predicted bound of KLD-resampling. This technique reduces the fluctuations of RSS samples in WSN leading to improve the location of object. It verifies that reduces the number of particles used when compared to traditional methods.

References

1. Schön, T.B.: Solving nonlinear state estimation problems using particle filters-an engineering perspective. Report No. 2953, Automatic Control at Linköping University, Linköping universitet, Sweden (2010)
2. Arulampalam, M.S., Maskell, S., Gordon, N., Clapp, T.: A tutorial on particle filters for online nonlinear/non-Gaussian Bayesian tracking. IEEE Trans. Sig. Process. **50**(2), 174–188 (2002)
3. Gordon, N.J., Salmond, D.J., Smith, A.F.: Novel approach to nonlinear/non-Gaussian Bayesian state estimation. In: IEE Proceedings F (Radar and Signal Processing), vol. 140, no. 2, pp. 107–113. IET Digital Library (1993)
4. Li, T., Sattar, T.P., Sun, S.: Deterministic resampling: unbiased sampling to avoid sample impoverishment in particle filters. Sig. Process. **92**(7), 1637–1645 (2012)
5. Li, T., Bolic, M., Djuric, P.: Resampling methods for particle filtering: classification, implementation, and strategies. IEEE Sig. Process. Mag. **32**(3), 70–86 (2015)
6. Li, T., Sun, S., Sattar, T.P.: Adapting sample size in particle filters through KLD-resampling. Electron. Lett. **49**(12), 740–742 (2013)
7. Fox, D.: Adapting the sample size in particle filters through KLD-sampling. Int. J. Robot. Res. **22**(12), 985–1003 (2003)
8. Ly-Tu, N., Le-Tien, T., Mai, L.: Performance of sampling/resampling-based particle filters applied to non-linear problems. REV J. Electron. Commun. **4**(3–4), 75–83 (2016)
9. Park, S.-H., Kim, Y.-J., Lee, H.-C., Lim, M.-T.: Improved adaptive particle filter using adjusted variance and gradient data. In: Proceedings of IEEE International Conference on Multisensor Fusion and Integration for Intelligent Systems, Seoul, Korea, pp. 650–655, August 2008
10. Zhao, Q., Wei, C., Qi, L., Yuan, W.: Adaptive double-resampling particle filter algorithm for target tracking. In: International Conference on Frontier Computing, 13 July 2016, pp. 777–787. Springer, Singapore (2016)
11. Dihua, S., Hao, Q., Min, Z., Senlin, C., Liangyi, Y.: Adaptive KLD sampling based Monte Carlo localization. In: IEEE Proceedings Chinese Control and Decision Conference (CCDC), pp. 4154–4159 (2018)
12. Wang, Z., Zhao, X., Qian, X.: The analysis of localization algorithm of unscented particle filter based on RSS for linear wireless sensor networks. In: IEEE Proceedings of the 32nd Chinese Control Conference, pp. 7499–7504, 26 July 2013 (2013)
13. Swamynathan, M.: Mastering Machine Learning with Python in Six Steps: A Practical Implementation Guide to Predictive Data Analytics Using Python. Apress, New York (2017)
14. Ly-Tu, N., Le-Tien, T., Mai, L.: A new resampling parameter algorithm for Kullback-Leibler distance with adjusted variance and gradient data based on particle filter. In: International Conference on Industrial Networks and Intelligent Systems, 4 September 2017, pp. 347–358. Springer, Cham (2017)
15. Ly-Tu, N., Le-Tien, T., Vo-Phu, Q., Huynh-Kha, T.: A new bound error based K-nearest neighbor for Kullback-Leibler distance particle filter in tracking. In: Proceedings National Conference, Thai Binh province, 28–29 June 2019, pp. 1–6 (2019). ISBN 978-604-67-1287-9
16. Tharwat, A., Gaber, T., Ibrahim, A., Hassanien, A.E.: Linear discriminant analysis: a detailed tutorial. AI Commun. **30**(2), 169–190 (2017)
17. Garg, A., Upadhyaya, S., Kwiat, K.: A user behavior monitoring and profiling scheme for masquerade detection. In: Handbook of Statistics, vol. 31, pp. 353–379. Elsevier (2013)
18. Ly-Tu, N., Le-Tien, T., Mai, L.: A modified particle filter through Kullback-Leibler distance based on received signal strength. In: IEEE Proceedings of the 3rd National Foundation for Science and Technology Development Conference on Information and Computer Science (NICS), pp. 229–233 (2016)

Stacking of SVMs for Classifying Intangible Cultural Heritage Images

Thanh-Nghi Do[1,2(✉)], The-Phi Pham[1], Nguyen-Khang Pham[1],
Huu-Hoa Nguyen[1], Karim Tabia[3], and Salem Benferhat[3]

[1] College of Information Technology,
Can Tho University, Can Tho 92000, Vietnam
dtnghi@cit.ctu.edu.vn
[2] UMI UMMISCO 209 (IRD/UPMC),
UPMC, Sorbonne University, Pierre and Marie Curie University, Paris 6, France
[3] CRIL UMR 8188, CRIL CNRS and Artois University, Arras, France

Abstract. Our investigation aims at classifying images of the intangible cultural heritage (ICH) in the Mekong Delta, Vietnam. We collect an images dataset of 17 ICH categories and manually annotate them. The comparative study of the ICH image classification is done by the support vector machines (SVM) and many popular vision approaches including the handcrafted features such as the scale-invariant feature transform (SIFT) and the bag-of-words (BoW) model, the histogram of oriented gradients (HOG), the GIST and the automated deep learning of invariant features like VGG19, ResNet50, Inception v3, Xception. The numerical test results on 17 ICH dataset show that SVM models learned from Inception v3 and Xception features give good accuracy of 61.54% and 62.89% respectively. We propose to stack SVM models using different visual features to improve the classification result performed by any single one. Triplets (SVM-Xception, SVM-Inception-v3, SVM-VGG19), (SVM-Xception, SVM-Inception-v3, SVM-SIFT-BoW) achieve 65.32% of the classification correctness.

Keywords: Images of the intangible cultural heritage in the Mekong Delta · Image classification · Visual features · Support vector machines · Stacking

1 Introduction

The Aniage project[1] focuses on high dimensional heterogeneous data based animation techniques for Southeast Asian Intangible Cultural Heritage (ICH) digital content. It aims to develop novel techniques and tools to reduce the production costs and improve the level of automation without sacrificing the control

---

[1] https://www.euh2020aniage.org.

© Springer Nature Switzerland AG 2020
H. A. Le Thi et al. (Eds.): ICCSAMA 2019, AISC 1121, pp. 186–196, 2020.
https://doi.org/10.1007/978-3-030-38364-0_17

from the artists, in order to preserve the performing art related ICHs of Southeast Asia. The classification of ICH images[2] is the work package in the AniAge project.

The main aim is to automatically classify the image into one of predefined ICH categories. It requires to collect high quality ICH images organized by their categories (classes/labels) and study vision approaches to classify ICH images. To pursue this goal, we build an images dataset of 17 ICH categories by querying a text-based web search engine of Google, followed which we manually annotate them. And then, we explore popular vision approaches to deal with the classification task of ICH images. The extraction of visual features are performed by three popular handcrafted features such as the scale-invariant feature transform (SIFT [15,16]) and the bag-of-words model (BoW [2,13,22]), the histogram of oriented gradients (HOG [7]), the GIST [18]. Recent pre-trained deep learning networks, including VGG19 [21], ResNet50 [10], Inception v3 [23], Xception [5] are used to extract invariant features from ICH images. And then, Support vector machines (SVM [24]) models are learned from visual features to classify ICH images. The numerical test results on 17 ICH dataset show that SVM models learned from Inception-v3 and Xception features give good accuracy of 61.54% and 62.89% respectively. We propose to stack SVM classifiers using different visual features to improve classification results given by any single one. Triplets (SVM-Xception, SVM-Inception-v3, SVM-VGG19), (SVM-Xception, SVM-Inception-v3, SVM-SIFT-BoW) achieve 65.32% of the classification correctness.

The paper is organized as follows. Section 2 describes how to collect a dataset of ICH images and how to build classification models from vision approaches. Section 3 shows the experimental results before conclusions and future works presented in Sect. 4.

2   Classification of Intangible Cultural Heritage Images

The classification system of ICH images in Fig. 1 follows the usual framework for the classification of images. It involves three steps: (1) building the high quality dataset of images, (2) extracting visual features from images and representing them, and (3) training SVM classifiers.

2.1   The Dataset of Intangible Cultural Heritage Images

Firstly, we need to build the dataset of ICH images in the Mekong Delta, Vietnam. Figure 2 shows an images sample of 17 ICH categories. Our proposal is to collect ICH images from Google due to the availability of this biggest public repository. It just does image search by textual query being key words related to 17 ICH categories and retrieve them. However, there are still noisy and irrelevant images. And then we do the manual post-processing stage and tagging images to obtain the high quality images organized by their ICH categories. Table 1 presents the dataset description with a total of 7409 images.

---

[2] http://aniage.ctu.edu.vn/myproj.

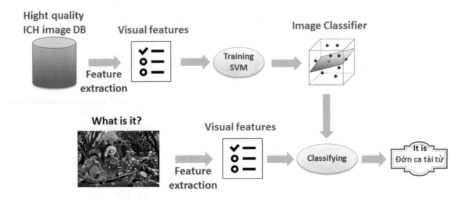

Fig. 1. Framework for classifying ICH images

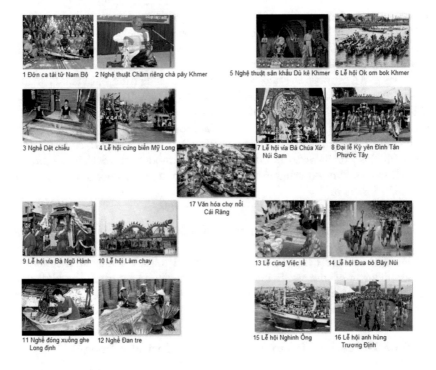

Fig. 2. Images of 17 ICH categories

2.2 Visual Approaches for Classifying Intangible Cultural Heritage Images

Visual approaches perform the classification task of ICH images via two key steps. The first one is to extract visual features from images and represent them. Followed which, the second one is to train SVM models to classify images.

Table 1. Dataset description of 17 ICH categories

No	Category	#Images
1	Đờn ca tài tử Nam Bộ	513
2	Nghệ thuật Chằm riêng chà pây Khmer	185
3	Nghề Dệt chiếu	642
4	Lễ hội cúng biển Mỹ Long	398
5	Nghệ thuật sân khấu Dù kê Khmer	404
6	Lễ hội Ok om bok Khmer	465
7	Lễ hội vía Bà Chúa Xứ Núi Sam	405
8	Đại lễ Kỳ yên Đình Tân Phước Tây	223
9	Lễ hội vía Bà Ngũ Hành	569
10	Lễ hội Làm chay	365
11	Nghề đóng xuồng ghe Long định	281
12	Nghề Đan tre	641
13	Lễ cúng Việc lề	447
14	Lễ hội Đua bò Bảy Núi	449
15	Lễ hội Nghinh Ông	523
16	Lễ hội anh hùng Trương Định	361
17	Văn hóa chợ nổi Cái Răng	538
	Total	7409

Three popular methods for handcrafted features include the scale-invariant feature transform (SIFT [15,16]) and the bag-of-words model (BoW [2,13,22]), the histogram of oriented gradients (HOG [7]), the GIST [18].

Scale-Invariant Feature Transform: The SIFT descriptors [15,16]) and the bag-of-words model (BoW) are the most commonly image represenation for tasks of images classification [2,13,22]. The SIFT method detects the appearance of the object at particular interest points, invariant to image scale, rotation, and also robust to changes in illumination, noise, and occlusion.

Histogram of Oriented Gradients: The HOG descriptors are used for human detection [7]. The HOG method computes the distribution of local intensity gradients or edge directions to describe local object appearance and shape within an image. The combined distributions form the image representation. The HOG descriptor is invariant to geometric and photometric transformations, except for object orientation.

GIST: The GIST descriptors proposed by [18] are used for images retrieval. The GIST method uses Gabor filters to extract the set of perceptual dimensions (naturalness, openness, roughness, expansion, ruggedness) that represent the spatial structure of a scene.

Recent deep learning networks such as VGG19 [21], ResNet50 [10], Inception v3 [23], Xception [5] are pre-trained on ImageNet dataset [8]. These deep learning networks are used to extract invariant features from ICH images.

VGG19: The VGG19 network architecture [21] consists of 19 weight layers for large scale image recognition. The VGG19 network uses only 3×3 convolutional layers stacked on top of each other to develop depth. The max pooling layers are used to reduce volume size. From the input layer to the last max pooling layer are used as features extraction of images.

ResNet50: The ResNet50 network architecture [10] is designed with 50 weight layers for image recognition. The ResNet50 develops extremely deep networks by proposed micro-architecture modules (called network-in-network). Furthermore, network layers try to fit a residual mapping instead of desired one. From the input layer to the last pooling layer or the last convolutional layer are used to extract image features.

Inception-V3: The "Inception" module proposed by [23] is to learn multi-level features for image classification. The main idea uses 1×1, 3×3 and 5×5 convolutions within the Inception module of the network. And then these Inception modules are stacked on top of each other. The reduction of volume size bases on 1×1 convolutions. From the input layer to the last pooling layer or the last convolutional layer are regarded as features extractor for images.

Xception: The "Xception" network proposed by [4] is an extension of the Inception architecture. The Xception replaces the standard depthwise separable convolution (the depthwise convolution followed by a pointwise convolution) by the new modified one without any intermediate activation being the pointwise convolution followed by a depthwise convolution. Features extraction for images is performed by layers from the input layer to the last pooling layer or the last convolutional layer.

Support Vector Machines: For a binary classification problem depicted in Fig. 3, the SVM algorithm proposed by [24] tries to find the best separating plane furthest from both class +1 and class −1. To pursue this aim, the training SVM algorithm simultaneously maximize the margin (or the distance) between the supporting planes for each class and minimize errors.

The binary SVM solver can be extended for dealing with the multi-class problems (c classes, $c \geq 3$). The main idea is to decompose multi-class into a series of binary SVMs, including One-Versus-All [24], One-Versus-One [12]. The One-Versus-All strategy (as illustrated in Fig. 4) builds c different binary SVM models where the ith one separates the ith class from the rest. The One-Versus-One strategy (as illustrated in Fig. 5) constructs $c(c-1)/2$ binary SVM models for all the binary pairwise combinations of the c classes. The class is then predicted with the largest distance vote. In practice, the One-Versus-All strategy is implemented in LIBLINEAR [9] and the One-Versus-One technique is also used in LibSVM [3].

SVM algorithms use different kernel functions [6] for dealing with non-linear classification tasks. The commonly non-linear kernel functions include a polynomial function of degree d, a radial basis function (RBF).

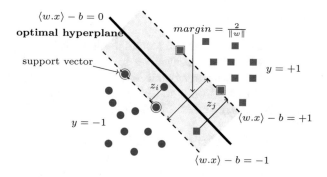

Fig. 3. Classification of the datapoints into two classes

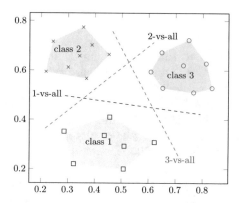

Fig. 4. Multi-class SVM (One-Versus-All)

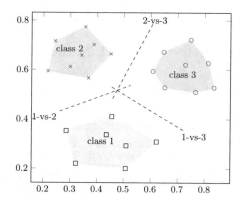

Fig. 5. Multi-class SVM (One-Versus-One)

3   Experimental Results

In this section, we present experimental results of different visual approaches for classifying ICH images. We implement them in Python using library Keras [4] with backend Tensorflow [1], library Scikit-learn [19] and library OpenCV [11]. All experiments are conducted on a machine Linux Fedora 23, Intel(R) Core i7-4790 CPU, 3.6 GHz, 4 cores and 32 GB main memory and the Nvidia GeForce GTX 960M 2 GB GPU.

The image dataset of 17 ICH categories in the Mekong Delta, Vietnam is randomly split into the trainset (6001 images), the validation set (667 images) and testset (749 images). We use the trainset to build visual classification models. Then, results are reported on the testset using the resulting visual classification models.

3.1   Tuning Parameters

We use the validation set to tune parameters for building visual classification models on the trainset. With methods for feature extractor and image representation, only handcrafted features SIFT and BoW model needs tuning the number of clusters (visual words) well-known as the parameter of $k$means algorithm [17]. We try to vary the number of visual words from 1000 to 5000 for finding the best experimental results. And then, the results are unchanged while increasing the number of visual words over 2000. Therefore, we use 2000 visual words for the BoW model.

With SVM models, we propose to use RBF kernel functions because it is general and efficient [14]. There is need to tune the hyper-parameter γ of RBF function $[K\langle x_i, x_j \rangle = exp(-\gamma||x_i - x_j||^2)]$ and the cost C (a trade-off between the margin size and the errors) to obtain the best correctness. Finally, we find out best parameters' SVM in Table 2 for visual classification models.

3.2   Classification Results for 17 ICH Categories

We obtain classification results of visual approaches in Table 3 and Fig. 6. The highest accuracy is bold-faced and the second one is in italic. In the comparison

Table 2. Hyper-parameters for training SVM models

No	Feature extraction method	γ	C
1	SIFT and BoW	0.0001	1000000
2	HOG	0.1	100000
3	GIST	5	1000000
4	VGG19	0.005	100000
5	ResNet50	0.005	100000
6	Inception-v3	0.005	100000
7	Xception	0.005	100000

among visual classification approaches, we can see that methods for handcrafted features extraction such as SIFT-BoW, HOG, GIST are not suited for classifying ICH images. Recent deep networks (excepting ResNet50) for extracting invariant features from ICH images are most accurate results. Typically, Xception and Inception v3 achieve 62.89% and 61.54% in terms of overall classification accuracy, respectively. We also try to tune these pre-trained deep networks by re-training about their 10% of layers from our image trainset but obtained results can not be improved even degraded.

Table 3. Overall classification accuracy for 17 ICH categories

No	Visual approach	Accuracy (%)
1	SVM-SIFT-BoW	33.87
2	SVM-HOG	32.93
3	SVM-GIST	37.25
4	SVM-VGG19	50.47
5	SVM-ResNet50	34.14
6	SVM-Inception-v3	*61.54*
7	SVM-Xception	**62.89**

3.3 Stacking of SVM Classifiers for Classifying 17 ICH Categories

We propose to use voting scheme [25] among visual models to improve classification correctness for ICH images. The main idea is to combine multiple visual classifiers learned for the classification task by weighted voting between the prediction of each visual classifier VC_i as illustrated in Eq. (1).

$$Majority-vote\{w_1*pred(x,VC_1)+w_2*pred(x,VC_2)+\cdots+w_k*pred(x,VC_k)\} \tag{1}$$

Voting schemes always use the visual classifier SVM-Xception because this model gives the best result. Followed which, other visual models are included in voting schemes with the hope that the models can complement one another in the classification. Table 4 and Fig. 7 show results obtained by weighted voting schemes.

The couple of SVM-Xception and SVM-Inception-v3 improves 1.62% and 2.97% of classification correctness against SVM-Xception and SVM-Inception-v3, respectively.

The improvements of the triplet SVM-Xception, SVM-Inception-v3 and SVM-VGG19 over each single visual classifier are 2.43%, 3.78% and 14.85%, respectively.

The triplet SVM-Xception, SVM-Inception-v3 and SVM-SIFT-BoW achieves the best accuracy as the triplet SVM-Xception, SVM-Inception-v3 and SVM-VGG19. It also improves 31.45% of the accuracy compared to SVM-SIFT-BoW.

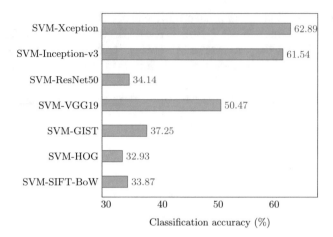

Fig. 6. Overall classification accuracy for 17 ICH categories

Table 4. Overall classification accuracy of voting schemes for 17 ICH categories

No	Voting scheme	Accuracy (%)
8	0.85*SVM-Xception + 0.15*SVM-SIFT-BoW	63.29
9	0.8*SVM-Xception + 0.2*SVM-HOG	63.02
10	0.8*SVM-Xception + 0.2*SVM-GIST	63.02
11	0.75*SVM-Xception + 0.25*SVM-VGG19	63.56
12	0.75*SVM-Xception + 0.25*SVM-ResNet50	63.02
13	0.65*SVM-Xception + 0.35*SVM-Inception-v3	64.51
14	0.55*SVM-Xception + 0.225*SVM-Inception-v3 + 0.225*SVM-VGG19	**65.32**
15	0.65*SVM-Xception + 0.22*SVM-Inception-v3 + 0.13*SVM-SIFT-BoW	**65.32**

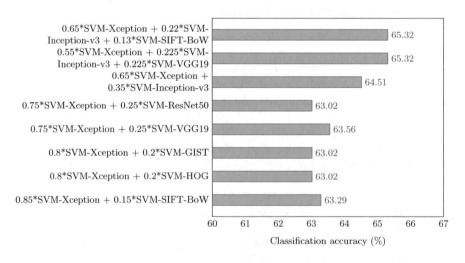

Fig. 7. Overall classification accuracy of voting schemes for 17 ICH categories

4   Conclusion and Future Works

We have presented visual approaches for classifying images of the intangible cultural heritage (ICH) in the Mekong Delta, Vietnam. We collect an images dataset of 17 ICH categories from Google and manually tagging images according to their categories. Visual approaches are used to deal with the ICH image classification. The feature extraction methods include three popular handcrafted features such as SIFT-BoW, HOG, GIST and four recent deep learning networks of invariant features like VGG19, ResNet50, Inception v3, Xception. Followed which SVM models are learned from these visual features to classify ICH images. The numerical results on 17 ICH dataset show that SVM-Xception and SVM-Inception-v3 give good accuracy of 61.54% and 62.89% respectively. We propose to use voting schemes between visual models to improve the classification result performed by any single one. Triplets (SVM-Xception, SVM-Inception-v3, SVM-VGG19), (SVM-Xception, SVM-Inception-v3, SVM-SIFT) achieve 65.32% of the classification correctness.

These visual approaches can be used to re-rank images retrieved from Google and then we select top-ranked images for automated organizing ICH images by their ICH categories. It allows us to build a large number of images for a specified ICH category. Another approach [20] for developing the images database size combines textual and visual features.

Acknowledgments. This work has received support from the European Project H2020 Marie Sklodowska-Curie Actions (MSCA), Research and Innovation Staff Exchange (RISE): Aniage project (High Dimensional Heterogeneous Data based Animation Techniques for Southeast Asian ICH Digital Content), No: 691215.

References

1. Abadi, M., Agarwal, A., Barham, P., Brevdo, E., Chen, Z., Citro, C., Corrado, G.S., Davis, A., Dean, J., Devin, M., Ghemawat, S., Goodfellow, I., Harp, A., Irving, G., Isard, M., Jia, Y., Jozefowicz, R., Kaiser, L., Kudlur, M., Levenberg, J., Mané, D., Monga, R., Moore, S., Murray, D., Olah, C., Schuster, M., Shlens, J., Steiner, B., Sutskever, I., Talwar, K., Tucker, P., Vanhoucke, V., Vasudevan, V., Viégas, F., Vinyals, O., Warden, P., Wattenberg, M., Wicke, M., Yu, Y., Zheng, X.: TensorFlow: large-scale machine learning on heterogeneous systems (2015). https://www.tensorflow.org/
2. Bosch, A., Zisserman, A., Munoz, X.: Scene classification via pLSA. In: Proceedings of the European Conference on Computer Vision, pp. 517–530 (2006)
3. Chang, C.C., Lin, C.J.: LIBSVM: a library for support vector machines. ACM Trans. Intell. Syst. Technol. **2**(27), 1–27 (2011)
4. Chollet, F., et al.: Keras. https://keras.io (2015)
5. Chollet, F.: Xception: deep learning with depthwise separable convolutions. CoRR abs/1610.02357 (2016)
6. Cristianini, N., Shawe-Taylor, J.: An Introduction to Support Vector Machines: and Other Kernel-based Learning Methods. Cambridge University Press, New York (2000)

7. Dalal, N., Triggs, B.: Histograms of oriented gradients for human detection. In: Proceedings of the 2005 IEEE Computer Society Conference on Computer Vision and Pattern Recognition (CVPR 2005) - vol. 1, pp. 886–893. IEEE Computer Society (2005)

8. Deng, J., Berg, A.C., Li, K., Li, F.: What does classifying more than 10,000 image categories tell us? In: Proceedings of Computer Vision - ECCV 2010 - 11th European Conference on Computer Vision, Heraklion, Crete, Greece, 5–11 September 2010, Part V, pp. 71–84 (2010)

9. Fan, R.E., Chang, K.W., Hsieh, C.J., Wang, X.R., Lin, C.J.: LIBLINEAR: a library for large linear classification. J. Mach. Learn. Res. **9**(4), 1871–1874 (2008)

10. He, K., Zhang, X., Ren, S., Sun, J.: Deep residual learning for image recognition. CoRR abs/1512.03385 (2015)

11. Itseez: Open source computer vision library (2015). https://github.com/itseez/opencv

12. Kreßel, U.H.G.: Pairwise classification and support vector machines. In: Schölkopf, B., Burges, C.J.C., Smola, A.J. (eds.) Advances in Kernel Methods, pp. 255–268. MIT Press, Cambridge (1999)

13. Li, F., Perona, P.: A bayesian hierarchical model for learning natural scene categories. In: 2005 IEEE Computer Society Conference on Computer Vision and Pattern Recognition (CVPR 2005), San Diego, CA, USA, pp. 524–531 (2005)

14. Lin, C.: A practical guide to support vector classification (2003)

15. Lowe, D.: Object recognition from local scale invariant features. In: Proceedings of the 7th International Conference on Computer Vision, pp. 1150–1157 (1999)

16. Lowe, D.: Distinctive image features from scale invariant keypoints. Int. J. Comput. Vis. **60**, 91–110 (2004)

17. MacQueen, J.: Some methods for classification and analysis of multivariate observations. In: Berkeley Symposium on Mathematical Statistics and Probability, University of California Press, vol. 1, pp. 281–297 (1967)

18. Oliva, A., Torralba, A.: Modeling the shape of the scene: a holistic representation of the spatial envelope. Int. J. Comput. Vis. **42**, 145–175 (2001)

19. Pedregosa, F., Varoquaux, G., Gramfort, A., Michel, V., Thirion, B., Grisel, O., Blondel, M., Prettenhofer, P., Weiss, R., Dubourg, V., Vanderplas, J., Passos, A., Cournapeau, D., Brucher, M., Perrot, M., Duchesnay, E.: Scikit-learn: machine learning in Python. J. Mach. Learn. Res. **12**, 2825–2830 (2011)

20. Schroff, F., Criminisi, A., Zisserman, A.: Harvesting image databases from the web. IEEE Trans. Pattern Anal. Mach. Intell. **33**(4), 754–766 (2011)

21. Simonyan, K., Zisserman, A.: Very deep convolutional networks for large-scale image recognition. CoRR abs/1409.1556 (2014)

22. Sivic, J., Zisserman, A.: Video Google: a text retrieval approach to object matching in videos. In: 9th IEEE International Conference on Computer Vision (ICCV 2003), Nice, France, 14–17 October 2003, pp. 1470–1477 (2003)

23. Szegedy, C., Vanhoucke, V., Ioffe, S., Shlens, J., Wojna, Z.: Rethinking the inception architecture for computer vision. CoRR abs/1512.00567 (2015)

24. Vapnik, V.: The Nature of Statistical Learning Theory. Springer, Berlin (1995)

25. Wolpert, D.: Stacked generalization. Neural Netw. **5**, 241–259 (1992)

Assessment of the Water Area in the Lowland Region of the Mekong River Using MODIS EVI Time Series

Chien Pham Van$^{(\boxtimes)}$ and Giang Nguyen-Van

Thuyloi University, 175 Tay Son, Dong Da district, Ha Noi, Vietnam
Pchientvct_tv@tlu.edu.vn

Abstract. This paper presents an application of reconstructing Moderate-Resolution Imaging Spectroradiometer (MODIS) Enhanced Vegetation Index (EVI) time series to extract the water area in the lowland region of the Mekong River. Firstly, MODIS13A1 EVI time series with land surface reflectance 16-day and 500 m spatial resolution is collected from 2000 to 2017, resulting total of 411 images. Then, these images are used for reconstructing EVI time-series by using the Whittaker smoother method on the Google Earth Engine. Next, the water area in each image is computed based on the smooth EVI value. The results showed that the extracting water areas in year 2000 was in line with the observed water elevation at Tan Chau station (for the Mekong Delta) and at Phnom Penh station (for the Cambodian region). The correlation coefficient between the extracting water area and water elevation equals to 0.885 for the Mekong Delta while its value is 0.924 for the Cambodian region. The extracting water area from MODIS13A1 EVI for the lowland region of the Mekong River can be used for assessment of inundated area resulting from different flow conditions as well as for studying inundation processes in the lowland region of the Mekong River when using hydrodynamic models.

Keywords: MODIS images · Mekong River · EVI

1 Introduction

In the last decade, satellite images such as Landsat, Sentinel, and MODIS have widely applied to investigate and manage water resources in river systems. This is because of satellite images allow to look in large spatial scale of the whole river system. Previous studies [1–4] also confirmed that data sources collected from satellite images at different times are extremely valuable data sources, making disaster management to be more effective. Indeed, these data sources can be combined with ground surface measurements to allow for accurate presentations of the field of interest.

In regards to satellite image processing, high performance computing systems are often applied with purposes of reducing computational time and of improving computing capacity. In this context, Google Earth Engine (GEE) has recently launched and allowed users in different fields and aspects to process quickly satellite images [4–6]. It should be noted that recent advances in high performance computing systems and parallel computing platform have allowed many useful and interesting processes to be

© Springer Nature Switzerland AG 2020
H. A. Le Thi et al. (Eds.): ICCSAMA 2019, AISC 1121, pp. 197–207, 2020.
https://doi.org/10.1007/978-3-030-38364-0_18

done in various topics when using satellite images. Here are some selected examples from the last decade, such as this load of generating annual map of open surface water body [6], estimating ecosystem production [3], producing crop yield and rice paddy maps [1, 4], and mapping inundated and flooding water extent as well [1].

The main objective of the paper is to apply the reconstructing global MODIS EVI time series, which is obtained by using MODIS13A1 EVI time series collected in the period from 2000 to 2017 and performed on the GEE, for extracting water area within lowland region of the Mekong River. Specific objectives of the study are to (i) investigate the spatial distribution and temporal change of inundated area within the Mekong Delta in the Vietnam and lowland area in the Cambodia for the whole year 2000 when using the smooth MODIS EVI time series and (ii) examine the relationship between the extracting water area and the observed water elevation at the hydrological stations in the study domain.

2 The Study Domain

The Mekong River is known as one of the largest rivers in the world. The river originates from the eastern Tibetan Plateau (China), and then flows through six countries of Southeast Asia (i.e. China, Laos PDR, Myanmar, Thailand, Cambodia, and Vietnam) before discharging into the East Sea of Vietnam (Fig. 1). The Mekong River meanders over the 4,900 km and its catchment area covers of about 795,000 km^2, with the mean annual river discharge at Kratie is 13,600 m^3/s [7]. The weather in the river basin is characterized by the Western North Pacific and Indian monsoons, with a wet season from June to November and a dry season from December to May. The mean monthly temperature (at Phnom Penh station) varies between 26 °C and 31 °C while annual evapotranspiration ranges from 1000 mm to 2000 mm, with high relative humidity. The mean annual rainfall ranges between 1200 mm and 3000 mm, with double peaks of rainfall in most lowland regions during wet years. Hydrological characteristics vary significantly within the river basin, especially in the lowland region of the river because of combined impacts of anthropogenic activities and natural disturbances such as significant variation of weather, El Nino, typhoons and tropical storms, hydropower development, climate change and sea level rise.

Regarding lowland region of the Mekong River, which extends from Kratie in Cambodia to the whole floodplains in Vietnam, the river length is about 700 km and the area is roughly about 130,000 km^2. In this region, the Tonle Sap Lake is an important water body, located in the Cambodian floodplain which contributes largely to the flow processes in the Mekong River. The Tonle Sap Lake consists of a permanent lake itself, twelve tributaries, extensive adjacent floodplains and the Tonle Sap river. The latter is linking the Tonle Sap Lake to the Mekong River. The Tonle Sap river with a length of approximately 120 km is situated at the Southeast end of the Tonle Sap Lake and joins the Mekong River at Chaktomuk confluence. At the Chaktomuk confluence, the Mekong River splits into the Bassac river in the West and the Mekong river in the East.

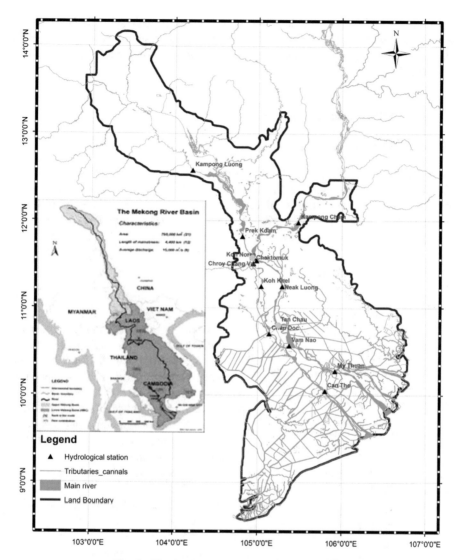

Fig. 1. The lowland region of the Mekong River

Note that the lowland region of the Mekong River, home to more than 60 million people, is the most important region in terms of agricultural and inland fishery production in both Cambodia and Vietnam. For example, Vietnamese Mekong Delta is farmed intensively with paddy rice (~ 1.9 million ha) and contributes about 50% (~ 23 million tons) of Vietnam's rice production in 2015 [9, 10]. However, the region is under threat from a combination of climate change, hydropower development, rising sea level, land surface subsidence and hydropower activities, resulting in various societal issues related to drought and flood disasters, water resources management and environmental protection. Therefore, management of water resources within this region under climate change and significant variation of anthropogenic activities is critical not only for understanding hydrological characteristics but also for developing suitable adaptation measures in the lowland region of the Mekong River while safeguarding its environment.

3   Material and Method

3.1   MODIS Imagery Data

The MODIS images (i.e. MODIS13A1 EVI) with land surface reflectance 16-day and 500 m spatial resolution are used in this study. Although MODIS13A1 EVI is composed by using the best available pixel value, cloud and cloud shadow still cover on most satellite images [5]. Thus, MODIS13A1 EVI images from 2000 to 2017 are collected and used for time-series reconstruction purposes.

3.2   Imagery Processing

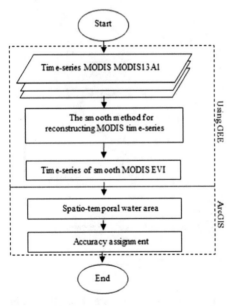

Fig. 2. General flowchart of the methodology to extract the water area

General procedure of the methodology to extract the water area in the domain of interest is shown in Fig. 2. Firstly, MODIS13A1 EVI images are collected in the period from 2000 to 2017. Then, the Whittaker smoother method is applied for reconstructing time-series EVI based on the computational platform of the GEE. Finally, spatio-temporal inundated water area and assessment of accuracy are done by using the ArcGIS.

Among different smoother methods, Whittaker smoother method [10] is selected and applied to reduce noises efficiently in time series of the MODIS13A1 EVI. Because this method is successfully applied to vegetation phenology extraction, land cover classification, hyperspectral remote sensing [5].

It should be noted that the Whittaker smoother method is performanced based on three indicators, i.e. balance fidelity S, roughness R, and difference matrix Q. These indicators are computed as [5, 9]:

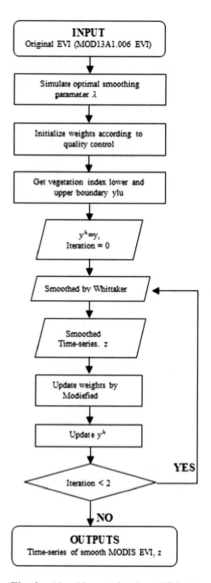

Fig. 3. Algorithm of the Whittaker smoother method (modified after [5])

$$S = |y - z|^2 = \sum_i (y_i - z_i)^2 \quad (1)$$

$$R = |Dz|^2 = \sum_i (z_i - 2z_{i-1} + z_{i-2})^2 \quad (2)$$

$$Q = S + \lambda R \quad (3)$$

where y and z are the original and smoothed time series, respectively; λ is the roughness parameter.

When weight w is taken into account, Eq. (1) is rewritten as:

$$S = |y - z|^2 = \sum_i w_i(y_i - z_i)^2$$
$$= (y - z)^T W(y - z) \quad (4)$$

with W is the diagonal matrix of w, w (0, 1].

The detailed algorithm of the Whittaker smoother method is given in Fig. 3.

3.3 Water Area Extraction

In order to extract the water area in the domain of interest, we use the Enhanced Vegetation Index (EVI) for identification and classification of non-water and full-water pixels in each image. The EVI indicator is calculated as:

$$EVI = 2.5 \times \frac{NIR - RED}{NIR + 6 \times RED - 7.5 \times BLUE + 1} \quad (5)$$

where *NIR*, *RED* and *BLUE* are the surface reflectance value of near infrared Band 2 (841–876 nm), visible Band 1 (*RED*, 620–670 nm) and visible Band 3 (*BLUE*, 459–479 nm), respectively. If *EVI* value of a pixel is smaller than the given value ε_0, this pixel is set as the fully water pixel. On contrast, if *EVI* value of a pixel is larger than the given value ε_0, the pixel is marked as the non-water pixel. A given value $\varepsilon_0 = 0.05$ is used in this study as in many previous studies [1, 5, 10, 11].

3.4   Water Area and Water Elevation

To analyze relationship between extracting water area and water elevation, a linear regression is used in order to render the examination of the relationship as simple as possible. This linear regression is given by

$$y = a \times x + b \tag{6}$$

where y is the water elevation at hydrological stations (i.e. Tan Chau in the Mekong Delta and Phnom Penh in the Cambodia region), x is the extracting water area, $a\psi$ and b are the regression coefficients.

4   Results and Discussion

4.1   Temporal Variation of Extracting Water Area

Figure 4 shows the time-series of extracting water and non-water areas in year 2000 for the Mekong Delta in the Vietnam and lowland region in the Cambodia. It is clearly observed that small water areas are obtained as expected in the dry season period from February to June in both sub-areas of the lowland region of the Mekong River. A large value of extracting water area is archived in the wet season period. Maximum extracting water area occurs in September (see the left panel in Fig. 4) because this period is the flood times in the lowland region of the Mekong River. On the other hand, a high value of non-water area is obtained in the dry season for both Mekong Delta and Cambodia region, while a small value of non-water area is archived in the wet season. These results reveal that the variation of extracting water and non-water areas in the domain of interest is in line with weather characteristics.

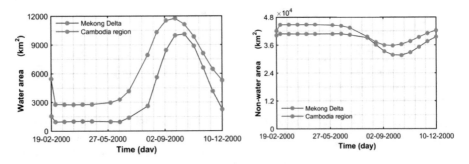

Fig. 4. Time-series of extracting: water area (left panel) and non-water area (right panel)

4.2   Spatial Distribution of Extracting Water Area

The spatial distribution of extracting water and non-water areas at different given dates in year 2000 is shown in Fig. 5. The results clearly show that the spatial distribution of extracting water area is consistent with bed elevation. For instance, water occurs and

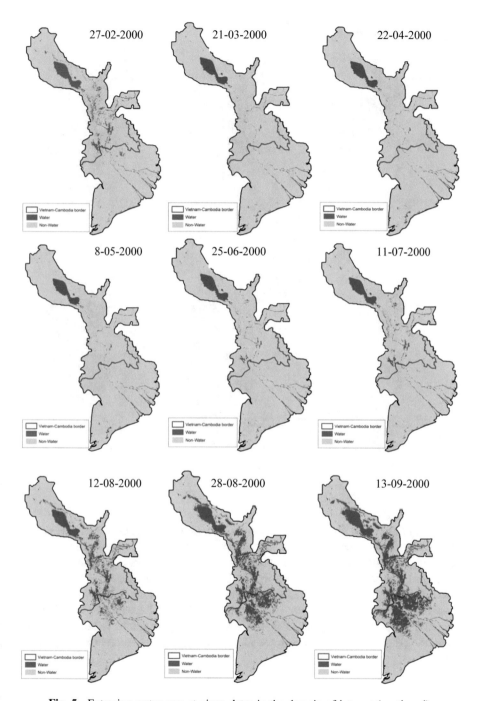

Fig. 5. Extracing water area at given dates in the domain of interest (continued)

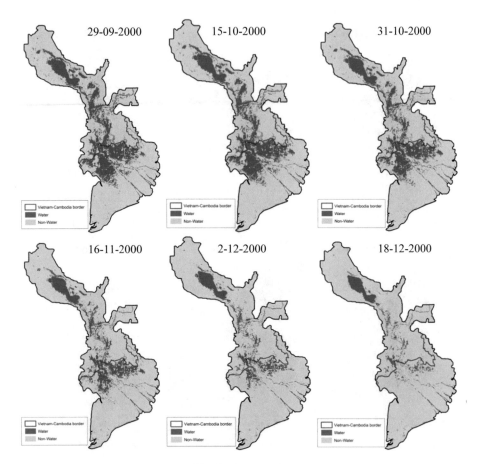

Fig. 5. (*continued*)

remains in the almost areas (where the bed elevation is low) such as main channel of the Mekong River, Tien river, Hau river as well as the Tonle Sap Lake in Cambodia. Regarding the Tonle Sap Lake, it is not supprised that water remains during the whole considered time and the water area in this lake varies depending on given instantaneous. This trend is also observed in the Long Xuyen Quadrangle and Plain of Reeds. Most areas in the Long Xuyen Quadrangle and Plain of Reeds are inundated in September and October of year 2000. These results are quite similar to results obtained when using three different indices (i.e. EVI, LSWI, and DVEL) reported in the previous study [1].

4.3   Relationship Between Extracting Water Area and Water Elevation

Figure 6 shows the time-series of extracting water area and water elevation for the Mekong Delta in the Vietnam and lowland area in the Cambodia. Note that observed water elevation at Tan Chau and Phnom Penh stations (Fig. 1) is used for investigating the relationship between extracting water area and water elevation in the Mekong Delta and Cambodia region, respectively. As can be seen, the higher water elevation the larger extracting water area.

Figure 7 depicts the extracting water area versus the water elevation at Tan Chau and Phnom Penh stations. A linear regression describing the extracting water area as a function of the water elevation is also presented. The correlation coefficient between the extracting water areas from MODIS EVI and the observed water elevation is 0.885 for the Mekong Delta in the Vietnam and 0.924 for the lowland area in the Cambodia. These results reveal that the variation of extracting water area from MODIS EVI within the domain of interest is in line with the change of observed water elevation at hydrological stations, namely Tan Chau and Phnom Penh.

In the lowland region of the Mekong River, extent of inundated area can vary depending on the water elevation and instantaneous, revealing that non-water, mixture and full-water pixels are inherently existed in each considered MODIS images. However, in order to render the performance procedure as simple as possible, we used only EVI indicator for identification and classification of full-water and non-water pixels within each image. Extracting water area resulting from that procedure shows a significant promise in relationship with the observed water elevation at hydrological stations (i.e. Tan Chau and Phnom Penh) even the mixture water pixels were disregarded. Further investigation of the mixture water pixels will be considered in the future investigation effort for accurate presentation of inundated area in the Mekong Delta and lowland area in the Cambodia.

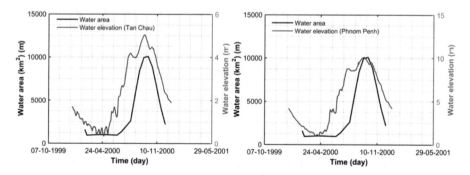

Fig. 6. Extracting water area and water elevation in the: Mekong Delta (left panel) and Cambodia region (right panel)

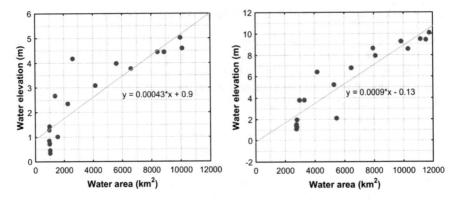

Fig. 7. Water area versus water elevation, for the Mekong Delta (left panel) and Cambodia region (right panel)

5 Conclusion

Using the reconstructing global MODIS EVI time-series, the extracting water area in the lowland region of the Mekong River was firstly carried out, and the results were correlate well with weather characteristics, observation of water elevation at Tan Chau and Phnom Penh stations for the high flow year 2000. The correlation coefficient between the extracting water area and observed water elevation was closely to the unity for both the Mekong Delta in the Vietnam and lowland area in the Cambodia. Secondly, the extracting water area from the reconstruction EVI time-series for the year 2000 was also consistent with the variation of the spatial distribution of the topography as well as the hydrological characteristics. Finally, the results obtained in the present study are believed to be useful for assessment of water extend caused by floods as well as for studying inundation processes in the lowland region of the Mekong River when using hydrodynamic models.

Acknowledgements. This research is funded by Vietnam National Foundation for Science and Technology Development (NAFOSTED) under grant number 105.06-2017.320.

References

1. Sakamoto, T., Van Nguyen, N., Kotera, A., Ohno, H., Ishitsuka, N., Yokozawa, M.: Detecting temporal changes in the extent of annual flooding within the Cambodia and the Vietnamese Mekong Delta from MODIS time-series imagery. Remote Sens. Environ. **109**, 295–313 (2007)
2. Shi, H., Li, L., Eamus, D., Huete, A., Cleverly, J., Tian, X., Yu, Q., Wang, S., Montagnani, L., Magliulo, V., Rotenberg, E., Pavelka, M., Carrara, A.: Assessing the ability of MODIS EVI to estimate terrestrial ecosystem gross primary production of multiple land cover types. Ecol. Indic. **72**, 153–164 (2017)

3. Zou, Z.H., Xiao, X.M., Dong, J.W., Qin, Y.W., Doughty, R.B., Menarguez, M.A., Zhang, G.L., Wang, J.: Divergent trends of open-surface water body area in the contiguous United States from 1984 to 2016. Proc. Natl. Acad. Sci. U.S.A. **115**, 3810–3815 (2018)
4. Gorelick, N., Matt, H., Mike, D., Simon, I., Thau, D., Moore, R.: Google Earth Engine: planetary-scale geospatial analysis for everyone. Remote Sens. Environ. **202**, 18–27 (2017)
5. Kong, D., Zhang, Y., Gu, X., Wang, D.: A robust method for reconstructing global MODIS EVI time series on the Google Earth Engine. ISPRS J. Photogramm. Remote. Sens. **155**, 13–24 (2019)
6. Wang, X., Xiao, X., Zou, Z., Chen, B., Ma, J., Dong, J., Doughty, R.B., Zhong, Q., Qin, Y., Dai, S., Li, X., Zhao, B., Li, B.: Tracking annual changes of coastal tidal flats in China during 1986–2016 through analyses of Landsat images with Google Earth Engine. Remote Sens. Environ. (2018)
7. Tuan, L.A., Chinvanno, S.: Climate change in the Mekong River Delta and key concerns on future climate threats. In: Advances in Global Change Research, vol. 45, pp. 207–217 (2011)
8. Demont, M., Rutsaert, P.: Restructuring the Vietnamese rice sector: towards increasing sustainability. Sustainability **9**(2), 325 (2017)
9. Mekong River Commission: Annual mekong flood report 2011, 72 p (2015)
10. Whittaker, E.T.: On a new method of graduation. Proc. Edinburgh Math. Soc. **41**, 63–75
11. Huete, A.R., Lui, H.Q., Batchily, K., van Leeuwen, W.: A comparison of vegetation indices over a Global set of IM images for EOS-MODIS. Remote Sens. Environ. **59**, 440–451 (1997)

Palmprint Recognition Using Discriminant Local Line Directional Representation

Hoang Thien Van[1](✉), Kiet Dang Hung[2], Giang Vu Van[3], Quynh Pham Thi[1],
and Thai Hoang Le[4]

[1] The Saigon International University (SIU), Ho Chi Minh City, Vietnam
vanthienhoang@siu.edu.vn, phamthiquynh.lamdong@gmail.com
[2] University of Information Technology (UIT), Vietnam National University,
Ho Chi Minh City, Vietnam
kietdh.13@grad.uit.edu.vn
[3] University of People's Security, Ho Chi Minh City, Vietnam
vuvangiang@gmail.com
[4] University of Sciences, Vietnam National University, Ho Chi Minh City, Vietnam
Lhthai@fit.hcmus.edu.vn

Abstract. Palmprint is a new biometric feature for personal identification with a high degree of privacy and security. In this paper, we propose the palmprint feature extraction method which combines the direction-based method (Local line direction pattern) and learning-based method (two-directional two-dimensional linear discriminant analysis ($(2D)^2LDA$)) to get the high discriminant direction based features, so-called Discriminant local line Directional Representation (DLLDR). First, the algorithm computes the LLDP features with two strategies of encoding multi-directions. Then, $(2D)^2LDA$ is applied to extract DLLDR features with higher discriminant and lower-dimensional from the LLDP matrix. The experimental results on the public databases of Hong Kong Polytechnic University demonstrate that our method is effective for palmprint recognition.

Keywords: Biometrics · 2DLDA · Palmprint · Discriminant local line directional pattern (DLLDP)

1 Introduction

Recently, palmprint has been increasingly studied and applied for personal recognition because of its advantages such as high performance, cost-effectiveness, user-friendliness, and etc. [1]. Low resolution Palmprint image could be used for recognition. The low resolution refers to 100 pixels per inch (PPI). Low resolution palmprint images contain features such as: principal lines (longest lines), wrinkles (weaker lines), and texture [2]. There are many methods exploiting these features, grouped into two categories such as subspace-based approaches and local feature based approachs [3–7].

Subspace-based approaches (PCA, LDA, ICA, …) project palmprint images from high dimensional space to a lower dimensional feature space. The subspace coefficients are considered as features [8]. To overcome illumination, contrast, and position changes,

© Springer Nature Switzerland AG 2020
H. A. Le Thi et al. (Eds.): ICCSAMA 2019, AISC 1121, pp. 208–217, 2020.
https://doi.org/10.1007/978-3-030-38364-0_19

inputs are directional images [9–11]. Rida et al. [12] used both 2D-PCA and 2D-LDA for feature extraction and identification.

Local feature representations commonly use dominant direction features and texture because that are insensitive to illumination changes [25]. Zhang et al. [13] used Gabor phase in a fixed orientation to encode the palmprint, called palmcode for palmprint recognition system. Kong and Zhang [14] proposed the dominant direction feature extraction method for palmprint recognition by using Gabor filters with different directions. Then, the improved dominant direction based methods are fusion code [15], DRCC code [16], and etc. Jia et al. proposed RLOC feature which is computed by using line-shape filter (MFRAT based filter) [17]. Sun et al. [18] used three grouped Gaussian filters with three orthogonal directions with the aim that describes multiline at each pixel. Fei et al. [20] computes a double orientation code by using two dominant orientations for palmprint recogntion. BOCV feature were proposed for palmprint recogntion, in which orientation features were encoded with all six orientations by using six Gabor filters [19]. The local line direction pattern representation jointly encoded by two optional directions is also proposed for palmprint recognition [21]. Zheng et al. [22] proposed DoN of palmprint to build the direction feature descriptor for recognition. Many orientation based methods were investigated in [23]. In general, the direction features are robust and discriminant for palmprint recognition [24]. Li et al. [29] proposed Local Microstructure Tetra Pattern (LMTrP) for extracting palmprint features. Moreover, the modern deep convolutional neural network is also studied for palmprint recognition [26–28]. Recently, the approaches combined direction features and subspace based methods are deemed to be promising methods because of the following reasons: (1) Direction feature could be stable and robust against illumination. (2) Subspace-based methods compute the global discriminative features with low dimensions. Hoang et al. [9–11] have proposed some methods of combining the global and local features using the local directional feature and the linear discriminant analysis method. However, the local line directional pattern is reported to be distinctive and gives higher accuracy than the dominant directional feature [1, 21].

This paper proposes a novel Discriminant local line Directional Representation for palmprint recognition by combining local line discriminant pattern (LLDP) and two-directional two-dimensional linear discriminant analysis ($(2D)^2LDA$), so-called DLLDR. First, the algorithm computes the LLDP features with two strategies of encoding multi-directions. Then, $(2D)^2LDA$ is applied to compute DLLDR features with lower-dimensional and higher discriminant. The experiments on the public databases from Hong Kong Polytechnic University demonstrate that DLLDR is a complete and robust direction representation for palmprint recognition.

In the following sections, we present in detail our proposed algorithm of palmprint recognition. Section 2 presents our proposed method. Experimental results are presented in Sect. 3. Finally, the paper conclusions are drawn in Sect. 4.

2 Our Proposed Method

The local line directional pattern (LLDP) is insensitive to illumination change and is more discriminative than the dominant directional code. However, this feature has two ways to represent the potential directional code: (1) using the direction of the brightest and darkest lines, (2) using the direction of the darkest line and the second darkest line. To exploit the distinctiveness of these two features, our proposed method first computes LLDP features with these two encoding strategies. Then, we apply the $(2D)^2$LDA method to reduce the dimensional number of two LLDP feature maps and eliminate the less distinctive information. Therefore, in this section, we present the proposed method in detail including LLDP, $(2D)^2$LDA method and combination scheme.

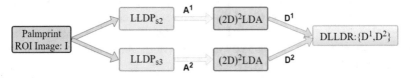

Fig. 1. The scheme of the proposal method.

2.1 LLDP

LLDP descriptor uses the index numbers of line directions to compute the feature code. There are three ways to build LLDP code [21].

S1: The directional bits of minimum k line magnitudes $\{m_i\}$, $(i = 0, 1, \ldots, K)$ are set to 1 and the remaining bits are set to 0, as:

$$LLDP_k = \sum_{i=0}^{K} b_i(m_i - m_j)/2^i, b_i(a) = \begin{cases} 0, a \geq 0 \\ 1, a < 0 \end{cases} \quad (1)$$

where m_k is the *k-th* minimum magnitude. K is the number of consider directions.
S2: The indexes of the first and the second minimum line magnitudes, t_{12} and t_{11} are used as:

$$LLDP = t_{12} \times K^1 + t_{11} \times K^0 \quad (2)$$

S3: The index numbers of the minimum line response t_{12} and the maximum line response t_1 are used as follows:

$$LLDP = t_{12} \times K^1 + t_1 \times K^0 \quad (3)$$

The lines could be computed by MFRAT or Gabor filter bank. In an image, given a square local area Z_p, whose size is $p \times p$, MFRAT computes magnitues of different lines $\{m_i\}$, $(i = 0, 1, ..K)$ at the pixel (x_0, y_0) as:

$$m_i = \sum_{x,y \in L_i} f(x, y) \quad (4)$$

$$L_i = \{(x, y) : y = S_i(x - x_0) + y_0, x \in Z_P\} \quad (5)$$

where $f(x, y)$ is gray value at (x, y), L_i is the set of points built a line on the Z_P, and i means the index number of a slope of S_i.

Given an image I, Gabor filter bank can be applied for detecting the lines $\{m_i\}$, $(i = 0, 1, \ldots, 12)$, located in (x, y) as follows:

$$
\begin{aligned}
m_i &= \langle I * G(x, y, \theta_i, \mu, \sigma) \rangle, \\
\theta_i &= \tfrac{\pi(i-1)}{12}, i = 1, 2, .., 12, \\
G(x, y, \theta, \mu, \sigma) &= \tfrac{1}{2\pi\sigma^2} exp\left\{-\tfrac{x^2+y^2}{2\sigma^2}\right\} f(x, y, \theta, \mu), \\
f(x, y, \theta, \mu) &= exp\{2\pi j(\mu x cos\theta + \mu y sin\theta)\}
\end{aligned}
\tag{6}
$$

where $j = \sqrt{-1}$, μ is the frequency of the sinusoidal wave, θ controls the orientation of the function and σ is the standard deviation of the Gaussian envelop.

LLDP has three strategies for coding line directional patterns. However, strategy 2 can represent strategy 1 with $k = 2$, so we only choose two strategies to get the candidate patterns in order to fully exploit the distinctiveness of the palm lines. That is strategy 2 and strategy 3. With these two strategies, LLDP created by the darkest, second darkest and least dark lines which are stable and clear lines, and effect to the accuracy of recognition.

2.2 (2D)²LDA

$(2D)^2LDA$ is applied for reducing the dimension of LLDP matrix. Suppose $\{A_k\}$, $k = 1 \ldots N$ are the LLDP matrices compute by formula (2) with strategy s2 (or s3) which belong to C classes, and the jth class C_i has n_i templates $\left(\sum_{i=1}^{C} n_i = N\right)$. Let $\overline{A}$ is the means of the registration set, and $\overline{A}_i$ is the means of ith class.

$$
A_k = \left[\left(A_k^{(1)}\right)^T, \left(A_k^{(2)}\right)^T, \ldots, \left(A_k^{(m)}\right)^T\right]^T, \overline{A}_i = \left[\left(\overline{A}_k^{(1)}\right)^T, \left(\overline{A}_k^{(2)}\right)^T, \ldots, \left(\overline{A}_k^{(m)}\right)^T\right]^T,
$$
$$
\overline{A} = \left[\left(\overline{A}^{(1)}\right)^T, \left(\overline{A}^{(2)}\right)^T, \ldots, \left(\overline{A}^{(m)}\right)^T\right]^T,
\tag{7}
$$

where $A_k^{(j)}, \overline{A}_k^{(j)}, \overline{A}^{(j)}$ are the jth row vector of A_k, $\overline{A}_k$ and $\overline{A}$, respectively. 2DLDA compute a set of optimal vectors to find the optimal projection matrices:

$$
X = \{x_1, x_2, \ldots, x_d\}
\tag{8}
$$

by maximizing the criterion as follows:

$$
J(X) = \frac{X^T G_b^X}{X^T G_w^X}
\tag{9}
$$

$$
G_b = \frac{1}{N} \sum_{i=1}^{C} n_i \sum_{j=1}^{m} \left(\overline{A}_i^{(j)} - \overline{A}^{(j)}\right)^T \left(\overline{A}_i^{(j)} - \overline{A}^{(j)}\right)
\tag{10}
$$

$$
G_w = \frac{1}{N} \sum_{i=1}^{C} \sum_{k \in c_i} \sum_{j=1}^{m} \left(A_i^{(j)} - \overline{A}_k^{(j)}\right)^T \left(A_i^{(j)} - \overline{A}^{(j)}\right)
\tag{11}
$$

where T is matrix transpose, G_b is between-class matrix, G_w is within-class scatter matrix. Therefore, X is the orthonormal eigenvectors of $G_w^{-1}G_b$ corresponding to the d largest eigenvalues $\lambda_1, \ldots, \lambda_d$. The value of d is selected based on the predefined threshold θ as follow:

$$\frac{\sum_{i=1}^{d} \lambda_i}{\sum_{i=1}^{n} \lambda} \geq \theta \tag{12}$$

2DLDA described above works in the row-wise direction to learn an optimal matrix X from a set of training LLDP matrices, and then project an LLDP matrix $A_{m \times n}$ onto X, yielding m by d matrix, i.e. $Y_{m \times d} = A_{m \times n}.X_{n \times d}$. Similarly, the alternative 2DLDA learns optimal projection matrix Z reflecting information between columns of LLDP matrices and then projects A onto Z, yielding a q by n matrix, i.e. $B_{q \times n} = Z_{m \times q}^{T}.A_{m \times n}$. Suppose we have obtained the projection matrices X and Z, projecting the LLDP matrix $A_{m \times n}$ onto X and Z simultaneously, yielding a q by d matrix D:

$$D = Z^{T}.A.X \tag{13}$$

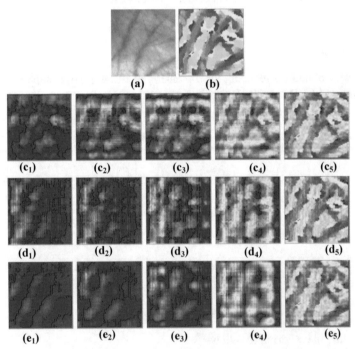

Fig. 2. Results of LLDP with strategy 2 and $(2D)^2$LDA: (a) original palmprint image, (b) LLDP image, (d1)–(d5), (e1–e5) some reconstructed images of the LLDP image with (c1)–(c5) d = 10, 15, 20, 25, 50 and q = 64, (d1)–(d5) d = 64, q = 10, 15, 20, 25, 50, (e1)–(e5) q = d=10, 15, 20, 25, 50, respectively.

The matrix D is also called the discriminant local line directional representation matrix (DLLDR) for recognition.

2.3 Discriminant Local Line Directional Representation

The input of our proposed algorithm is palmprint ROI image: I. Figure 1 demonstrates the proposed the method. The processing steps for extracting DLLDR feature are summarized as follows:

- Step 1: Compute the LLDP with strategies 2 to get A^1 matrix by using formula (2).
- Step 2: Compute the LLDP with strategies 3 to get A^2 matrix by using formula (3).
- Step 3: Based on $(2D)^2LDA$, compute the DLLDR feature D^1 by applying Eq. (13) to the A^1 feature matrix to get D^1.
- Step 4: Based on $(2D)^2LDA$, compute the DLLDR feature D^2 by applying Eq. (13) to the A^2 feature matrix to get D^2.
- Step 5: The combined feature matrix $\{D^1, D^2\}$ is DLLDR of the input image: I.

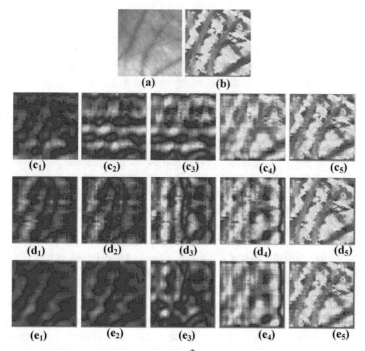

Fig. 3. Results of LLDP with strategy 3 and $(2D)^2LDA$: (a) original palmprint image, (b) LLDP image, (d1)–(d5), (e1–e5) some reconstructed images of the LLDP image with (c1)–(c5) d = 10, 15, 20, 25, 50 and q = 64, (d1)–(d5) d = 64, q = 10, 15, 20, 25, 50, (e1)–(e5) q = d=10, 15, 20, 25, 50, respectively.

Given a query image I, apply the proposed method to get DLLDR feature $D:\{D^1, D^2\}$, and apply our method to all the training images to get the DLLDR feature matrix $D_k(k = 1, 2, \ldots, N)$. The Euclidean distance is used to compare two features.

The distance between D and D_k is defined by:

$$d(D, D_k) = \| D - D_k \|$$
$$score(D, D_k) = 1 - d(D, D_k)$$

(14)

The $d(D, D_k)$ is between 0 and 1. The score of perfect match is 1.

3 Experimental Results

We evaluate the proposed method in comparison some methods (LLDP [21], RDORIC [10]) on the PolyU 3D database of Hong Kong Polytechnic University [30]. These methods were implemented using C# on a PC with a CPU Intel(R) Core(TM) i3-3110 M @ 2.4 GHz and Windows 7 Professional. In the PolyU 3D database, there are *400* different palms. The twenty images from each of palms were captured in each session. The time interval between the two sessions is about 30 days. Each sample contains a 3D ROI (region of interest) and 2D ROI at a resolution of 128 × 128 pixels. In our experiments, we use 2D-ROI database in which the resolution of these ROI images is *64 × 64* pixels. In identification, a query compares to all templates in training set to select the most similar template as result. In verification, each image in the query set is compared with all templates in the registered set to generate incorrect scores and correct scores. The correct score is the maximum of the scores created by the query and templates from the same registered palm. Similarly, the incorrect score is the maximum of the scores created by the query and all templates of the different registered palms. If the query does not have any registered images, we only obtain the incorrect score. If we have N queries of registered palms and M queries of unregistered palms, we obtain N correct scores and $N + M$ incorrect score. We get the verification results: the receiver operating characteristic (ROC) curve. Similar to the number of employees in small and medium-sized companies, we set up two experiments with dataset 1 and dataset 2 with N = 100 and 200. In dataset 1, the training database contains *500* templates from *100* random different palms, where each palm has five templates. The testing database contains *1000* templates from *200* different registered palms. In dataset 2, the training database contains *1000* templates from *200* palms registered. The testing database contains *1000* templates from *200* registered palms. Therefore, there are *500, 1000* correct identification distances and *1000, 2000* incorrect identification scores for $N = 100, 200$, respectively. None of samples in the testing datasets is contained in any of the training datasets. Table 1 presents these parameters of our experiments. Table 2 represents the recognition accuracy of our method in comparison with others. The ROC curve illustrating the verification performances of our method and others are shown in Fig. 4. From this group of figures and tables, we can see that the recognition accuracy rate of our method is higher than the state of art methods (RDORIC [11], LLDP [21]).

Table 1. Parameters of databases in recognition experiments.

Dataset	Each class			All class		Number of identification	
	Training set	Testing set		Training set	Testing set	Correct distance	Incorrect distance
		Registration set	Unregistration set				
1	5	5	5	500	500 + 500 = 1000	500	500 + 500 = 1000
2	5	5	5	1000	1000 + 1000 = 2000	1000	1000 + 1000 = 2000

Table 2. Rank one identification in our experiments.

Method	Dataset 1		Dataset 2	
	Recognition rate (%)	Test time for a query image (ms)	Recognition rate (%)	Test time for a query image (ms)
RDORIC [10]	97.80	147	97.67	204
LLDP [21]	98.80	352	98.70	526
Our method (DLLDR)	99.60	153	99.30	275

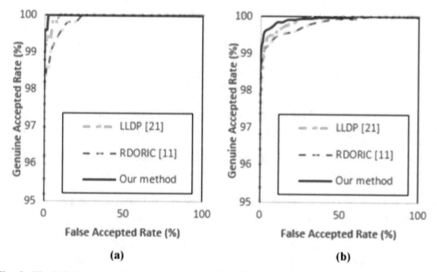

Fig. 4. The ROC curves of our proposed method and other methods with dataset 1 (a), dataset 2 (b), respectively.

4 Conclusion

This paper proposes a novel technique called Discriminant local line Directional Representation for palmprint recognition which combines LLDP and $(2D)^2LDA$. First, the algorithm computes the LLDP features with two strategies of encoding multi-directions. Then, $(2D)^2LDA$ is applie to extract the DLLDR features with lower dimension and higher discriminant. Experimental results on two palmprint datasets of 3D PolyU database show that the proposed method achieves the best results in comparison to the state of art methods.

Acknowledgments. The authors would like to thank the Saigon International University (SIU) for funding this project.

References

1. Fei, L., Zhang, B., Xu, Y., Guo, Z., Wen, J., Jia, W.: Learning discriminant direction binary palmprint descriptor. IEEE Trans. Image Process. **28**(8), 3808–3820 (2019)
2. Genovese, A., Piuri, V., Plataniotis, K.N., Scotti, F.: PalmNet: Gabor-PCA convolutional networks for touchless palmprint recognition. IEEE Trans. Inf. Forensics Secur. **14**(12), 3160–3174 (2019)
3. Zhang, D.: Advanced pattern recognition technologies with applications to biometrics (2009). Medical Information Science Reference
4. Jain, A.K., Nandakumar, K., Ross, A.: 50 years of biometric research: accomplishments, challenges, and opportunities. Pattern Recogn. Lett. **79**, 80–105 (2016)
5. Jaswal, G., Kaul, A., Nath, R.: Knuckle print biometrics and fusion schemes-overview, challenges, and solutions. ACM Comput. Surv. **49**, 34–79 (2016)
6. Huang, D., Shan, C., Ardabilian, M.: Local binary patterns and its applications to facial image analysis: a survey. IEEE Trans. Syst. Man Cybern. Part C Appl. Rev. **41**(6), 765–781 (2011)
7. Fei, L., Lu, G., Teng, S., Zhang, D.: Feature extraction methods for palmprint recognition: a survey and evaluation. IEEE Trans. Syst. Man Cybern. Syst. **49**(2), 346–363 (2019)
8. Hu, D., Feng, G., Zhou, Z.: Two-dimensional locality preserving projections (2DLPP) with its application to palmprint recognition. Pattern Recogn. **40**, 339–342 (2007)
9. Van, H.T., Le, T.H.: GridLDA of Gabor wavelet features for palmprint identification. In: SoICT 2012 Proceedings of the Third Symposium on Information and Communication Technology, pp. 1–10 (2012)
10. Van, H.T., Thanh, V.P., Le, T.H.: Discriminant orientation feature for palmprint recognition. In: 13th International Conference on Computational Science and Its Applications, pp. 1–7 (2013)
11. Van, H.T., Le, T.H.: A palmprint identification system using robust discriminant orientation code. VNU J. Sci. Comput. Sci. Commun. Eng. **30**(4), 1–13 (2014)
12. Rida, I., Herault, R., Marcialis, G., Gasso, G.: Palmprint recognition with an efficient data driven ensemble classifier. Pattern Recogn. Lett. **126**, 21–30 (2019)
13. Zhang, D., Kong, W.-K., You, J., Wong, M.: Online palmprint identification. IEEE Trans. Pattern Anal. Mach. Intell. **25**(9), 1041–1050 (2003)
14. Kong, A., Zhang, D.: Competitive coding scheme for palmprint verification. In: Proceeding of International Conference on Pattern Recognition, pp. 520–523 (2004)
15. Kong, A., Zhang, D., Kamal, M.: Palmprint identification using feature level fusion. Pattern Recogn. **39**(3), 478–487 (2008)
16. Xu, Y., Fei, L., Wen, J., Zhang, D.: Discriminative and robust competitive code for palmprint recognition. IEEE Trans. Syst. Man Cybern. Syst. **48**(2), 232–241 (2018)
17. Jia, W., Huang, D., Zhang, D.: Palmprint verification based on robust line orientation code. Pattern Recogn. **41**(5), 1504–1513 (2008)

18. Sun, Z., Tan, T., Wang, Y., Li, S.: Ordinal palmprint representation for personal identification. In: Proceeding of International Conference on Computer Vision and Pattern Recognition, pp. 279–284 (2005)
19. Guo, Z., Zhang, D., Zhang, L., Zuo, W.: Palmprint verification using binary orientation co-occurrence vector. Pattern Recogn. Lett. **30**(13), 1219–1227 (2009)
20. Fei, L., Xu, Y., Tang, W., Zhang, D.: Double-orientation code and nonlinear matching scheme for palmprint recognition. Pattern Recogn. **49**, 89–101 (2016)
21. Luo, Y., Zhao, L., Zhang, B., Jia, W., Xue, F., Lu, J., Zhu, Y., Xu, B.: Local line directional pattern for palmprint recognition. Pattern Recogn. **50**, 26–44 (2016)
22. Zheng, Q., Kumar, A., Pan, G.: A 3D feature descriptor recovered from a single 2D palmprint image. IEEE Trans. Pattern Anal. Mach. Intell. **38**(6), 1272–1279 (2016)
23. Jia, W., Hu, R.X., Lei, Y.K.: Histogram of oriented lines for palmprint recognition. IEEE Trans. Syst. Man Cybern. Syst. **44**(3), 385–395 (2014)
24. Jia, W., Zhang, B., Lu, J., Zhu, Y., Zhao, Y., Zuo, W., Ling, H.: Palmprint recognition based on complete direction representation. IEEE Trans. Image Process. **26**(9), 4483–4498 (2017)
25. Fei, L., Zhang, B., Xu, Y., Huang, D., Jia, W., Wen, J.: Local discriminant direction binary pattern for palmprint representation and recognition. IEEE Trans. Circ. Syst. Video Technol. 1–14 (2018)
26. Zhao, D., Pan, X., Luo, X., Gao, X.: Palmprint recognition based on deep learning. In: ICWMMN, pp. 214–216 (2015)
27. Minaee, S., Wang, Y.: Palmprint recognition using deep scattering convolutional network. arXiv preprint arXiv:1603.09027 (2016)
28. Izadpanahkakhk, M., Razavi, S., Gorjikolaie, M., Zahiri, S., Uncini, A.: Deep region of interest and feature extraction models for palmprint verification using convolutional neural networks transfer learning. Appl. Sci. **8**(7), 1210 (2018)
29. Li, G., Kim, J.: Palmprint recognition with local micro-structure tetra pattern. Pattern Recogn. **61**, 29–46 (2017)
30. Poly U 3D palmprint database http://www.comp.polyu.edu.hk/~biometrics/2D_3D_Palmprint.htm

Speech Assessment Based on Entropy and Similarity Measures

Michele Della Ventura$^{(\boxtimes)}$

Department of Music Technology, Music Academy "Studio Musica", Treviso, Italy
michele.dellaventura@tin.it

Abstract. Identifying similarities between audio signals is an important goal for the Speech Recognition Systems. This paper proposes a new method for the assessment of the similarity between two audio signals. Based on the entropy theory, the audio signals are segmented and compared on the basis of the information value carried by each segment: this value is obtained considering an alphabet derived from the trend lines of each audio signal. It is an experimental method that permits to represent and analyse the results through an isometric diagram, for a better interpretation of them.

Keywords: Entropy · Information theory · Sound similarity · Speech assessment

1 Introduction

The voice recognition system is a part of everyday life, so that it can be considered a constant presence in any activity.

Speech recognition is a "young" technology and currently under development and improvement. The challenges that this technology presents are numerous but are gradually being reduced thanks to numerous research in this field. Listening to and understanding what a person says is so much more than hearing the words the person speaks. Each person is different from another person and pronounces the same words differently: it depends on many factors such as health or emotional state. Machines are learning to "listen" to accents, emotions and inflections [1, 2], but there is still work to be done in this area and improvements still need to be made.

There are different methods used for the speech recognition and these include:

- Hidden Markov Model method [3–6] in which the observation is linked to the state through a probabilistic function;
- Neural Network method [7–10] which uses systems consisting of interconnected computational nodes working somewhat similarly to human neurons.

The research has invested and is still investing in this field in order to improve voice assistants and guarantee better services to people. The everyday situations in which a person usually uses the voice assistant are many and different, and it is normal that in similar situations the person, who may be engaged in other activities (for example, he

© Springer Nature Switzerland AG 2020
H. A. Le Thi et al. (Eds.): ICCSAMA 2019, AISC 1121, pp. 218–227, 2020.
https://doi.org/10.1007/978-3-030-38364-0_20

is driving or, more generally, has hands busy), pronounces the words incorrectly or too hastily. It may also happen that if the person is in particularly noisy environments the recorded audio has imperfections and discontinuities, or it may happen that the person who uses the speech recognition system has speech or language defects. The main task of a speech recognition system is not limited to the simple detection of what has been pronounced, but, to guarantee an optimal functioning, it carries out a series of analysis on the voice, on the speech and on all the recorded audio in order to better interpret and eventually correct what was said. One of the essential concepts to be considered in speech recognition is the concept of similarity: only in this way is it possible to go beyond the problems mentioned above.

This paper illustrates a method to analyse and evaluate the similarity between two voice patterns considering them not as separate entities but as belonging to the same pattern. The method is based on the calculation of the information content carried by each voice pattern so as to assess the similarity between two sound waves and, consequently, be able to detect phoneme mispronunciations.

This paper is organized as follows.

A brief introduction to the concept of similarity is presented in Sect. 2. Section 3 explains the method used to assess the similarity, and the information theory (as well as the properties related to the information quantification). In Sect. 4, the effectiveness of the method is illustrated through some experimental results Finally, Sect. 5 concludes this paper with the brief discussion.

2 What Is Similarity?

Every day, a lot of information reaches each person, but what is the use of reading, listening or watching the media if we do not systematically proceed with an analysis of the news? How to understand and choose this flow of information?

The analysis becomes a tool that allows a defence against such a vast flow of information: it allows to filter the information [11]. Information takes on meaning from the comparison with something already known by a person. The comparison is an intuitive operation, almost automatic for a person: for example, while listening to a piece of music, the person continually makes a comparison between what he is listening to and what he has already heard. The analysis is able to define the structural elements of a message only through a comparison operation.

It is in this context that the concept of similarity assumes importance.

This concept has been studied from different perspective, such as cosine coefficient [12], information content [13–15], mutual information [16–18] and distance-based measurements [19–21]. A common feature of all these methods is that each of them is related to a particular context or assumes a particular domain model.

The difference between the concept of similarity and the concept of identity is very subtle: generally, two entities (A and B) can be defined as similar if they have some features in common, but not necessarily all [11]. Therefore, the similarity between A and B depend on:

- their common characteristics (the more they have and the more they are similar),
- their differences (the more they have and the less they are similar).

On the basis of these considerations it emerges that to compare two entities (A and B) it is necessary to carry out a segmentation of these entities in order to identify common characteristics or dissimilarities (Fig. 1).

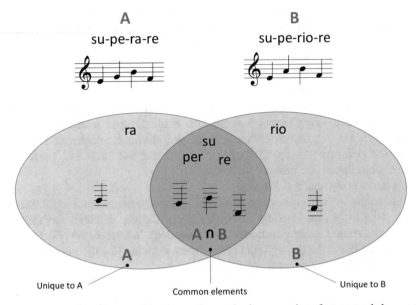

Fig. 1. Representation of two entities that each contains its own unique features and also contains common features.

3 The Method

The method presented in this paper aims to assess the similarity between two speech audio messages (finding common characteristics or dissimilarities) using the information content method. In the following examples two audio files have been considered, with the pronunciation of the word "SUPERARE" (the Italian word for "overtake") by two different people: a 6-year-old child and a 10-year-old child.

To compare the audio messages between them, each audio message is segmented into syllables (using KALDI an open-source toolkit [22]) and then the entropy of each syllable is calculated. This calculation necessarily implies the reference to a specific alphabet. The alphabet is a set of symbols that characterize a language [23] and are associated with it [24–27]: their different combination permits you to construct a message. Thus, in the transmission of a message the meaning changes according to the probability that some symbols are transmitted, as it may be immediately inferred from the formula (1).

$$H(X) = E[I(x_i)] = \sum_{i=1}^{n} I(x_i) \bullet P(x_i) = \sum_{i=1}^{n} P(x_i) \bullet \log_2 \frac{1}{P(x_i)} \quad (1)$$

The following procedure is used to obtain the alphabet from each syllable of the audio message:

(1) extraction (from a Waveform Audio File) of the values of the single samples (in csv format),
(2) elimination of negative values,
(3) representation of the Line Chart of the remaining positive values,
(4) visualization of the 6th order polynomial trend line for the positive values (Fig. 2),

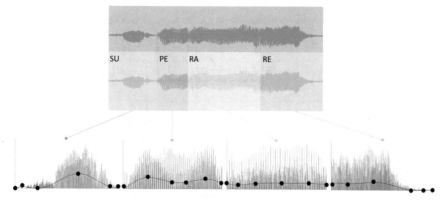

Fig. 2. 6th order polynomial trend line.

(5) compilation of the alphabet table (Table 1), considering the trend of the line (A = Ascending, D = Descending, S = Stable if the curve variation is less than 10%) and the height difference between two consecutive points; the height of each point is defined proportionally with respect to the maximum value 10.

Table 1. Example of Alphabet Table: the first column indicates the height of the point, the second column indicates the trend of the graph line and the third column shows the number of trends of this specific type identified in the audio message (considering the trend line).

Height	Trend	Number
0	A	
	D	
1	A	
	D	
2	A	
	D	
≈	≈	≈
	S	

After defining the alphabet, a transition matrix is created (Table 2) so as to calculate the entropy of each syllable: it is necessary to take into account the manner in which the trend of the Line Chart (between two points) succeeds one another within the audio message. Whit the aim of doing this, the Markov's stochastic process is used: continuous sequence of states of a process in which the probability of passing from one state to another in a unitary time depends probabilistically only on the state immediately preceding it and not on the overall "history" of the system.

Table 2. Example of transition matrix for the word represented in Fig. 2.

		0A	0D	1A	1D	2A	2D	3A	3D	4A	4D	5A	5D	6A	6D	7A	7D	8A	8D	9A	9D	10A	10D	SA	SD
0	A							1																	
	D									1								1							2
1	A			2																					
	D									1															4
2	A			1																					
	D							1																	
3	A					1																			1
	D		1					1		1															2
4	A		1							2	1														
	D																								
5	A									1															
	D																								
6	A									2															
	D																								
7	A																								
	D																								
8	A					1																			
	D																								
9	A																								
	D																								
10	A																								
	D																								
S				1		1	1	1		1		1		2											3

Only one transition matrix is created for both the two speech audio.

By means of the alphabet table and the transition matrix it is possible to calculate the informative value of each syllable and define the information slots, each representing a specific range of values.

Figure 3 shows an example of analysis of two audio messages and the respective representation with the isometric diagram. In the upper part of the diagram is represented (through colours) the information value of each single syllable of the first audio message and in the lower part of the diagram the information value of each syllable of the second audio message. The colour representing a specific information value will have a darker or lighter shade depending on whether it is a greater or lesser value within one of the information ranges. In the event that (case where) the two syllables compared have the same information value, on the diagram there will be a column with a single colour, corresponding to one of the information intervals previously defined (see Table 3). On the contrary, if the information values of the two syllables are different, on the diagram there will be a column with a colour that changes, passing from the colour of the first syllable to the colour of the second syllable.

Table 3. Information value of the syllables.

Syllable	Model	Sample
SU	0,216096405	0,216096405
	0,528320834	0,528320834
	0	0
	0,528320834	0,507650812
	0,5	0,5
PE	0,216096405	0,216096405
	0,528771238	0,528771238
	0,430827083	0,430827083
	0	0,528320834
	0,464385619	0,5
RA	0,133048202	0,410544839
	0,464385619	0,5
	0,298746875	0
	0,5	0,430827083
	0,5	0,464385619
RE	0,375	0,375
	0,528771238	0,528771238
	0,464385619	0,430827083
	0,430827083	0,5
	0,5	0,5
Range1	Range 2	Range 3
0 – 0,2	0.2 – 0,4	0.4 – 0,6

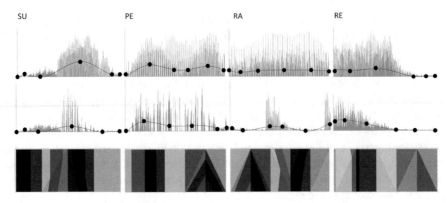

Fig. 3. Isometric analysis of two audio messages.

4 Experimental Results

The analysis method described in this article was tested using a specially developed algorithm, which has no parameters to set for analysis.

The speech audios were recorded considering some technical characteristics:

- audio: mono signal;
- sample rate: more than or equal to 44.100 (for a superior description of the audio signal);
- audio file format: wave or aiff (uncompressed audio formats).

Before starting the tests, a list of (30) specific words was created (to be used for testing): words with high recognition error rates and composed of (at least) three syllables.

Table 4. Excerpt of the word used for testing.

Word	Number of syllables	Language
Ag-nos-tic	3	English
En-vi-ron-ment	4	English
Ham-bur-ger	3	English Italian
Ol-tre-ma-re	4	Italian (word for "overseas")
Ir-reg-u-lar	4	English
Scen-de-re	3	Italian (word for "descend")
Sig-na-ture	3	English
Sti-ra-re	3	Italian (word for "iron")
Stre-go-ne	3	Italian (word for "shaman")
…	…	…

The choice was made on Italian and English words (15 Italian words and 15 English words), considering words that require a different articulation of the vocal organ (see Table 4).

The tests involved 2 different speakers, for the British pronunciation and the Italian pronunciation, and 8 people aged between 12 and 20.

First of all, the words spoken by the speakers were recorded; then, participants were tested one at a time. They sat in front of a microphone and listened to the auditory stimuli on headphones (where it was reproduced the recording realized with the two speakers): after listening three times a word, they had to repeat the same word.

The results of the analysis of the recordings supplied important information:

(1) the difference in intensity in the pronunciation of the same word by two people is not an important discriminant: as we can see in the example of Fig. 3, the first syllable "SU" has two different intensities in the two recordings (see the trend between points 3 and 4), and in the corresponding isographic diagram the columns show small inflections of colour, not very evident;
(2) words that begin with "h" show a high degree of discrimination: the "h" sound is considered like the noise made by the air when it passes through the glottis, without the vocal cords stretched;
(3) words containing the letter "r" show a high degree of discrimination: the lack of ability, or difficulty in, pronouncing the sound "r" (Rhotacism) influences the recognition of the syllables of a word and consequently the analysis. In cases where the words have a maximum of three syllables, often the registration is recognized as different and not similar.

Out of a total of 30 subjects, the algorithm successfully discriminated 87 cases, which means a matching rate of 72%.

5 Discussion and Conclusions

The overall goal of this study is to realize an automatic system able to detect pronunciation defects based on voice audio recordings. The method requires to consider two audio recordings of the same word: the first one with the correct pronunciation and the second one with a generic pronunciation. Using the information theory method, it is possible to highlight, through an isometric diagram, the points in which there is a pronunciation defect.

It is important that this study be viewed more as a demonstration of an approach than as the presentation of an absolute method. The trend-line technique is very promising as a tool for assessing phonetic similarity. One of the main advantages of this technique is that it allows the analysis of words in different languages (Italian, English, French, German,…): algorithm corrections are not required. At the same time, the results expressed through the isometric diagram allow an easy reading and identification of the gravity of the pronunciation defect (on the base of the colour). This method could be a useful tool for a speech therapy analysis, for the analysis of the young children pronunciation.

However, there are some limitations to consider: the need for words to be recorded in the absence of noise and the need to approximate the interpretation of the results due to the poor precision of automatic spelling systems.

Concerning future work, in addition to trying to improve the spelling of audio messages, the research field should consider the distance between two points in a trend line. From this point of view, the results presented in this paper are only preliminary.

References

1. Herman, K., Niesler Thomas, R.: The impact of accent identification errors on speech recognition of South African English. S. Afr. J. Sci. **110**(1–2), 63–68 (2014)
2. Hanani, A., Russell, M.J., Carey, M.J.: Human and computer recognition of regional accents and ethnic groups from British English speech. Comput. Speech Lang. **27**(1), 59–74 (2013)
3. Wydra, S.: Recognition quality improvement in automatic speech recognition system for Polish, pp. 218–223. Eurocon, Warsaw (2007)
4. Gales, M., Young, S.: The application of hidden Markov models in speech recognition. Found. Trends Sig. Process. **1**(3), 195–304 (2007)
5. Torbati, A.H.H.N., Picone, J.: A doubly hierarchical Dirichlet process hidden Markov model with a non-ergodic structure. IEEE-ACM Trans. Audio Speech Lang. Process. **24**(1), 174–184 (2016)
6. Cuiling, L.: English speech recognition method based on hidden Markov model. In: International Conference on Smart Grid and Electrical Automation (ICSGEA), pp. 94–97 (2016)
7. Mohamad, A., Al-Marghilani, A., Akram, A.: Automatic detection technique for speech recognition based on neural networks inter-disciplinary. Int. J. Adv. Comput. Sci. Appl. **9**, 179–184 (2018)
8. Palaz, D., Doss, M., Collobert, R.: Convolutional neural networks based continuous speech recognition using raw speech signal. In: Acoustics, Speech and Signal Processing (2015)
9. Maas, A., Xie, Z., Jurafsky, D., Ng, A.: Lexicon-free conversational speech recognition with neural networks. In: Proceedings of the Conference of the North American Chapter of the Association for Computational Linguistics. Human Language Technologies (2015)
10. Cui, X., Goel, V., Kingsbury, B.: Data augmentation for deep neural network acoustic modeling. In: Proceedings of ICASSP, pp. 5582–5586 (2014)
11. Della Ventura, M.: Similarity measures for music information retrieval: In: Proceedings of the 5th International Conference on Computer Science, Applied Mathematics and Applications. Springer, Germany (2017)
12. Frakes, W.B., Baeza-Yates, R.: Information Retrieval. Data Structure and Algorithms. Prentice Hall, Upper Saddle River (1992)
13. Birjali, M., Beni-Hssane, A., Erritali, M., Madani, Y.: Information content measures of semantic similarity between documents based on Hadoop system. In: International Conference on Wireless Networks and Mobile Communications (WINCOM), Fez, pp. 187–192 (2016)
14. Sanchez, D., Batet, M.: A semantic similarity method based on information content exploiting multiple ontologies. Expert Syst. Appl. **40**(4), 1393–1399 (2013)
15. Resnik, P.: Using information content to evaluate semantic similarity in a taxonomy. In: Proceedings of IJCAI-95, Montreal, Canada, pp. 448–453 (1995)
16. Chen, H.: Mutual information: a similarity measure for intensity based image registration. In: Advanced Image Processing Techniques for Remotely Sensed Hyperspectral Data. Springer, Heidelberg (2004)

17. Russakoff, D., Tomasi, C., Rohlfing, T., Maurer, C.R.: Image similarity using mutual information of regions. In: European Conference on Computer Vision, pp. 596–607 (2004)
18. Rogelj, P., Kovačič, D.S.: Point similarity measure based on mutual information. In: 2nd International Workshop on Biomedical Image Registration, USA (2003)
19. Lee, J.H., Kim, M.H., Lee, Y.J.: Information retrieval based on conceptual distance in IS-A hierarchies. J. Doc. **49**(2), 188–207 (1989)
20. Rada, R., Mili, H., Bicknell, E., Blettner, M.: Development and application of a metric on semantic nets. IEEE Trans. Syst. Man Cybern. **19**(1), 17–30 (1989)
21. Eidenberger, H.: New perspective on visual information retrieval. In: Proceedings of SPIE 5307, Storage and Retrieval Methods and Applications for Multimedia (2004)
22. https://kaldi-asr.org/doc/index.html
23. Weaver, W., Shannon, C.: The Mathematical Theory of Information. Illinois Press, Urbana (1964)
24. Angeleri, E.: Information, meaning and universalit. UTET, Turin (2000)
25. Abraham, M.: Teorie de l'information et Perception esthetique. Paris, Flammarion Editeur (1958)
26. Lerdhal, F., Jackendoff, R.: A Grammatical Parallel Between Music and Language. Plenum Press, New York (1982)
27. Della Ventura, M.: Analysis of algorithms' implementation for melodic operators in symbolical textual segmentation and connected evaluation of musical entropy. In: Proceedings of 1st Models and Methods in Applied Sciences, Drobeta Turnu Severin, pp. 66–73 (2011)

Machine Learning Methods
and Applications

Deep Clustering with Spherical Distance in Latent Space

Bach Tran[✉] and Hoai An Le Thi

Computer Science and Applications Department, LGIPM,
University of Lorraine, Metz, France
{bach.tran,hoai-an.le-thi}univ-lorraine.fr

Abstract. This paper studies the problem of deep joint-clustering using auto-encoder. For this task, most algorithms solve a multi-objective optimization problem, where it is then transformed into a sing-objective problem by linear scalarization techniques. However, it introduces the scaling problem in latent space in a class of algorithms. We propose an extension to solve this problem by using scale invariance distance functions. The advantage of this extension is demonstrated for a particular case of joint-clustering with MSSC (minimizing sum-of-squares clustering). Numerical experiments on several benchmark datasets illustrate the superiority of our extension over state-of-the-art algorithms with respect to clustering accuracy.

Keywords: Clustering · Deep learning · Auto-encoder · Spherical distance

1 Introduction

Clustering is an important task in data mining with the aim of segmenting data-points into groups which pose similarity. Despite decades of research, clustering high-dimensional datasets is still a difficult problem due to the *"curse of dimensionality"* phenomenon. In general terms, it is a *"widely observed phenomenon that data analysis techniques (including clustering), which work well at lower dimensions, often perform poorly as the dimensionality of analyzed data increases"* [23]. Using dimension reduction techniques is a popular way to overcome this problem, where the original features are presented by a smaller but informative set of features. Among them, auto-encoders are recently considered and are often referred to as "deep clustering" in the literature. They have demonstrated a significant improvement in clustering high-dimensional data, which are currently the state-of-the-art methods for clustering.

There are several works focus on developing deep joint-clustering algorithm with different clustering techniques such as K-means [10,22,25,28,30], sub-space clustering [13,31,32], robust continuous clustering [20], soft-assignment clustering [10,11,27], hierarchical clustering [29], etc. They can be classified into

© Springer Nature Switzerland AG 2020
H. A. Le Thi et al. (Eds.): ICCSAMA 2019, AISC 1121, pp. 231–242, 2020.
https://doi.org/10.1007/978-3-030-38364-0_21

two main types: (1) simultaneously minimizing the auto-encoder reconstruction and clustering in a joint framework by linear scalarization techniques (i.e. min $F^{joint} = F^{AE} + \lambda F^{clustering}$) [10,13,20,22,25,28,30–32], or (2) progressively updating neural network mapping and clustering assignment in order to match a target distribution/pseudo label, which is updated during the optimization process [4,8,10,11,27,29]. Generally speaking, it can be viewed as solving an optimization problem or a sequence of problems where each has the form similar to (1).

However, the derived problem in several works is not well-defined. Since clustering is applied to data's representation in the latent space, the clustering objective is affected by the scaling whereas the reconstruction objective is not. This affects deep clustering algorithm that realized on Euclidean distance or ℓ_p norm such as K-means (MSSC) [10,15,22,25,28,30], latent space clustering [13,31,32]; or integrating additional tasks on the latent space depends on Euclidean distance/ℓ_p norm [5,6,12,21]. To solve the mentioned problem, one could consider regularization techniques for neural networks such as regularization for $\|\theta^{(l-1)}\|_p$ or $\|f_E(\theta_E, x_i)\|_p$ [12]. However, it introduces another trade-off parameter that needs to be tuned/searched, which is not encouraged in the unsupervised setting. Hence, instead of using Euclidean distance (or other ℓ_p norm instead of ℓ_2), we could choose another distance function that is invariance to scaling. In this direction, cosine distance has been extensively used, especially for clustering document clustering [7]. Similarly, [1] introduces ℓ_2-layer to project data points' representation onto a unit-hypersphere. From the context of deep clustering, this approach is roughly similar to use the cosine distance directly in the latent space of the auto-encoder, which improves clustering accuracy significantly over other regularization methods (such as Batch Norm and Layer Norm). Hence, motivated by the success of cosine distance in deep clustering, this paper considers two variants of cosine distance.

Our contributions in this work are to:

– Propose a solution to the scaling problem in deep joint-clustering, and instantiate a specific algorithm for the deep joint-clustering using MSSC.
– Conduct numerical experiments for high-dimensional large-scale datasets with several recent clustering methods to study the quality of the proposed algorithms.

The rest of this paper is organized as follows. Section 2 introduces the limitation of existing deep joint-clustering problems by auto-encoder. Section 3 and outlines our proposed solution, with a specific application for the deep joint-clustering by MSSC (Sect. 3.2). The numerical experiment on real-world datasets is reported in Sect. 4.

2 Scaling Problem in a Class of Deep Clustering Algorithms

2.1 Auto-Encoder

The standard stacked auto-encoder consists of two components: an encoder and a decoder. An encoder f^E (reps. decoder f^D) is a neural network parametrized by parameter θ_E (reps. θ_D), maps data from $\mathbb{R}^m \rightarrow \mathbb{R}^d$ (reps. $\mathbb{R}^d \rightarrow \mathbb{R}^m$). Given a raw input data point $x_i \in \mathbb{R}^m$ (i.e. an image, a document, etc.), the encoder first produces code $z_i \in \mathbb{R}^d$, then the decoder reconstructs $\hat{x}_i$ of x_i only from code z_i (Fig. 1 illustrates the corresponding auto-encoder). By reconstructing the input, the network can learn the under-complete but informative representation from the raw input data. Mathematically, given a data-set $\{x_i\}_{i=1,\dots,n}$ where $x_i \in \mathbb{R}^m$, the problem of auto-encoder reconstruction can be defined as

$$\min_{\theta_E,\theta_D} \left\{ F^{AE}(\theta_E, \theta_D) = \sum_{i=1}^n \ell(x_i, f^D(\theta_D, f^E(\theta_E, x_i))) \right\}, \qquad (1)$$

where ℓ measures the reconstruction error. Common choices for ℓ are mean squared error, binary cross-entropy, ℓ_1 norm, etc. In this work, we consider the square error $\ell(x,y) = \|x - y\|_2^2$.

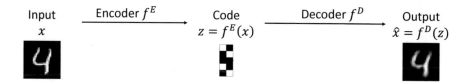

Fig. 1. Illustration of an auto-encoder.

The autoencoder $f^{AE} = f^D \circ f^E$ can be seen as a neural network of L layers, where $f^D = f^{(L)} \circ \cdots \circ f^{(l)}$ and $f^E = f^{(l)} \circ \cdots \circ f^{(1)}$. Each function $f^{(i)}$ represents a layer of the neural network, which maps the output of the previous layer $z^{(i-1)}$ into new code $z^{(i)} = h(z, \theta^{(i-1)})$ where h is the activation function. A typical choice for h is linear function ($h_{\text{linear}}(z, \theta) = \phi(z, \theta)$) or non-linear function such as ReLU [18] ($h_{\text{ReLU}}(z, \theta) = \max(\phi(z, \theta), 0)$, where max is a element-wise operator), where $\phi(z, \theta)$ is the matrix multiplication operators (dense layer in neural network) or convolution operation (convolution layer in convolution neural network).

2.2 Scaling Problem of Joint-Clustering by Auto-Encoder

Let us consider the problem of deep joint-clustering by MSSC. Several works [10, 15, 22, 25, 28, 30] combines auto-encoder with MSSC into an optimization problem by using the linear scalarization technique of multi-objective programming. Formally, this problem is defined as

$$\min_{\theta_E, \theta_D, u, s} F^{\text{J-MSSC}}(\theta_E, \theta_D, u, s) := F^{AE}(\theta_E, \theta_D) + \lambda F^{\text{IP-MSSC}}(\theta_E, u, s) \quad (2)$$

$$= \sum_{i=1}^{n} \|x_i - f^D(\theta_D, f^E(\theta_E, x_i))\|_2^2 + \lambda \|us - f^E(\theta_E, x_i)\|_2^2$$

$$\text{s.t.} \quad s_{j,i} \in \{0, 1\} \quad \text{for} \quad i = 1, \ldots, n \text{ and } j = 1, \ldots, k,$$

$$\sum_j s_{i,j} = 1 \quad \text{for} \quad i = 1, \ldots, n,$$

or

$$\min_{\theta_E, \theta_D, u} F^{\text{J-MSSC}}(\theta_E, \theta_D, u, s) := F^{AE}(\theta_E, \theta_D) + \lambda F^{\text{Bi-MSSC}}(\theta_E, u) \quad (3)$$

$$= \sum_{i=1}^{n} \left(\|x_i - f^D(\theta_D, f^E(\theta_E, x_i))\|_2^2 + \lambda \min_{l=1\ldots k} \|u_l - f^E(\theta_E, x_i)\|_2^2 \right),$$

where λ controls the trade-off between two terms; $u \in \mathbb{R}^{k \times m}$ are k centroids in latent space.

However, problem (2) and (3) both suffer the scaling problem in latent space, which is explained as following. Let l is the bottle-neck layer of the L-layer auto-encoder, i.e. $f^E = f^{(l)} \circ f^{(l-1)} \circ \cdots \circ f^{(1)}$ and $f^D = f^{(L)} \circ \cdots \circ f^{(l+1)}$, where $f^{(i)}$ is the transformation function of i-th layer/block. In modern auto-encoder in general and in deep clustering in particular, $f^{(l)}$ and $f^{(l+1)}$ are linear function or sometimes are nonlinear activation function (such as ReLU):

$$\text{ReLU:} \quad f^{(i)}(\theta^{(i)}, z^{(i-1)}) = \text{ReLU}(\theta^{(i)} z^{(i-1)}),$$
$$\text{Linear activation:} \quad f^{(i)}(\theta^{(i)}, z^{(i-1)}) = \theta^{(i)} z^{(i-1)},$$

for $i \in \{l, l-1\}$, and $z^{(i)}$ is the output of the i-th layer. For these networks, the first terms F^{AE} is invariance to the scaling of $z_i^{(l)} = f^E(\theta_E, x_i)$, but the second term F^{MSSC} scale quadratically to $z_i^{(l)}$. This problem affects deep clustering algorithm that realized on Euclidean distance or ℓ_p norm such as K-means (MSSC) [10,15,22,25,28,30], latent space clustering [13,31,32]; or integrating additional task on the latent space depends on Euclidean distance/ℓ_p norm [5,6,12,21].

3   Proposed Solution

3.1   Spherical Distance

Instead of computing the Euclidean distance on $\mathbb{R}^m$ space, we measure the distance on the projection of data points on the surface of the unit hypersphere $\mathbb{S}^{m-1}$ instead. By only considering the distance between projections, we eliminate the magnitude of z_i but only its direction. Among them, the cosine distance function is a popular measurement, which has been used in several clustering algorithms, notably by Spherical K-means [7]. It is defined as

$$d_{\text{cosine}}(z_i, z_j) = 1 - \left\langle \frac{z_i}{\|z_i\|_2}, \frac{z_j}{\|z_j\|_2} \right\rangle.$$

This distance function has recently been used in deep clustering problem [2], and have demonstrated its effectiveness in comparison with other regularization methods such as Batch Normalization and Layer Normalization. However, the cosine distance is not a metric because it violates the triangular inequality. In addition, consider the case where comparing the distance between two projects on the surface of the hypersphere $\mathbb{S}^{m-1}$. Let us consider the example in Fig. 2. Since $\bar{x}_i$ and $\bar{x}_j$ are the projection of x_i and x_j onto the hypersphere, measuring the arc length between $\bar{x}_i$ and $\bar{x}_j$ (length of the dashed orange arc) is more suitable than the Euclidean distance between them (length of the solid blue line segment). Hence, instead of using the d_{cosine}, we employ the $d_{\text{spherical}}$ as

$$d_{\text{spherical}}(z_i, z_j) = \frac{\arccos(s_{\text{cosine}}(z_i, z_j))}{\pi} = \frac{1}{\pi} \arccos \left\langle \frac{z_i}{\|z_i\|_2}, \frac{z_j}{\|z_j\|_2} \right\rangle,$$

where $\arccos(\alpha)$ is the inverse cosine 1function for $\alpha \in [-1, 1]$. To avoid numerical problem of $\|z_i\|_2 = 0$, we slightly modify $d_{\text{spherical}}$ to

$$d_{\text{spherical}}^{\epsilon}(z_i, z_j) = \frac{1}{\pi} \arccos \left\langle \frac{z_i}{\|z_i\|_2 + \epsilon}, \frac{z_j}{\|z_j\|_2 + \epsilon} \right\rangle, \tag{4}$$

for a very small value of ϵ.

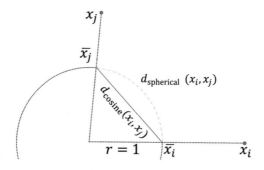

Fig. 2. Illustration of the spherical distance. The solid blue line segment (reps. dashed orange arc) represents the Euclidean distance between $\bar{x}_i$ and $\bar{x}_i$ (reps. spherical distance), which are the projection of x_i and x_j onto the hypersphere (i.e. $\bar{x} = \frac{x}{\|x\|_2}$).

3.2   Application for Deep Clustering with MSSC

In this section, we applied the proposed extension in Sect. 2.2 for the specific problem (3). Modifying the distance in problem (3) from squared Euclidean distance to $d^\epsilon_{\text{spherical}}$ leads to the

$$\min_{\theta_E,\theta_D,u} F^{\text{J-Spherical}_\epsilon}(\theta_E,\theta_D,u,s) := F^{AE}(\theta_E,\theta_D) + \lambda F^{\text{Spherical}_\epsilon}(\theta_E,u) \tag{5}$$

$$= \sum_{i=1}^{n} \|x_i - f^D(\theta_D, f^E(\theta_E,x_i))\|_2^2 + \lambda \sum_{i=1}^{n} \min_{l=1,\dots,k} d^\epsilon_{\text{spherical}}(f^E(\theta_E,x_i),u_l),$$

where λ is the trade-off parameter.

Motivated by the success of Adam algorithm [14] in deep learning, especially for solving the first term $F^{AE}(\theta_E,\theta_D)$ in (5), we adopt Adam to solve our problem. The problem (5) is difficult due to the non-differentiable of the second term $F^{\text{Spherical}_\epsilon}$. We apply the following smoothing technique for min function with $\alpha_s > 0$,

$$\min_{l=1,\dots,k} r_l \approx \text{LSE}_{\alpha_s}(r) = -\alpha_s \log \sum_{l=1}^{k} \exp(-\alpha_s r_l),$$

which turns problem (5) into following optimization problem

$$\min_{\theta_E,\theta_D,u} F^{\text{J-Spherical}_\epsilon}(\theta_E,\theta_D,u,s) := F^{AE}(\theta_E,\theta_D) + \lambda F^{\text{Smooth-Spherical}_\epsilon}(\theta_E,u)$$
$$\tag{6}$$

$$= \sum_{i=1}^{n} \|x_i - f^D(\theta_D, f^E(\theta_E,x_i))\|_2^2 + \sum_{i=1}^{n} \frac{-\lambda\alpha}{\pi} \log \sum_{l=1}^{k} \exp -\alpha d^\epsilon_{\text{spherical}}(f^E(\theta_E,x_i),u_l).$$

Each iteration of Adam for solving the problem (6) requires computing gradient $\nabla_{\theta_E,\theta_D,u}F^J$. The computation of $\nabla_{\theta_E,\theta_D}F^J$ can be calculated by the back-propagation algorithm [19], and the computation for $\nabla_u F^J$ is computed by

$$\frac{\partial F^J(u)}{\partial u_l} = \frac{\alpha^2\lambda}{\pi} \sum_{i=1}^{n} \text{LSE}_1(\alpha t_i) \, \text{softmax}(\alpha t_i) \frac{\frac{1}{\|u_l\|_2+\epsilon}\mathbb{I} - \frac{1}{\|u_l\|_2(\|u_l\|_2+\epsilon)^2}u_l u_l^T}{\sqrt{1 - \left\langle z_i, \frac{u_l}{\|u_l\|_2+\epsilon}\right\rangle}} z_i, \tag{7}$$

where $\text{softmax}(\bar{t}) = \frac{\exp(-\bar{t})}{\sum_l \exp(-t_l)}$, $t_i = (t_l^{(i)})_{l=1,\dots,k} = \left(\arccos\left\langle z_i, \frac{u_l}{\|u_l\|_2}\right\rangle\right)_{l=1,\dots,k}$ and $z_i = \frac{F^E(\theta_E,x_i)}{\|F^E(\theta_E^0,x_i)\|}$, $i=1,\dots,n$.

The Adam applied for problem (5) is outlined in Algorithm 1.

Similar to MSSC-JAE-Sphere, MSSC-JAE-Cosine solves the joint-clustering problem of auto-encoder with MSSC using the cosine distance (problem (5) with d_{cosine} instead of $d^\epsilon_{\text{spherical}}$) by Adam. The difference is the step of computing $\nabla_u F^{\text{J-Spherical}_\epsilon}(\theta_E^t,\theta_D^t,u^t)$, which is done automatically by autograd[1].

---

[1]   https://pytorch.org/docs/stable/autograd.html.

Algorithm 1. `MSSC-JAE-Sphere`: Adam applied for problem (5)

Input: Data x, number of clusters k, trade-off parameter λ, smooth parameter α, batch-size b, Adam's parameter $(\alpha, \beta_1, \beta_2)$

Initialization:

    Initialize θ_E, θ_D (by pre-train of by random initialization).

    Initialize u (by random initialization or by clustering on $z = f^E(\theta_E, x)$).

repeat

    Sample a batch x^t.

    Compute $G^t = \nabla_{\theta_E, \theta_D, u} F^{\text{J-Spherical}}(\theta_E^t, \theta_D^t, u^t)$ with data x^t where

        $\nabla_{\theta_E, \theta_D} F^{\text{J-Spherical}_\epsilon}(\theta_E^t, \theta_D^t, u^t)$ is computed by back-propagation and

        $\nabla_u F^{\text{J-Spherical}_\epsilon}(\theta_E^t, \theta_D^t, u^t)$ is computed by (7).

    Update $(\theta_E, \theta_D, \mu)$ using Adam with G^t.

    $t \leftarrow t + 1$.

until Stopping condition.

---

4 Numerical Experiment

4.1 Datasets

To study the performances of clustering algorithms, we consider the following image and text dataset(s), which are all widely used to benchmark deep clustering algorithms:

- `mnist`: The `mnist` dataset [16] consists of 70000 gray-scale 28×28 images over 10 classes of handwritten digits.
- `fashion`: The `fashion` dataset [26] consists of 70000 gray-scale 28×28 images. `fashion` contains 10 classes of clothing items (shirt, dress, shoe, bag, etc.).
- `usps`: Similar to `mnist` dataset, the `usps` dataset consists of 9298 gray-scale 16×16 images over 10 classes of handwritten digits.
- `rcv1`: Similar to [27,28], we used a subset of 10000 documents from the full RCV1–v2 corpus (Reuters Corpus Volume 1 Version 2) of the four largest classes. Following the procedure in [27], we represent each document by a tf-idf vector of the 2000 most frequently occurring words.

4.2 Comparative Algorithms

We compare the proposed methods (`MSSC-JAE-Sphere` and `MSSC-JAE-Cosine`) with the following baseline:

- `MSSC`: MSSC clustering for raw data.
- `AE-MSSC`: The 2-step approach, which an auto-encoder is first applied for dimensionality reduction, followed by MSSC for clustering
- `DC-Kmeans` [25] solves an alternative of problem (2) by Alternating Direction of Multiplier Method (ADMM).
- `DCN` [28] solves the problem (2) by their proposed alternating stochastic gradient algorithm.

Auto-Encoder Settings: For a fair comparison, we following the setting from [28]. For all algorithms, we follow the standard architecture for clustering with auto-encoder: the number of node in the encoder is $m-500-500-2000-k$, where m is the number of features in the dataset and k is the number of clusters; and the decoder is the mirror of the encoder. The activation function of embedding layer and the last layer are linear, whereas the rest are ReLU. Unless specified otherwise, the auto-encoder is initialized follow the "Xavier Uniform" scheme [9] and pre-train by Adam with learning rate $\alpha = 10^{-3}$ and $(\beta_1, \beta_2) = (0.9, 0.999)$ for 50 epochs with batch-size of 256.

Setting for `MSSC-JAE-Cosine`**:** We use the Adam optimizer with learning rate $\alpha = 3\times10^{-4}$, $(\beta_1, \beta_2) = (0.9, 0.999)$ and batch-size of 256. The smooth parameter $\alpha_s = 16$. The algorithm stop when either of the following criteria is met: (1) 500 epochs or (2) convergence ($\frac{F^J(\theta^k,u^k)-F^J(\theta^{k-1},u^{k-1})}{F^J(\theta^{k-1},u^{k-1})} < \epsilon$ or $\|(\theta,u)^k-(\theta,u)^{k-1}\|_2 < \epsilon$). For initial point, the weighted $\theta^0 = (\theta_E^0, \theta_D^0)$ is obtained by the procedure above. u^0 is initialized as the best result (by objective value) among 10 runs of K-means on extracted feature $\left\{ \frac{F^E(\theta_E^0,x_i)}{\|F^E(\theta_E^0,x_i)\|} \right\}_{i=1,...,n}$. The activation in auto-encoder is Soft Plus - a smooth approximation of ReLU: $Softplus(x) = \frac{1}{\beta}\log(1+\exp(\beta x))$, with $\beta = 256$.

Setting for `MSSC-JAE-Sphere`**:** The setting is the same as `MSSC-JAE-Cosine`'s except u^0. We set $\epsilon = 10^{-4}$ for the $d^\epsilon_{spherical}$. For u^0, we solve the `MSSC-Sphere` problem (MSSC with $d^\epsilon_{spherical}$ instead of $d_{Squared\ Euclidean}$) by Adam optimizer (default parameters) for 10 runs for extracted data $\left\{F^E(\theta_E^0,x_i)\right\}_{i=1,...,n}$. u^0 is selected as the result whose objective value is smallest. The algorithm is implemented in PyTorch[2].

Setting for `DC-Kmeans`**:** We use the authors' implementation[3]. For training the auto-encoder, we use Matlab's neural network toolbox. We also notice that the initial point scheme for K-means from the author's implementation often produces bad results, so we use to the same procedure for initial point as `MSSC-JAE-Cosine` but with extracted feature $\{F^E(\theta_E^0,x_i)\}_{i=1,...,n}$. For `DC-Kmeans`, the authors set $\lambda = 1$.

Setting for `DCN`**:** We use the source code available at[4].

4.3 Experiment Setting

Evaluation criteria: Given an input x_i with ground-truth label l_i; and p_i is the assignment label from clustering algorithm, we measure the following criteria to evaluate experimental results:

---

[2] https://pytorch.org/.
[3] https://github.com/JennyQQL/DeepClusterADMM-Release.
[4] https://github.com/MaziarMF/deep-k-means.

– **Clustering Accuracy (ACC** [3]**):** ACC is calculated as $ACC(l, p) = \frac{1}{n} \sum_{i=1}^{n} 1_{m(p_i)=l_i}$, where $m(x_i)$ is the function which maps each clustering assign in p_i to the equivalent label l_i. In our case, we use the mapping by using the Kuhn-Munkres algorithm [17].

– **Normalized mutual information (NMI** [24]**):** The NMI criterion is calculated as $\text{NMI}(l, p) = \frac{I(l,p)}{\sqrt{H(l)H(p)}}$, where $I(l, p)$ is the mutual information of l and p.

All deep clustering algorithms in our experiment (except MSSC and AE+KM) has a hyper parameter λ controls the trade-off between the auto-encoder and the clustering. For choosing the λ, we performs a grid search $\lambda \in \{10^{-7}, 10^{-6}, \ldots, 10^2\}$. We repeat the experiment 10 times, select the results which has the highest accuracy among 10 runs, and report the average with standard deviation of each criterion.

All experiments are conducted on a Intel(R) Xeon (R) CPU E5-2630 v4 @2.20 GHz with 32 GB of RAM and a GTX 1080 GPU.

4.4   Experiment Results

The results for the evaluation of deep clustering alogrithms on different datasets are reported in in Table 1.

The results show that reducing the number of dimension by auto-encoder facilitates the clustering process: the increase in accuracy by using the 2-step approach (AE+MSSC) over clustering on raw data (MSSC) is significant. The gap in clustering accuracy is up to 32.56% as in *mnist* dataset. However, the final result of both MSSC and AE+MSSC do not exceed the joint-clustering approach (DCN, MSSC-JAE-Cosine, and MSSC-JAE-Sphere). DCN further improves the accuracy of AE+MSSC among all 4 datasets, range from 0.33% to 2.68%.

The proposed methods (MSSC-JAE-Cosine and MSSC-JAE-Sphere) further improve the clustering quality over DCN. The increase in terms of clustering accuracy is consistent, ranging from 1.21% to 7.63% (reps. from 2.48% to 8.17%) for MSSC-JAE-Cosine (reps. MSSC-JAE-Sphere). Both MSSC-JAE-Cosine and MSSC-JAE-Sphere produces better results than DCN. The NMI of both results of MSSC-JAE-Cosine and MSSC-JAE-Sphere is higher than DCN's, up to 3.90% and 6.72% respectively. This result demonstrates the importance of regularization on the latent space, which is achieved in this case by projection the data point's representation in the latent space onto the ℓ_2 ball. In addition, among two methods that utilize the ℓ_2 projection in the latent space, MSSC-JAE-Sphere is undoubtedly better than MSSC-JAE-Cosine: the gap in clustering accuracy is from 0.22% (*usps* dataset) to 2.5% (*mnist* dataset). This increase indicates the importance of using the appropriate distance measure in the latent space.

In conclusion, both MSSC-JAE-Cosine and MSSC-JAE-Sphere are the improvement over DCN and DC-Kmeans. In addition, MSSC-JAE-Sphere is better than MSSC-JAE-Cosine, which demonstrates the importance of using the appropriate distance measure.

Table 1. Comparative result between Auto-encoder-based joint-clustering algorithms. Bold values correspond to best results for each dataset. NA means that the algorithm fails to furnish a result.

Dataset	Algorithms	NMI		Accuracy	
		Average	STD	Average	STD
usps	MSSC	61.35%	0.01%	67.26%	0.05%
$n = 9298$	AE+MSSC	65.41%	1.09%	69.14%	0.52%
$d = 256$	DC-Kmeans	55.26%	0.11%	60.55%	0.16%
$k = 10$	DCN	69.68%	0.60%	71.21%	0.29%
	MSSC-JAE-Cosine	**70.59%**	**1.73%**	73.46%	0.62%
	MSSC-JAE-Sphere	69.98%	1.11%	**73.68%**	**0.79%**
rcv1	MSSC	31.30%	5.40%	50.80%	2.90%
$n = 10000$	AE+MSSC	35.99%	5.47%	55.36%	4.70%
$d = 2000$	DC-Kmeans	NA	NA	NA	NA
$k = 4$	DCN	31.54%	4.58%	58.05%	4.74%
	MSSC-JAE-Cosine	34.80%	10.26%	61.69%	9.27%
	MSSC-JAE-Sphere	**38.27%**	**5.55%**	**64.19%**	**6.09%**
mnist	MSSC	44.25%	0.02%	48.24%	0.05%
$n = 70000$	AE+MSSC	75.63%	0.54%	81.23%	1.83%
$d = 784$	DC-Kmeans	75.65%	0.02%	78.04%	0.02%
$k = 10$	DCN	76.96%	0.70%	83.83%	1.31%
	MSSC-JAE-Cosine	80.86%	0.75%	85.04%	2.30%
	MSSC-JAE-Sphere	**82.81%**	**2.04%**	**86.85%**	**6.44%**
fashion	MSSC	51.24%	0.01%	53.99%	0.07%
$n = 70000$	AE+MSSC	55.48%	0.72%	53.04%	1.99%
$d = 784$	DC-Kmeans	51.64%	2.96%	47.61%	2.44%
$k = 10$	DCN	56.51%	0.58%	53.37%	1.18%
	MSSC-JAE-Cosine	60.21%	1.15%	61.01%	2.58%
	MSSC-JAE-Sphere	**61.34%**	**0.67%**	**61.54%**	**3.25%**

5   Conclusion

We have studied the scaling problem in the latent space for a class of deep clustering algorithm. We proposed an extension by using cosine and spherical distance measure, which is applicable when the derived optimization problems suffer from the scaling of data's representation in the latent space. Both distance measures are invariance to scaling since they compute the distance between projections of data points onto the surface of the unit hypersphere $\mathbb{S}^{m-1}$ instead of between the data points in $\mathbb{R}^m$. As an application, we considered the specific problem of deep joint-clustering with MSSC and proposed two algorithms (MSSC-JAE-Cosine and MSSC-JAE-Sphere) to solve the mentioned problem. The numerical results present the effectiveness of proposed algorithms in comparison with state-of-the-art algorithms in joint-clustering by K-means: the clustering accuracy is higher than the second-best method (from 3.90% to 6.72%). Among the proposed extensions, `MSSC-JAE-Sphere` improves the clustering accuracy `MSSC-JAE-Cosine` by up to 2.5%. It demonstrates the importance of using the appropriate distance measure in the latent space for a class of deep clustering algorithms.

References

1. Affeldt, S., Labiod, L., Nadif, M.: Spectral clustering via ensemble deep autoencoder learning (SC-EDAE). arXiv:1901.02291 [cs, stat], January 2019
2. Aytekin, C., Ni, X., Cricri, F., Aksu, E.: Clustering and unsupervised anomaly detection with ℓ_2 normalized deep auto-encoder representations. arXiv:1802.00187 [cs], February 2018
3. Cai, D., He, X., Han, J.: Locally consistent concept factorization for document clustering. IEEE Trans. Knowl. Data Eng. **23**(6), 902–913 (2011)
4. Caron, M., Bojanowski, P., Joulin, A., Douze, M.: Deep clustering for unsupervised learning of visual features. In: Ferrari, V., Hebert, M., Sminchisescu, C., Weiss, Y. (eds.) Computer Vision – ECCV 2018. Lecture Notes in Computer Science, pp. 139–156. Springer, Cham (2018)
5. Chen, D., Lv, J., Zhang, Y.: Unsupervised multi-manifold clustering by learning deep representation. In: Workshops at the Thirty-First AAAI Conference on Artificial Intelligence, March 2017
6. Das, D., Ghosh, R., Bhowmick, B.: Deep representation learning characterized by inter-class separation for image clustering. In: 2019 IEEE Winter Conference on Applications of Computer Vision (WACV), pp. 628–637, January 2019
7. Dhillon, I.S., Modha, D.S.: Concept decompositions for large sparse text data using clustering. Mach. Learn. **42**(1), 143–175 (2001)
8. Ghasedi Dizaji, K., Herandi, A., Deng, C., Cai, W., Huang, H.: Deep clustering via joint convolutional autoencoder embedding and relative entropy minimization. In: Proceedings of the IEEE International Conference on Computer Vision, pp. 5736–5745 (2017)
9. Glorot, X., Bengio, Y.: Understanding the difficulty of training deep feedforward neural networks. In: Proceedings of the Thirteenth International Conference on Artificial Intelligence and Statistics, pp. 249–256, March 2010
10. Guo, X.: Deep embedded clustering with data augmentation. In: ACML, p. 16 (2018)
11. Guo, X., Gao, L., Liu, X., Yin, J.: Improved deep embedded clustering with local structure preservation. In: International Joint Conference on Artificial Intelligence (IJCAI-17), pp. 1753–1759 (2017)
12. Huang, P., Huang, Y., Wang, W., Wang, L.: Deep embedding network for clustering. In: 2014 22nd International Conference on Pattern Recognition, pp. 1532–1537, August 2014
13. Ji, P., Zhang, T., Li, H., Salzmann, M., Reid, I.: Deep subspace clustering networks. In: NIPS, September 2017
14. Kingma, D.P., Ba, J.: Adam: a method for stochastic optimization. arXiv:1412.6980 [cs], December 2014

15. Le Tan, D.K., Le, H., Hoang, T., Do, T.T., Cheung, N.M.: DeepVQ: a deep network architecture for vector quantization. In: Proceedings of the IEEE Conference on Computer Vision and Pattern Recognition Workshops, pp. 2579–2582 (2018)
16. Lecun, Y., Bottou, L., Bengio, Y., Haffner, P.: Gradient-based learning applied to document recognition. Proc. IEEE **86**(11), 2278–2324 (1998)
17. Lovász, L., Plummer, M.D.: Matching Theory. American Mathematical Society, Providence (2009)
18. Nair, V., Hinton, G.E.: Rectified linear units improve restricted Boltzmann machines. In: Proceedings of the 27th International Conference on International Conference on Machine Learning, ICML 2010, pp. 807–814. Omnipress, USA (2010)
19. Rumelhart, D.E., Hinton, G.E., Williams, R.J.: Learning representations by back-propagating errors. Nature **323**(6088), 533 (1986)
20. Shah, S.A., Koltun, V.: Deep continuous clustering. arXiv:1803.01449 [cs], March 2018
21. Shaol, X., Ge, K., Su, H., Luo, L., Peng, B., Li, D.: Deep discriminative clustering network. In: 2018 International Joint Conference on Neural Networks (IJCNN), pp. 1–7. IEEE, Rio de Janeiro, July 2018
22. Song, C., Liu, F., Huang, Y., Wang, L., Tan, T.: Auto-encoder based data clustering. In: Progress in Pattern Recognition. Image Analysis, Computer Vision, and Applications, pp. 117–124. Lecture Notes in Computer Science, Springer, Heidelberg, November 2013
23. Steinbach, M., Ertöz, L., Kumar, V.: The challenges of clustering high dimensional data. In: Wille, L.T. (ed.) New Directions in Statistical Physics: Econophysics, Bioinformatics, and Pattern Recognition, pp. 273–309. Springer, Heidelberg (2004)
24. Strehl, A., Ghosh, J.: Cluster ensembles - a knowledge reuse framework for combining multiple partitions. J. Mach. Learn. Res. **3**, 583–617 (2002). http://www.jmlr.org/papers/v3/strehl02a.html
25. Tian, K., Zhou, S., Guan, J.: DeepCluster: a general clustering framework based on deep learning. In: Joint European Conference on Machine Learning and Knowledge Discovery in Databases, pp. 809–825. Springer (2017)
26. Xiao, H., Rasul, K., Vollgraf, R.: Fashion-MNIST: a novel image dataset for benchmarking machine learning algorithms. arXiv:1708.07747 [cs, stat], August 2017
27. Xie, J., Girshick, R., Farhadi, A.: Unsupervised deep embedding for clustering analysis. In: International Conference on Machine Learning, pp. 478–487 (2016)
28. Yang, B., Fu, X., Sidiropoulos, N.D., Hong, M.: Towards k-means-friendly spaces: simultaneous deep learning and clustering. In: International Conference on Machine Learning, pp. 3861–3870, 17 July 2017. http://proceedings.mlr.press/v70/yang17b.html
29. Yang, J., Parikh, D., Batra, D.: Joint unsupervised learning of deep representations and image clusters. In: 2016 IEEE Conference on Computer Vision and Pattern Recognition (CVPR), pp. 5147–5156, June 2016
30. Zhang, P., Gong, M., Zhang, H., Liu, J.: DRLNet: deep difference representation learning network and an unsupervised optimization framework. In: IJCAI (2017)
31. Zhang, T., Ji, P., Harandi, M., Hartley, R., Reid, I.: Scalable deep k-subspace clustering. In: Jawahar, C., Li, H., Mori, G., Schindler, K. (eds.) Computer Vision – ACCV 2018. Lecture Notes in Computer Science, pp. 466–481. Springer International Publishing, Cham (2019)
32. Zhang, T., Ji, P., Harandi, M., Huang, W., Li, H.: Neural collaborative subspace clustering. In: Chaudhuri, K., Salakhutdinov, R. (eds.) Proceedings of the 36th International Conference on Machine Learning. Proceedings of Machine Learning Research, vol. 97, pp. 7384–7393. PMLR, June 2019

A Channel-Pruned and Weight-Binarized Convolutional Neural Network for Keyword Spotting

Jiancheng Lyu[1(✉)] and Spencer Sheen[2]

[1] UC Irvine, Irvine, CA 92697, USA
jianchel@uci.edu
[2] UC San Diego, La Jolla, CA 92093, USA
spsheen97@gmail.com

Abstract. We study channel number reduction in combination with weight binarization (1-bit weight precision) to trim a convolutional neural network for a keyword spotting (classification) task. We adopt a group-wise splitting method based on the group Lasso penalty to achieve over 50% channel sparsity while maintaining the network performance within 0.25% accuracy loss. We show an effective three-stage procedure to balance accuracy and sparsity in network training.

Keywords: Convolutional neural network · Channel pruning · Weight binarization · Classification

1 Introduction

Reducing complexity of neural networks while maintaining their performance is both fundamental and practical for resource limited platforms such as mobile phones. In this paper, we integrate two methods, namely channel pruning and weight quantization, to trim down the number of parameters for a keyword spotting convolutional neural network (CNN, [4]).

Channel pruning aims to lower the number of convolutional channels, which is a group sparse optimization problem. Though group Lasso penalty [8] is known in statistics, and has been applied directly in gradient decent training of CNNs [7] earlier, we found that the direct approach is not effective to realize sparsity for the keyword CNN [4,6]. Instead, we adopt a group version of a recent relaxed variable splitting method [2]. This relaxed group-wise splitting method (RGSM, see [10] for its first study on deep image networks) accomplished over 50% sparsity while keeping accuracy loss at a moderate level. In the next stage (II), the original network accuracy is recovered with a retraining of float precision weights while leaving out the pruned channels in stage I. In the last stage (III), the network weights are binarized into 1-bit precision with a warm start training based on stage II. At the end of stage III, a channel pruned (over 50%) and weight binarized slim CNN is created with validation accuracy within 0.25 % of that of the original CNN.

© Springer Nature Switzerland AG 2020
H. A. Le Thi et al. (Eds.): ICCSAMA 2019, AISC 1121, pp. 243–254, 2020.
https://doi.org/10.1007/978-3-030-38364-0_22

The rest of the paper is organized as follows. In Sect. 2, we review the network architecture of keyword spotting CNN [4,6]. In Sect. 3, we introduce the proximal operator of group Lasso, RGSM, and its convergence theorem where an equilibrium condition is stated for the limit. We also outline binarization, the BinaryConnect (BC) training algorithm [1] and its blended version [11] to be used in our experiment. Through a comparison of BC and RGSM, we derive a hybrid algorithm (group sparse BC) which is of independent interest. In Sect. 4, we describe our three stage training results, which indicate that RGSM is the most effective method and produces two slim CNN models for implementation. Concluding remarks are in Sect. 4.

2   Network Architecture

Let us briefly describe the architecture of keyword CNN [4,6] to classify a one second audio clip as either silence, an unknown word, 'yes', 'no', 'up', 'down', 'left', 'right', 'on', 'off', 'stop', or 'go'. After pre-processing by windowed Fourier transform, the input becomes a single-channel image (a spectrogram) of size $t \times f$, same as a vector $v \in \mathbb{R}^{t \times f}$, where t and f are the input feature dimension in time and frequency respectively. Next is a convolution layer that operates as follows. A weight tensor $W \in \mathbb{R}^{(m \times r) \times 1 \times n}$ is convolved with the input v. The weight tensor is a local time-frequency patch of size $m \times r$, where $m \leq t$ and $r \leq f$. The weight tensor has n hidden units (feature maps), and may down-sample (stride) by a factor s in time and u in frequency. The output of the convolution layer is n feature maps of size $(t - m + 1)/s \times (f - r + 1)/u$. Afterward, a max-pooling operation replaces each $p \times q$ feature patch in time-frequency domain by the maximum value, which helps remove feature variability due to speaking styles, distortions etc. After pooling, we have n feature maps of size $(t - m + 1)/(s\,p) \times (f - r + 1)/(u\,q)$. An illustration is in Fig. 1. The keyword CNN has two convolutional (conv) layers and a fully connected layer. There is 1 channel in the first conv. layer and there are 64 channels in the second. The weights in the second conv. layer form a 4-D tensor $W^{(2)} \in \mathbb{R}^{W \times H \times C \times N}$, where (W, H, C, N) are dimensions of spatial width, spatial height, channels and filters, $C = 64$.

3   Complexity Reduction and Training Algorithms

3.1   Group Sparsity and Channel Pruning

Our first step is to trim the 64 channels in the second conv. layer to a smaller number while maintaining network performance. Let weights in each channel form a group, then this becomes a group sparsity problem for which group Lasso (GL) has been a classical differentiable penalty [8]. Let vector

$$w = (w_1, \cdots, w_g, \cdots, w_G), \ w_g \in \mathbb{R}^d, \ w \in \mathbb{R}^{d \times G},$$

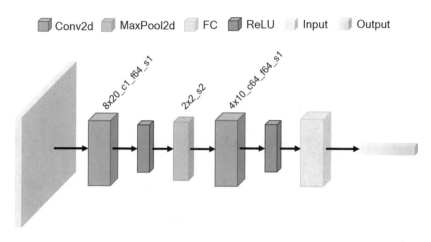

Fig. 1. A convolutional model with single-channel input. We use a simple notation to indicate conv. and max-pooling layers. For instance, $8 \times 20\_c1\_f64\_s1$ indicates a conv. layer with kernel size 8×20, channel number 1, filter number 64 and stride 1.

where G is the number of groups. Let I_g be the indices of w in group g. The group-Lasso penalty is [8]:

$$\|w\|_{GL} := \sum_{g=1}^{G} \|w_g\|_2. \tag{3.1}$$

It is easy to implement GL as an additive penalty term for deep neural network training [7] by minimizing a penalized objective function of the form:

$$f(w) := \ell(w) + \mu\, P(w), \quad \mu > 0, \tag{3.2}$$

where $\ell(w)$ is a standard loss function on data such as cross entropy [12], and $P(w)$ is a penalty function equal to sum of weight decay (ℓ_2 norm all network weights) and GL. For ease of notation, we merge the weight decay term with ℓ and take P as GL below.

In a case study of training CNN with un-structured weight sparsity [9], a direct minimization of ℓ_1 type penalty as an additive term in the training objective function provides less sparsity and accuracy (Table 4 of [9]) than the Relaxed Variable Splitting Method (RVSM [2]). In the group sparsity setting here, we shall see that the direct minimization of GL in (3.2) is also not efficient. Instead, we adopt a group version of RVSM [2], which minimizes the following Lagrangian function of (u, w) alternately:

$$\mathcal{L}_\beta(u, w) = \ell(w) + \mu\, P(u) + \frac{\beta}{2}\|w - u\|_2^2, \tag{3.3}$$

for a parameter $\beta > 0$.

The u-minimization is in closed form for GL. To see this, consider finding the GL proximal (projection) operator by solving:

$$y^* = \mathrm{argmin}_y \ \frac{1}{2}\,\|y - w\|^2 + \lambda\,\|y\|_{GL}, \tag{3.4}$$

for parameter $\lambda > 0$ or group-wise:

$$y_g^* = \mathrm{argmin}_{y_g} \ \lambda\,\|y_g\| + \frac{1}{2}\sum_{i \in I_g}\|y_{g,i} - w_{g,i}\|^2. \tag{3.5}$$

If $y_g^* \neq 0$, the objective function of (3.5) is differentiable and setting gradient to zero gives:

$$y_{g,i} - w_{g,i} + \lambda y_{g,i}/\|y_g\| = 0,$$

or:

$$(1 + \lambda/\|y_g\|)y_{g,i} = w_{g,i}, \ \forall i \in I_g,$$

implying:

$$(1 + \lambda/\|y_g\|)\,\|y_g\| = \|w_g\|,$$

or:

$$\|y_g^*\| = \|w_g\| - \lambda, \ \text{if } \|w_g\| > \lambda. \tag{3.6}$$

Otherwise, the critical equation does not hold and $y_g^* = 0$. The minimal point formula is:

$$y_{g,i}^* = w_{g,i}(1 + \lambda/(\|w_g\| - \lambda))^{-1} = w_{g,i}(\|w_g\| - \lambda)/\|w_g\|, \ \text{if } \|w_g\| > \lambda;$$

otherwise, $y_g^* = 0$. The result can be written as a soft-thresholding operation:

$$y_g^* = \mathrm{Prox}_{GL,\lambda}(w_g) := w_g \max(\|w_g\| - \lambda, 0)/\|w_g\| \tag{3.7}$$

The w minimization is by gradient descent, implemented in practice as stochastic gradient descent (SGD). Combining the u and w updates, we have the Relaxed Group-wise Splitting Method (RGSM):

$$u_g^t = \mathrm{Prox}_{GL,\lambda}(w_g^t), \ g = 1, \cdots, G,$$
$$w^{t+1} = w^t - \eta\,\nabla\ell(w^t) - \eta\,\beta\,(w^t - u^t), \tag{3.8}$$

where η is the learning rate.

3.2   Theoretical Aspects

The main theorem of [2] guarantees the convergence of RVSM algorithm under some conditions on the parameters (λ, β, η) and initial weights in case of one convolution layer network and Gaussian input data. The latter conditions are used to prove that the loss function ℓ obeys Lipschitz gradient inequality on the iterations. Assuming that the Lipschitz gradient condition holds for ℓ, we adapt the main result of [2] into:

Theorem 1. *Suppose that ℓ is bounded from below, and satisfies the Lipschitz gradient inequality: $\|\nabla\ell(x) - \nabla\ell(y)\| \le L\,\|x - y\|$, $\forall(x, y)$, for some positive constant L. Then there exists a positive constant $\eta_0 = \eta_0(L, \beta) \in (0, 1)$ so that if $\eta < \eta_0$, the Lagrangian function $\mathcal{L}_\beta(u^t, w^t)$ is descending and converging in t, with (u^t, w^t) of RGSM algorithm satisfying $\|(u^{t+1}, w^{t+1}) - (u^t, w^t)\| \to 0$ as $t \to +\infty$, and subsequentially approaching a limit point $(\bar{u}, \bar{w})$. The limit point $(\bar{u}, \bar{w})$ satisfies the equilibrium system of equations:*

$$\bar{u}_g = \mathrm{Prox}_{GL,\lambda}(\bar{w}_g),\ g = 1, \cdots, G,$$
$$\nabla\ell(\bar{w}) = \beta\,(\bar{u} - \bar{w}). \tag{3.9}$$

Remark 1. The system (3.9) serves as a "critical point condition". The $\bar{u}$ is the desired weight vector with group sparsity that network training aims to reach.

Remark 2. The group-ℓ_0 penalty is:

$$\|w\|_{GL0} := \sum_{g=1}^{G} 1_{(w_g : \|w_g\|_2 \neq 0)} \tag{3.10}$$

Then the GL proximal problem (3.5) is replaced by:

$$y_g^* = \mathrm{argmin}_{y_g}\ \lambda\,1_{\|y_g\| \neq 0} + \frac{1}{2}\sum_{i \in I_g} \|y_{g,i} - w_{g,i}\|^2. \tag{3.11}$$

If $y_g = 0$, the objective equals $\|w_g\|_2^2/2$. So if $\lambda \ge \|w_g\|^2/2$, $y_g = 0$ is a minimal point. If $\lambda < \|w_g\|^2/2$, $y_g = w_g$ gives minimal value λ. Hence the thresholding formula is:

$$y_g^* := \mathrm{Prox}_{GL0,\lambda}(w_g) = w_g\,1_{\|w_g\|_2 > \sqrt{2\lambda}}. \tag{3.12}$$

Theorem 1 remains true with (3.9) modified where $\mathrm{Prox}_{GL,\lambda}$ is replaced by $\mathrm{Prox}_{CL0,\lambda}$.

3.3   Weight Binarization

The CNN computation can speed up a lot if the weights are in the binary vector form: float precision scalar times a sign vector $(\cdots, \pm1, \pm1, \cdots)$, see [3]. For the keyword CNN, such weight binarization alone doubles the speed of an Android app that runs on Samsung Galaxy J7 cellular phone [5] with standard tensorflow functions such as 'conv2d' and 'matmul'.

  Weight binarized network training involves a projection operator or the solution of finding the closest binary vector to a given real vector w. The projection is written as $\mathrm{proj}_\mathbb{Q}\,w$, for $w \in \mathbb{R}^D$, $\mathbb{Q} = \mathbb{R}_+ \times \{\pm1\}^D$. When the distance is Euclidean (in the sense of ℓ_2 norm $\|\cdot\|$), the problem:

$$\mathrm{proj}_{\mathbb{Q},a}(w) := \mathrm{argmin}_{z \in \mathbb{Q}}\ \|z - w\| \tag{3.13}$$

has exact solution [3]:

$$\text{proj}_{\mathbb{Q},a}(w) = \frac{\sum_{j=1}^{D} |w_j|}{D} \, \text{sgn}(w) \tag{3.14}$$

where $\text{sgn}(w) = (q_1, \cdots, q_j, \cdots, q_D)$, and

$$q_j = \begin{cases} 1 & \text{if } w_j \geq 0 \\ -1 & \text{otherwise.} \end{cases}$$

The projection is simply the sgn function of w times the arithmetic average of the absolute values of the components of w.

The standard training algorithm for binarized weight network is BinaryConnect [1]:

$$\mathbf{w}_f^{t+1} = \mathbf{w}_f^t - \eta \, \nabla \ell(\mathbf{w}^t), \quad \mathbf{w}^{t+1} = \text{proj}_{\mathbb{Q},a}(\mathbf{w}_f^{t+1}), \tag{3.15}$$

where $\{\mathbf{w}^t\}$ denotes the sequence of binarized weights, and $\{\mathbf{w}_f^t\}$ is an auxiliary sequence of floating weights (32 bit). Here we use the blended version [11]:

$$\mathbf{w}_f^{t+1} = (1 - \rho)\, \mathbf{w}_f^t + \rho\, \mathbf{w}^t - \eta \, \nabla \ell(\mathbf{w}^t), \quad \mathbf{w}^{t+1} = \text{proj}_{\mathbb{Q},a}(\mathbf{w}_f^{t+1}), \tag{3.16}$$

for $0 < \rho \ll 1$. The algorithm (3.16) becomes the classical projected gradient descent at $\rho = 1$, which suffers from weight stagnation due the discreteness of $\mathbf{w}^t$ however. The blending in (3.16) leads to a better theoretical property [11] that the sufficient descent inequality holds if the loss function ℓ has Lipschitz gradient.

Remark 3. In view of (3.8) and (3.16), we see an interesting connection that both involve a projection step, as Prox is a projection in essence. The difference is that $\nabla \ell$ in BC is evaluated at the projected weight $\mathbf{w}^t$. If we mimic such a BC-gradient, and evaluate the gradient of Lagrangian in w at u^t instead of w^t, then (3.8) becomes:

$$\begin{aligned} u_g^t &= \text{Prox}_{GL,\lambda}(w_g^t), \quad g = 1, \cdots, G, \\ w^{t+1} &= w^t - \eta \, \nabla \ell(u^t). \end{aligned} \tag{3.17}$$

We shall call (3.17) a Group Sparsity BinaryConnect (GSBC) algorithm and compare it with RGSM in our experiment.

4   Experimental Results

In this section, we show training results of channel pruned and weight binarized audio CNN based on GL, RGSM, and GSBC. We assume that the objective function under gradient descent is $\ell(\cdot) + \mu \|\cdot\|_{GL}$, with a threshold parameter λ. For GL, $\mu > 0$, $\lambda = 0$, $\beta = 0$. For RGSM, $\mu = 0$, $\lambda > 0$, $\beta = 1$. For GSBC, $\mu = 0$, $\lambda > 0$, $\beta = 0$. The experiment was conducted in TensorFlow on a single GPU machine with NVIDIA GeForce GTX 1080. The overall architecture

[6] consists of two convolutional layers, one fully-connected layer followed by a softmax function to output class probabilities. The training loss $\ell(\cdot)$ is the standard cross entropy function. The learning rate begins at $\eta = 0.001$, and is reduced by a factor of 10 in the late training phase. The training proceeds in 3 stages:

- Stage I: channel pruning with a suitable choice of μ or λ so that sparsity emerges at a moderate accuracy loss.
- Stage II: retrain float precision (32 bit) weights in the un-pruned channels at the fixed channel sparsity of Stage I, aiming to recover the lost accuracy in Stage I.
- Stage III: binarize the weights in each layer with warm start from the pruned network of Stage II, aiming to nearly maintain the accuracy in Stage II.

Stage I begins with random (cold) start and performs 18000 iterations (default, about 50 epochs). Figure 2 shows the validation accuracy of RGSM at $(\lambda, \beta) = (0.05, 1)$ vs. epoch number. The accuracy climbs to a peak value above 80% at epoch 20, then comes down and ends at 59.84%. The accuracy slide agrees with channel sparsity gain beginning at epoch 20 and steadily increasing to nearly 56% at the last epoch seen in Fig. 3. The bar graph in Fig. 4 shows the pruning pattern and the remaining channels (bars of unit height). At $(\lambda, \beta) = (0.04, 1)$, RGSM stage I training yields a higher validation accuracy 76.6% with a slightly lower channel sparsity 51.6%. At the same (λ, β) values, GSBC gives an even higher validation accuracy 80.9% but much lower channel sparsity of 26.6%. The GL method produces minimal channel sparsity in the range $\mu \in (0, 1)$ covering the corresponding λ value where sparsity emerges in RGSM. The reason appears to be that the network has certain internal constraints that prevent the GL penalty from getting too small. Our experiments show that even with the cross-entropy loss $\ell(\cdot)$ removed from the training objective, the GL penalty cannot be minimized below some positive level. The Stage-I results are tabulated in Table 1 with a GL case at $\mu = 0.6$. It is clear that RGSM is the best method to go forward with to stage II.

Table 1. Validation accuracy (%) and channel (ch.) sparsity (%) after Stage I (ch. pruning).

Model	β	λ	μ	Accuracy	Ch. Sparsity
Original Audio-CNN	0	0	0	88.5	0
GL Ch-pruning	0	0	0.6	66.8	0
RGSM Ch-pruning	1	4.e−2	0	76.6	51.6
RGSM Ch-pruning	1	5.e−2	0	59.8	56.3
GSBC Ch-pruning	0	4.e−2	0	80.9	26.6

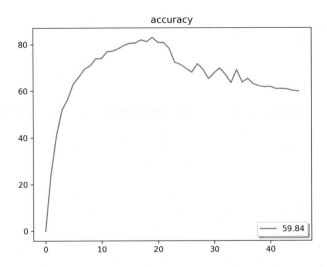

Fig. 2. Validation accuracy vs. number of epochs in Stage-1 training by RGSM at $(\lambda, \beta) = (0.05, 1)$. The accuracy at the last epoch is 59.84%.

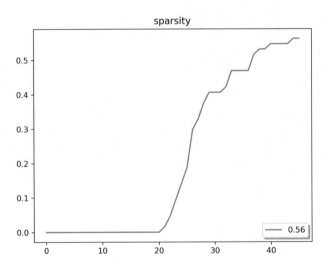

Fig. 3. Channel sparsity vs. number of epochs in Stage-1 training by RGSM at $(\lambda, \beta) = (0.05, 1)$. The sparsity at the last epoch is 56.3%.

In Stage II, we mask out the pruned channels to keep sparsity invariant (Fig. 5), and retrain float precision weights in the complementary part of the network. Figure 7 shows that with a dozen epochs of retraining, the accuracy of the RGSM pruned model at $\lambda = 0.04$ (0.05) in Stage I reaches 89.2% (87.9%), at the level of the original audio CNN (Table 2).

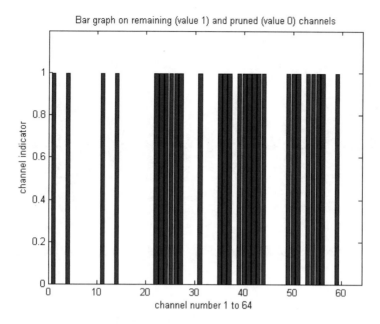

Fig. 4. Remaining channels illustrated by bars vs. channel number (1 to 64) after Stage-1 training by RGSM at $(\lambda, \beta) = (0.05, 1)$. The sparsity (% of 0's) is 56.3%.

Table 2. Validation accuracy (%) and channel (ch.) sparsity (%) after Stage II (float precision weight retraining).

Model	β	λ	μ	Accuracy	Ch. Sparsity
Original Audio-CNN	0	0	0	88.5	0
RGSM Ch-pruning + Float weight retrain	1	4.e−2	0	89.2	51.6
RGSM Ch-pruning + Float weight retrain	1	5.e−2	0	87.9	56.3

In Stage III, with blending parameter $\rho = 1.\mathrm{e}{-5}$, the weights in the network modulo the masked channels are binarized with validation accuracy 88.3% at channel sparsity 51.6%, and 87% at channel sparsity 56.3%, see Fig. 6 and Table 3.

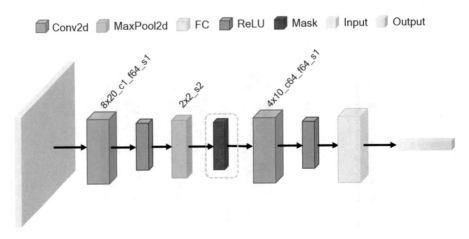

Fig. 5. The CNN model with pruned channels masked out (the masking layer in red).

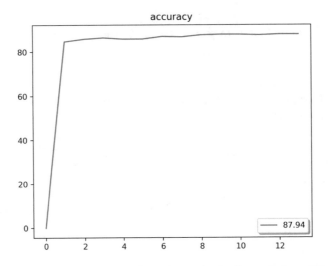

Fig. 6. Validation accuracy vs. number of epochs in Stage-2 float (32 bit) weight retraining. The accuracy at the last epoch is 87.94%. Channel sparsity is 56.3%.

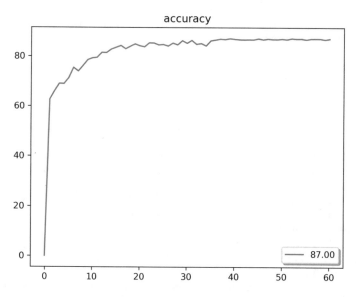

Fig. 7. Validation accuracy vs. number of epochs in Stage-3 binary (1-bit) weight training. The accuracy at the last epoch is 87%. Channel sparsity is 56.3%.

Table 3. Validation accuracy (%) and channel (ch.) sparsity (%) after Stage III (weight binarization training).

Model	β	λ	μ	Accuracy	Ch. Sparsity
Original Audio-CNN	0	0	0	88.5	0
RGSM Ch-pruning + Weight binarization	1	4.e−2	0	88.3	51.6
RGSM Ch-pruning + Weight binarization	1	5.e−2	0	87.0	56.3

5   Conclusion and Future Work

We successfully integrated a group-wise splitting method (RGSM) for channel pruning, float weight retraining and weight binarization to arrive at a slim yet almost equally performing CNN for keyword spotting. Since channel pruning involves architecture change, there is additional work to speed up a hardware implementation. Preliminary test on a MacBook Air with a CPU version of Tensorflow shows as much as 28.87% speed up by the network structure with float precision weight in Fig. 5. An efficient way to implement the masking layer without resorting to an element-wise tensor multiplication (especially on a mobile phone) is worthwhile for our future work.

We also plan to study other penalties [2] such as group-ℓ_0 (transformed-ℓ_1) in the RGSM framework as outlined in Remark 2, and extend the three stage process developed here to multi-level complexity reduction on larger CNNs and other applications in the future.

Acknowledgements. The work was supported in part by NSF grants IIS-1632935 and DMS-1854434 at UC Irvine.

References

1. Courbariaux, M., Bengio, Y., David, J.: BinaryConnect: training deep neural networks with binary weights during propagations. In: Conference on Neural Information Processing Systems (NIPS), pp. 3123–3131 (2015)
2. Dinh, T., Xin, J.: Convergence of a relaxed variable splitting method for learning sparse neural networks via ℓ_1, ℓ_0, and transformed-ℓ_1 penalties, arXiv preprint arXiv:1812.05719
3. Rastegari, M., Ordonez, V., Redmon, J., Farhadi, A.: XNOR-Net: ImageNet classification using binary convolutional neural networks. In: European Conference on Computer Vision (ECCV) (2016)
4. Sainath, T., Parada, C.: Convolutional neural networks for small-footprint keyword spotting. In: INTERSPEECH 2015, Dresden, Germany, 6–10 September 2015, pp. 1478–1482 (2015)
5. Sheen, S., Lyu, J.: Median binary-connect method and a binary weight convolutional neural network for word recognition. In: IEEE Data Compression Conference (DCC) (2019). arXiv:1811.02784, https://doi.org/10.1109/DCC.2019.00116
6. Simple audio recognition tutorial. tensorflow.org. Accessed 10 Aug 2019
7. Wen, W., Wu, C., Wang, Y., Chen, Y., Li, H.: Learning structured sparsity in deep neural networks. In: NIPS (2016)
8. Yuan, M., Lin, Y.: Model selection and estimation in regression with grouped variables. J. Roy. Stat. Soc. B **68**(1), 49–67 (2007)
9. Xue, F., Xin, J.: Learning sparse neural networks via L0 and TL1 by a relaxed variable splitting method with application to multi-scale curve classification. In: Proceedings of the World Congress Global Optimization, Metz, France, July 2019. arXiv preprint arXiv: 1902.07419. https://doi.org/10.1007/978-3-030-21803-4_80.
10. Yang, B., Lyu, J., Zhang, S., Qi, Y-Y., Xin, J.: Channel pruning for deep neural networks via a relaxed group-wise splitting method. In: Proceedings of 2nd International Conference on AI for Industries, Laguna Hills, CA, 25–27 September 2019
11. Yin, P., Zhang, S., Lyu, J., Osher, S., Qi, Y-Y., Xin, J.: Blended coarse gradient descent for full quantization of deep neural networks. Res. Math. Sci. **6**(1), 14 (2019). https://doi.org/10.1007/s40687-018-0177-6, arXiv: 1808.05240
12. Yu, D., Deng, L.: Automatic speech recognition: a deep learning approach. In: Signals and Communication Technology. Springer, New York (2015)

Fusing of Deep Learning, Transfer Learning and GAN for Breast Cancer Histopathological Image Classification

Mai Bui Huynh Thuy and Vinh Truong Hoang$^{(\boxtimes)}$

Faculty of Information Technology, Ho Chi Minh City Open University,
Ho Chi Minh City, Vietnam
{maibht.178i,vinh.th}@ou.edu.vn

Abstract. Biomedical image classification often deals with limited training sample due to the cost of labeling data. In this paper, we propose to combine deep learning, transfer learning and generative adversarial network to improve the classification performance. Fine-tuning on VGG16 and VGG19 network are used to extract the good discriminated cancer features from histopathological image before feeding into neuron network for classification. Experimental results show that the proposed approaches outperform the previous works in the state-of-the-art on breast cancer images dataset (BreaKHis).

Keywords: Deep learning · Transfer learning · BreaKHis dataset · Breast cancer · Histopathological image classification · GAN

1 Introduction

Breast cancer is the most common invasive cancer in women and have a significant impact to 2.1 million people yearly. In 2018, the World Health Organization (WHO) estimated 627,000 death cases because of breast cancer, be getting 15% death causes. Early cancer detection might help to treat and increase survival rate for patients. WHO finds that there are the effective diagnostic methods such as X-ray, Clinical Breast Exam but it needs to have the professional physicians or experts. In fact, the diagnostic result is not always 100% accuracy because of some reasons such as subjective experiments, expertise, emotional state. There are several applications of computer vision for Computer-Aided Diagnosis (CADx) have been proposed and implemented [6,7]. The breast cancer can be diagnosed via histopathological microscopy imaging, for which image analysis can aid physicians and technical expert effectively [7,12].

Moreover, the CADx system for breast cancer diagnosis is still challenging until now due to the complexity of the histopathological images. In the last decade, many works have been proposed to enhance the recognition performance of breast cancer image. They can be categorized into three groups:

H. A. Le Thi et al. (Eds.): ICCSAMA 2019, AISC 1121, pp. 255–266, 2020.
https://doi.org/10.1007/978-3-030-38364-0_23

- **Handcrafted-feature or deep feature**: Spanhol and Badejo [3,28] compare several handcrafted features extracted from Local Binary Patterns, Local Phase Quantization, Gray Level Co-Occurrence Matrices, Free Threshold Adjacency Statistic, Oriented FAST and Rotated BRIEF based on 1-NN, SVM and Random forest classifiers. Alom et al. [2] combine the strength of Inception, ResNet and Recurrent Convolutional Neural Network with and without augmentation for 4 magnification factors. Zhang et al. [34] propose a method to use skip connection in Resnet in order to solve the optimization issues when network becomes deeper. Roy et al. [21] propose a patch-based classifier using CNN network consisting of 6CONV-5POOL-3FC.
- **Transfer learning approach**: Weiss et al. [32] evaluate different features extracted from VGG, ResNet and Xception with a limited training samples and achieved a good result in the state-of-the-art on BACH dataset. This method downsized BACH image into 1024×768 in order to build the classification model. Vo et al. [31] apply the augmentation techniques as rotate, cut, transform image to increase the training data before extracting deep feature from Inception-ResNet-v2 model in order to avoid the over-fitting. Vo trained the model with multi-scale input images 600×600, 450×450, 300×300 to extract local and global feature. Then Gradient Boosting Trees model again was trained to detect breast cancer. Fusion model will vote the higher accuracy classifier. The accuracy rate archived to 93.8%–96.9% at low cost computation. Murtaza et al. [18] use Alexnet as feature extraction hierarchical classification model by combination of 6 classifiers to reduce the feature space and increase the performance.
- **Generative Adversarial Network (GAN) method**: Shin et al. [24] apply Image-to-Image Conditional GAN mode (pix2pix) to generate synthesis data and discriminate T1 brain tumor class on ADNI dataset. They then use this model on other dataset namely, BRATS to classify T1 brain tumor. This GAN model can increase accuracy compared to train on the real image dataset. Iqbal et al. [8] propose a new GAN model for Medical Imaging (MI-GAN) to generate synthetic retinal vessel images for STARE and DRIVE dataset. This method generated precise segmented image better than existing techniques. Author declared that synthetic image contained the content and structure from original images. Senaras et al. [22] employ a conditional GAN (cGAN) to generate synthetic histopathological breast cancer images. Six readers (three pathologists and image analysts) tried to differentiate 15 real from 15 synthetic images and the probability that average reader would be able to correctly classify an image as synthetic or real more than 50% of the time was only 44.7%. Mahapatra et al [15] propose a P-GANs network to generate a high-resolution image of defined scaling factors from a low-resolution image.

Both handcrafted and deep feature demonstrate the good cancer detection capability. Various researches combine numerous color features and local texture descriptors to improve the performance [1,16]. Modak et al. [16] did comparative analysis of several multi-biometric fusions consisting levels of feature-mostly feature concatenation, score or rules/algorithms level. Authors statistically analyzed that fusion approach represents many advantages than single mode such as accuracy improvement, noise data and spoof attack reduction, more convenience. Fotso Kamga Guy et al. [1] exploited the powerful transfer-learning technique from popular models such as Alexnet, VGGNet-16, VGGNet-19, GoogleNet and ResNet to design the fusion schema at feature level for satellite images classification. It is said that fusion from many ConvNet layers are better than feature extracted from single layer. Features extracted from CNN network is less effected by different conditions such as edge of view, color space; it is an invariant feature and getting the better generalization. Thus data augmentation methods might affect the accuracy if it is applied inadequately. In order to save low computation cost from scratch, transfer learning technique can be considered to employ in medical field. It needs to be retrained or fine-tuning in some layers so that these networks can detect the cancer features. Furthermore, GAN is the effective data augmentation method in computer vision but GAN training process is still a difficult problem. These method have been investigated intensively for common data and rarely for medical data. To overcome this limitation, we propose a composition method of three techniques to be boosting the breast cancer classification accuracy in a limited training data.

The rest of this paper is organized as follows. Section 2 introduces our proposed approach by combining three methods such as transfer learning, deep learning and GAN. The experimental results are then introduced in Sect. 3. Finally, the conclusion is given in Sect. 4.

2   Proposed Approach

In the recent years, Convolutional Neural Network (CNN) proved as an efficient approach in computer vision and have significantly improved in cancer classification. Both VGG16 and VGG19 are proven to be a good candidate in transfer-learning technique. To get the discriminated benign and malignant from the tumor features, the base networks have to retrained on BreaKHis dataset and then be used as an input for CNN network.

A combination of different feature extraction methods can increase the classification accuracy. This work uses VGG16 network and then both VGG16 & VGG19 to extract the features. The proposed architecture is summarized in Fig. 1 and can be described in the following steps:

- **Input layer**: the input layer has three channels of 256×256 pixels which normalized from RGB patch images.
- **Fine-tuning VGG16 and VGG16 & VGG19 feature extraction**: the first 17 layers of VGG16 and VGG19 has primitive low-level spatial characteristic learned on ImageNet dataset which can be transferred to medical dataset. To later higher convolutional layer, they are trained according to BreaKHis dataset.
- **Batch normalization**: layer to normalize a number of activations in combination layer of VGG16 & VGG19's output layer to reduce overfitting from ImageNet's original weight.

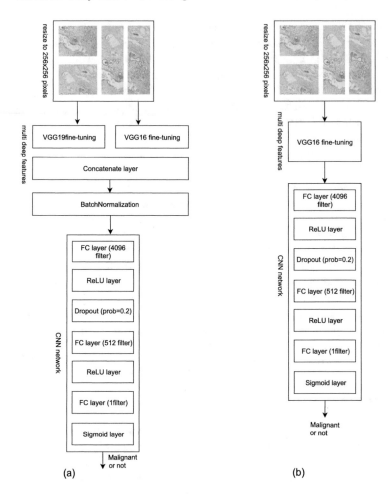

Fig. 1. (a) Fine-tuning VGG16 and CNN, (b) Fine-tuning VGG16 & VGG19 and CNN

- **Full connected layer**: all neurons in this layer have full connections to previous layer's neurons.
- **Rectified Linear Units (ReLU) layer**: ReLU activation layer

$$f(x) = \max(0, x) \tag{1}$$

will output previous layer value if it is positive, otherwise it will output zero. So ReLU layer is used many in deep learning because it helps the network to be trained easily and achieve the better performance.

- **Dropout layer**: is a regularization technique which removes some neurons randomly out network with probability 0.2 during forward or backward propagation process.

– **Output layer**: the layer uses a non-linear activation - sigmoid function.

$$h_\theta(x) = \frac{1}{1 + e^{-\theta^T x}} \qquad (2)$$

Furthermore, three voting methods are applied to compute the model accuracy based on the patch image for two malignant or benign class. We define the so called method A is to select a majority predicted accuracy of the 4 patch images as final result of orginal image. Method B is a similar to A however, if 2 patch images is correctly predicted and 2 patch images is wrongly predicted, the final results of original image will be assigned as correct. Otherwise, method C is defined as at least one patch image is correct, orginal image is predicted as correct.

3   Experiments

3.1   Dataset Description

We propose to evaluate the proposed approach on one real histopathological image database (BreaKHis) and two generated databases from BreaKHis by GAN. The following subsection describes theses datasets.

The BreaKHis Dataset. [28] is a recent benchmark database proposed by Spanhol et al. to study the automated classification problem for breast cancer. This dataset contains 7,909 images (see Fig. 2) of 82 patients using 4 magnifying factors (40×, 100×, 200×, 400×). It is divided into 2 main groups: benign and malignant tumors, 8 sub cancer type as well totally size is 4 GB. It is publicly available from https://web.inf.ufpr.br/vri/databases/breast-cancer-histopathological-database-breakhis

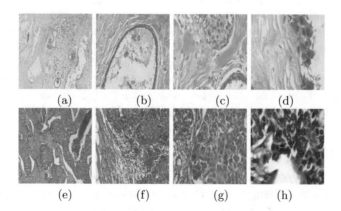

(a)   (b)   (c)   (d)

(e)   (f)   (g)   (h)

Fig. 2. Illustration of BreaKHis database at different magnification factors of benign cell 40× (a), 100× (b), 200× (c), 400× (d) and malignant cell 40× (e), 100× (f), 200× (g), 400× (h).

The Fake BreaKHis. images generated from StyleGAN transfers [11] the style image to input latent space z by using mapping network f to create an immediate feature space w. The adaptive instance normalization (AdaIN) technique is applied to control the style transferred image. We use StylgeGAN to generate the fake benign and malignant image for each scale of $40\times$, $100\times$, $200\times$, $400\times$ (Fig. 3). StyleGAN is trained with 256×256 BreaKHis image for the independent scale and type on a PC with NVIDIA Tesla P100 1GPU during 8 h.

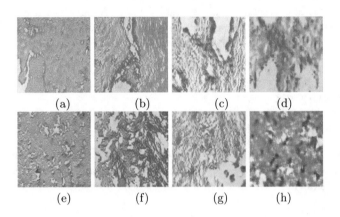

(a)   (b)   (c)   (d)

(e)   (f)   (g)   (h)

Fig. 3. Illustration of generated database by StyleGAN at different magnification factors of benign cell $40\times$ (a), $100\times$ (b), $200\times$ (c), $400\times$ (d) and malignant cell $40\times$ (e), $100\times$ (f), $200\times$ (g), $400\times$ (h).

The Fake BreaKHis. generated by Pix2Pix which is a conditional GAN network proposed by Isola et al. [9]. This framework applies U-Net model and skip connector technique as proposed generator network and discriminator architecture from PatchGAN to penalize structure at patch scale. To synthesize cancer image at each rate, we trained Pix2Pix network by using conditional image as the generated magnification rate image and the rest of magnification rates as input image. Benign $40\times$ rate image will be conditional image and Begnign $100\times$, $200\times$, $400\times$ rate images will be used as input image. Because of complex cancer structure, most of latent space from other magnification rate images can be transferred to the target image and might maintain original feature (Fig. 4).

3.2   Experimental Setup

The accuracy was estimated by a cross validation method through 5 iterations while the ratio of training and testing set ratio of each class are 70% and 30%, respectively. The reason that we choose this ration because it is the most common decomposition (be applied in more than 20 papers) in the literature on BreaKHis dataset. We train the proposed approach with BreaKHis dataset mentioned in a previous section. Firstly, the histopathological image will be divided into 2

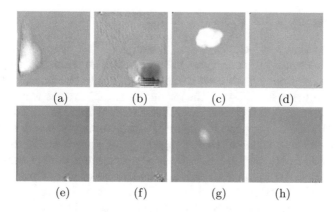

Fig. 4. Magnification factor of fake benign cell 40× (a), 100× (b), 200× (c), 400× (d) and fake malignant cell 40× (e), 100× (f), 200× (g), 400× (h) from Pix2Pix model.

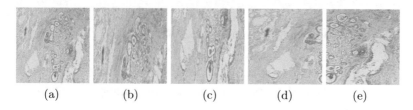

Fig. 5. Magnification factor of 40× benign image (a), top half (b), a bottom half (c), a left half (d) a right half (e)

patches by horizontal (resulting in Fig. 5b and c) and vertical direction (resulting in Fig. 5d and e).

The image patch size is 700×230 pixels in horizontal direction and 350×460 pixels in vertical direction. In stead of extracting small patch size as 32×32 pixels or 64×64 pixels, the approach can keep not only the textural and geometrical features but increase data's complexity and dimension. Most of discriminated features are twice stronger if it is at a central of images. After extracting all patch images needed, image pixel in each channel is normalized to the range of $[0, 1]$ in order to decrease the colored intensive rate. Then patch image is resized to 256×256 pixels, using the bilinear interpolation method. Each image in train comprises the 4 patches of an original image so that our network can learn the multi deep features and increase the performance.

Secondly, the discriminated features extracted from fine-tuning VGG16 and concatenated of fine-tuning VGG16 & VGG19 transfer learning is classified by our novel approach. In this work, all layers before 17th layer of VGG16 & VGG19 is freezed and the rest of layers is re-trained. The loss function is a binary cross-entropy and the Adam optimizer is applied. All experiments are implemented in TensorFlow-GPU version 2 on 16 CPU, 64 GB RAM Tesla P4.

3.3   Results

Table 1 shows that the concatenation of many transfer learning features can increase the recognition accuracy of breast cancer. To train the deep networks efficiently, a large enough dataset is needed so apply the transfer learning is nominated approach nowadays. This technique shared the low feature space but have many differences about textural and geometrical features between ImageNet and BreaKHis. So our approach suggest to train some top layers of VGG16 & VGG19 network and achieved the averaged accuracy from 91.7% to 95.0%.

Both of evaluation method B & C get the average accuracy from 94.9% to 99.2% which can be applied to quickly detect the cancer if patients present any potential signs before doing many costly medical examinations. In order to compare our results, we carefully select the works (Table 1) in the state-of-the-art with the same decomposition and experimental condition. We can observe that the proposed approach clearly outperforms all the previous works. Additionally, the local image descriptors based approach does not give a good results compared with deep learning based method. Our work is "a plus" since we apply GAN to generate more medical images and apply deep learning method to classify images (Table 2).

Table 1. The experimental results of two proposed approaches on BreaKHis dataset.

Model	Evaluation method	40×	100×	200×	400×	Average
VGG16 ft + CNN	Method C	97.5±1.6	98.3±0.8	97.3±1.7	96.8±1.5	97.5±1.4
	Method B	95.0±1.5	95.6±1.8	95.4±1.8	94.0±1.6	95.0±1.6
	Method A	91.6±2.4	92.2±2.6	92.7±2.2	89.6±2.2	91.6±2.2
VGG16 ft + CNN + StyleGAN	Method C	97.3±1.3	98.0±1.3	97.4±1.2	95.3±2.1	97.1±1.4
	Method B	94.7±1.9	95.7±1.9	95.0±1.9	93.0±2.4	94.6±1.9
	Method A	90.9±2.0	92.0±2.0	92.2±1.7	89.2±1.5	91.1±1.7
VGG16 ft + CNN + Pix2Pix	Method C	97.5±1.5	98.5±1.0	97.4±2.1	95.3±1.5	97.2±1.5
	Method B	94.9±2.9	96.2±1.6	95.4±2.2	92.8±1.9	94.9±2.1
	Method A	91.4±3.4	92.9±1.8	92.8±2.5	89.3±1.9	91.7±2.3
VGG16 &VGG19 ft + CNN	Method C	99.2±1.0	99.5±0.6	99.2±1.1	99.1±1.3	99.2±1.0
	Method B	98.2±1.6	98.3±1.3	98.2±1.3	97.5±2.1	98.1±1.5
	Method A	95.1±3.0	95.2±2.4	95.2±1.7	94.6±2.9	95.0±2.4
VGG16 &VGG19 ft + CNN + StyleGAN	Method C	98.6±0.8	99.0±1.3	99.0±1.0	98.1±1.8	98.7±1.2
	Method B	96.7±0.8	97.9±1.8	97.8±1.9	96.1±2.5	97.1±2.0
	Method A	93.5±3.2	95.2±3.0	94.4±2.7	92.6±3.5	94.0±3.0
VGG16 &VGG19 ft + CNN + Pix2Pix	Method C	98.8±1.4	98.8±1.4	98.7±1.6	97.8±1.7	98.6±1.5
	Method B	97.0±2.6	97.3±2.3	97.3±2.0	95.5±2.0	96.8±2.2
	Method A	93.8±3.4	94.4±3.1	94.2±2.7	91.8±2.8	93.6±2.9

Table 2. Comparison of the proposed approach with previous works in the state-of-the-art on BreaKHis dataset.

Ref,Year	Method	40x	100x	200x	400x	2 classes
[2] 2019	IRRCNN + augmentation	97.9	97.5	97.3	97.4	-
[31] 2019	Inception & Boosting & Fusion	95.1	96.3	96.9	93.8	-
[34] 2019	ResNet50 + CBAM	91.2	91.7	92.6	88.9	-
[23] 2018	VGG16 (finetuning) + LR	-	-	-	-	91.7
[19] 2018	Active learning	89.4	90.9	91.6	90.4	-
[25] 2018	CSE (Fish vector)	87.5	88.6	85.5	85.0	-
[26] 2017	Intra-embedding algorithm	87.7	87.6	86.5	83.9	-
[30] 2019	Non parametric	87.8	85.6	80.8	82.9	-
[27] 2017	DeCaf feature	84.6	84.8	84.2	81.6	-
[29] 2016	CNN	85.6	83.5	83.1	80.8	-
[28] 2016	PFTAS	83.8	82.1	85.1	82.3	-
[18]2019	BMIC Net	-	-	-	-	95.5
[5] 2018	DMAE	89.8	88.0	91.5	89.2	-
[10] 2018	MVPNet+NuView data	-	-	-	-	92.2
[3] 2018	Texture Descriptor	91.1	90.7	87.2	87	-
[13] 2018	CNN	82.0	86.2	84.6	84.0	-
[17] 2018	PCANet	96.1	97.4	90.9	85.9	-
[14] 2018	Multi-task deep learning	94.8	94.0	93.8	90.7	-
[4] 2018	Deep VGG16 & Reduction	86.3	84.9	84.7	81.0	-
[33] 2018	Domain Knowledge	-	-	-	-	81.2
[20] 2018	CNN + Over-sampling	-	-	-	-	86.8
Our - A	VGG16 & VGG19 & CNN	95.1	95.2	95.2	94.6	95.0
Our - B	VGG16 & VGG19 & CNN	**98.2**	**98.3**	**98.2**	**97.5**	**98.1**

4   Conclusion

We proposed a composition method of three techniques, transfer learning, deep learning and GAN to be boosting the breast cancer classification accuracy in a limited training dataset. We studied two GAN models such as StyleGAN and Pix2Pix to boost the medical train dataset. At each training iteration, we combine the additional fake images of 4,800 generated StyleGAN and 2,912 generated Pix2Pix images. The experiments show that GAN images created much noise and effected to classification accuracy. Although GAN network can not generate the similar structure as original images but it can synthesize some features from medical images which proved not to be different accuracy. The future of this work is to adjust the U-Net generator in Pix2Pix network to increase a volumes of training set and improve the classification performance.

References

1. Fotso Kamga Guy, A., Akram, T., Laurent, B., Naqvi, S.R., Alex, M.M., Muhammad, N.: A deep heterogeneous feature fusion approach for automatic land-use classification. Inf. Sci. **467**, 199–218 (2018)
2. Alom, M.Z., Yakopcic, C., Nasrin, M.S., Taha, T.M., Asari, V.K.: Breast cancer classification from histopathological images with inception recurrent residual convolutional neural network. J. Digital Imaging, 1–15 (2019)
3. Badejo, J.A., Adetiba, E., Akinrinmade, A., Akanle, M.B.: Medical image classification with hand-designed or machine-designed texture descriptors: a performance evaluation. In: Rojas, I., Ortuño, F. (eds.) Bioinformatics and Biomedical Engineering, vol. 10814, pp. 266–275. Springer, Cham (2018)
4. Cascianelli, S., Bello-Cerezo, R., Bianconi, F., Fravolini, M.L., Belal, M., Palumbo, B., Kather, J.N.: Dimensionality reduction strategies for CNN-based classification of histopathological images. In: De Pietro, G., Gallo, L., Howlett, R.J., Jain, L.C. (eds.) Intelligent Interactive Multimedia Systems and Services 2017, vol. 76, pp. 21–30. Springer, Cham (2018)
5. Feng, Y., Zhang, L., Mo, J.: Deep manifold preserving autoencoder for classifying breast cancer histopathological images. IEEE/ACM Trans. Comput. Biol. Bioinf. 1 (2018)
6. Guillén-Rondon, P., Robinson, M., Ebalunode, J.: Breast cancer classification: a deep learning approach for digital pathology. In: Meneses, E., Castro, H., Hernández, C.J.B., Ramos-Pollan, R. (eds.) High Performance Computing, vol. 979, pp. 33–40. Springer, Cham (2019)
7. Zilong, H., Tang, J., Wang, Z., Zhang, K., Zhang, L., Sun, Q.: Deep learning for image-based cancer detection and diagnosis - a survey. Pattern Recogn. **83**, 134–149 (2018)
8. Iqbal, T., Ali, H.: Generative adversarial network for medical images (MI-GAN). J. Med. Syst. **42**(11), 231 (2018)
9. Isola, P., Zhu, J.-Y., Zhou, T., Efros, A.A.: Image-to-Image Translation with Conditional Adversarial Networks. arXiv:1611.07004 [cs]
10. Jonnalagedda, P., Schmolze, D., Bhanu, B.: [Regular Paper] MVPNets: multiviewing path deep learning neural networks for magnification invariant diagnosis in breast cancer. In: 2018 IEEE 18th International Conference on Bioinformatics and Bioengineering (BIBE), pp. 189–194. IEEE, Taichung, October 2018
11. Karras, T., Laine, S., Aila, T.: A Style-Based Generator Architecture for Generative Adversarial Networks. arXiv:1812.04948 [cs, stat], December 2018
12. Komura, D., Ishikawa, S.: Machine learning methods for histopathological image analysis. Comput. Struct. Biotechnol. J. **16**, 34–42 (2018)
13. Kumar, K., Rao, A.C.S.: Breast cancer classification of image using convolutional neural network. In: 2018 4th International Conference on Recent Advances in Information Technology (RAIT), pp. 1–6. IEEE, Dhanbad, March 2018
14. Li, L., Pan, X., Yang, H., Liu, Z., He, Y., Li, Z., Fan, Y., Cao, Z., Zhang, L.: Multi-task deep learning for fine-grained classification and grading in breast cancer histopathological images. In: Multimedia Tools and Applications, December 2018
15. Mahapatra, D., Bozorgtabar, B., Garnavi, R.: Image super-resolution using progressive generative adversarial networks for medical image analysis. Comput. Med. Imaging Graph. **71**, 30–39 (2019)
16. Modak, S.K.S., Jha, V.K.: Multibiometric fusion strategy and its applications: a review. Inf. Fusion **49**, 174–204 (2019)

17. Mukkamala, R., Neeraja, P.S., Pamidi, S., Babu, T., Singh, T.: Deep PCANet framework for the binary categorization of breast histopathology images. In: 2018 International Conference on Advances in Computing, Communications and Informatics (ICACCI), pp. 105–110. IEEE, Bangalore, September 2018
18. Murtaza, G., Shuib, L., Mujtaba, G., Raza, G.: Breast cancer multi-classification through deep neural network and hierarchical classification approach. Multimed. Tools Appl. 1–31 (2019)
19. Qi, Q., Li, Y., Wang, J., Zheng, H., Huang, Y., Ding, X., Rohde, G.: Label-efficient breast cancer histopathological image classification. IEEE J. Biomed. Health Inf. **23**, 1 (2018)
20. Reza, M.S., Ma, J.: Imbalanced histopathological breast cancer image classification with convolutional neural network. In: 2018 14th IEEE International Conference on Signal Processing (ICSP), pp. 619–624. IEEE, Beijing, China, August 2018
21. Roy, K., Banik, D., Bhattacharjee, D., Nasipuri, M.: Patch-based system for classification of breast histology images using deep learning. Comput. Med. Imaging Graph. **71**, 90–103 (2019)
22. Senaras, C., Niazi, M.K.K., Sahiner, B., Pennell, M.P., Tozbikian, G., Lozanski, G., Gurcan, M.N.: Optimized generation of high-resolution phantom images using cGAN: application to quantification of Ki67 breast cancer images. PLoS ONE **13**(5), e0196846 (2018)
23. Shallu, Mehra, R.: Breast cancer histology images classification: training from scratch or transfer learning? ICT Express **4**(4), 247–254 (2018)
24. Shin, H.-C., Tenenholtz, N.A., Rogers, J.K., Schwarz, C.G., Senjem, M.L., Gunter, J.L., Andriole, K.P., Michalski, M.: Medical image synthesis for data augmentation and anonymization using generative adversarial networks. In: Gooya, A., Goksel, O., Oguz, I., Burgos, N. (eds.) Simulation and Synthesis in Medical Imaging, vol. 11037, pp. 1–11. Springer, Cham (2018)
25. Song, Y., Chang, H., Gao, Y., Liu, S., Zhang, D., Yao, J., Chrzanowski, W., Cai, W.: Feature learning with component selective encoding for histopathology image classification. In: 2018 IEEE 15th International Symposium on Biomedical Imaging (ISBI 2018) (2018)
26. Song, Y., Chang, H., Huang, H., Cai, W.: Supervised Intra-embedding of fisher vectors for histopathology image classification. In: Han, Y. (ed.) Physics and Engineering of Metallic Materials, vol. 217, pp. 99–106. Springer, Singapore (2017)
27. Spanhol, F.A., Oliveira, L.S., Cavalin, P.R., Petitjean, C., Heutte, L.: Deep features for breast cancer histopathological image classification. In: 2017 IEEE International Conference on Systems, Man, and Cybernetics (SMC), pp. 1868–1873. IEEE, Banff, AB, October 2017
28. Spanhol, F.A., Oliveira, L.S., Petitjean, C., Heutte, L.: A dataset for breast cancer histopathological image classification. IEEE Trans. Biomed. Eng. **63**(7), 1455–1462 (2016)
29. Spanhol, F.A., Oliveira, L.S., Petitjean, C., Heutte, L.: Breast cancer histopathological image classification using Convolutional Neural Networks. In: 2016 International Joint Conference on Neural Networks (IJCNN), pp. 2560–2567. IEEE, Vancouver, BC, Canada, July 2016
30. Sudharshan, P.J., Petitjean, C., Spanhol, F., Oliveira, L.E., Heutte, L., Honeine, P.: Multiple instance learning for histopathological breast cancer image classification. Expert Syst. Appl. **117**, 103–111 (2019)
31. Vo, D.M., Nguyen, N.-Q., Lee, S.-W.: Classification of breast cancer histology images using incremental boosting convolution networks. Inf. Sci. **482**, 123–138 (2019)

32. Weiss, N., Kost, H., Homeyer, A.: Towards interactive breast tumor classification using transfer learning. In: Campilho, A., Karray, F., Romeny, B.H. (eds.) Image Analysis and Recognition, vol. 10882, pp. 727–736. Springer, Cham (2018)
33. Zhang, G., Xiao, M., Huang,Y.-H.: Histopathological image recognition with domain knowledge based deep features. In: Huang, D.-S., Gromiha, M.M., Han, K., Hussain, A. (eds.), Intelligent Computing Methodologies, vol. 10956, pp. 349–359. Springer, Cham (2018)
34. Zhang, X., Zhang, Y., Qian, B., Liu, X., Li, X., Wang, X., Yin, C., Lv, X., Song, L., Wang, L.: Classifying breast cancer histopathological images using a robust artificial neural network architecture. In: Rojas, I., Valenzuela, O., Rojas, F., Ortuño, F. (eds.) Bioinformatics and Biomedical Engineering, vol. 11466, pp. 204–215. Springer, Cham (2019)

Attentive biLSTMs for Understanding Students' Learning Experiences

Tran Thi Oanh$^{(\boxtimes)}$ (iD)

International School, Vietnam National University, Hanoi,
144 Xuan Thuy, Cau Giay, Hanoi, Vietnam
oanhtt@isvnu.vn

Abstract. Understanding students' learning experiences on social media is an important task in educational data mining. Since it provides more complete and in-depth insights to help educational managers get necessary information in a timely fashion and make more informed decisions. Current systems still rely on traditional machine learning methods with hand-crafted features. One more challenge is that important information can appear in any position of the posts/sentences. In this paper, we propose an attentive biLSTMs method to deal with these problems. This model utilizes neural attention mechanism with biLSTMs to automatically extract and capture the most critical semantic features in students' posts in regard to the current learning experience. We perform experiments on a Vietnamese benchmark dataset and results indicate that our model achieves state-of-the-art performance on this task. We achieved 63.5% in the micro-average F1 score and 59.7% in the macro-average F1 score for this multi-label prediction.

Keywords: Attention mechanism · biLSTMs · Students' learning experience · Social media

1 Introduction

Students' learning experience refers to the feelings/thoughts of students in the process of getting knowledge or skills from studying in academic environments. It is considered to be one of the most relevant indicator of education quality in schools/universities [17]. Getting to understand this is an effective and important way to improve educational quality in schools/universities.

Learning experiences can vary dramatically for students. To determine students' learning experiences, the widespread used methods is to undertake a number of surveys, direct interviews or observations that provide important opportunities for educators to obtain student feedback and identify key areas for action. Unfortunately, these traditional methods usually cost time, thus cannot be frequently repeated. Moreover, they also raise the question of accuracy and validity of data collected because they do not accurately reflect on what students were thinking or doing something at the time the problems/issues happened.

© Springer Nature Switzerland AG 2020
H. A. Le Thi et al. (Eds.): ICCSAMA 2019, AISC 1121, pp. 267–278, 2020.
https://doi.org/10.1007/978-3-030-38364-0_24

Another drawback is that the selection of the standards of educational practice and student behavior implied in the questions is also criticized in the surveys [5].

Nowadays, social sites such as Facebook, forums, blog, etc. provide great venues for students to express their opinions, concerns and emotions about the learning process. When students post on these sites, they usually write about their feelings/thoughts at that moment. Therefore, the textual data collected from on-line conversations may be more authentic and unfiltered than responses to formal research surveys. These public data sets provide vast amount of insights for educators to understand students' experiences besides the above traditional methods. For mining these datasets, there existed several work done for English using traditional machine learning classifiers with hand-crafted features. Some typical classifiers used in mining various problems in students' learning process are Decision Tree [13], Naive Bayes [6], SVM [8], Memetic [2], etc. In Vietnamese, not much effort has been spent to mine such data so far. Tran and Nguyen [14] presented the first work towards mining social media to get insights from Vietnamese students' posts. They developed a framework using Naive Bayes and Decision Tree to automatically detect students' issues and problems in their study at universities.

Recently, deep neural network approaches provide an effective way of reducing the number of hand crafted features. Specifically, neural networks have been proved to improving the performance of many tasks ranging from question generation [18], machine translations [7], relation classification [19], etc. Hence, in this paper, we propose a novel architecture exploiting a neural network called attention-based biLSTMs for mining students' learning experiences. This model doesn't use any features derived from knowledge resources or Natural Language Processing (NLP) systems. We perform experiments on a benchmark dataset, and achieve 63.5% in the micro-average F1 score and 59.7% in the macro-average F1 score, higher than the existing methods in the literature for this critical task.

The rest of this paper is organized as follows: Sect. 2 presents related work. In Sect. 3, we show a proposed method using attention-based biLSTM to deal the task. Section 4 shows experimental setups, evaluation metrics, experimental results and some findings of this work on a dataset benchmark for Vietnamese. Finally, we summarize the paper in Sect. 5 and discuss some on-going work for the future.

2   Related Work

Social media has risen to be not only a communication media for personal purposes, but also a media to share opinions about products and services or even political issues among its users. Many researches from diverse fields have developed tools to formally represent, measure, model, and mine meaningful patterns (knowledge) from large scale social for the concerned domains. In healthcare, many researches, e.g Sue et al. [12] has shown that social media can be used to reveal lots of health information about its users, or to provide online social support for anyone with health problems [16]. In the marketing field, researchers mine the social data to recommend friends or items (e.g. online courses, videos, beauty product, research papers, search keywords, social tags, and other products in general.) on social media sites, etc.

Recently, research on mining web-based conversations in informal ways on social media (e.g., Facebook, forum, etc.) has started emerging. From these sites there are huge amount of textual data are generated which contain important data about students. There existed many researches proposing different techniques to process such data to better know about students and their learning environments. This information will be valuable to institutions/universities to make informed decisions related to students' learning. For example, Chen et al. [3] firstly provided a framework for analyzing these kind of data using Twitters' posts for educational goals. Takle et al. [13] did a detailed study to make comparison of different classification techniques such as Iterative dichotomiser (ID3), Naive Bayes Multi-label Classifier and Memetic Classifier using common dataset to analyze and get the information related to students in order to enhance the higher education system, etc. Blessy et al. [2] developed a framework to use both qualitative analysis and big data mining techniques using Naive Bayes Multi-label Classifier algorithm and Memetic classifier to categorize tweets presenting students' problems. Pande et al. [8] exploited the SVM method to determine Many issues like stress, suicide, sleepy problems, and anxiety in students' posts. Patil et al. [9] showed that the way students indicate their feelings via social media sites and which posts are in which category using Memetic algorithm. Jessiepriscilla et al. [6] built a sentiment analyzer tool for analyzing tweets which can be used to accomplish the goal of determining the student learning experiences using Navie Bayes multilabel classifier. All of these researches were done using traditional machine learning methods.

While most work has focused on English, a few attempts have been done for Vietnamese so far. Specifically, Tran and Nguyen [14] presented the first work towards mining social media to get insights from engineering students' posts. They developed a framework to automatically detect students' issues and problems in their study at universities. Similar to other work in English, the authors also exploited traditional machine-learning methods which are Naive Bayes and Decision Tree to build the prediction models. This work also contributed the first benchmark dataset on this field in Vietnamese. The experimental results were just the preliminary step and need more effort to enhance the performance of the methods.

As can be seen that, previous work mostly exploited traditional machine learning methods which require hand-crafted features. Designing these features is commonly time-consuming and requires experts' knowledge. Another challenge is that in a post, some words play more important roles in deciding its main meanings. Especially, when one students' post may contain more than one meaning. In recent years, deep neural network methods give us an effective way to make the quantity of hand crafted features less in size. It also does not use extra knowledge and NLP systems. Therefore, this research proposes a novel architecture exploiting attentive biLSTM for the task of mining students' learning experiences on social media. Specifically, we convert the multi-label classification into binary classification problems and then exploit the attentive biLSTM to build the corresponding models for these problems. The effective of the proposed method is verified on a Vietnamese benchmark dataset through extensive experiments.

3   An Attention-Based biLSTM for Understanding Students' Learning Experiences

In formal statement, multi-label learning problem can be seen as the problem of looking for a method that converts inputs x to binary vectors y. These binary vectors are not scalar outputs as in the single-label classification problem. Learning from multi-label classification problem can be solved by transformation techniques. This technique turns the problem into some single-label classification problems. This work uses the techniques called *binary relevance*. Specifically, assume that we have p labels, this method creates p new data sets, one dataset for each label. This binary relevance method then trains single-label classifiers for each of these new data sets. Each single-label classifier only classifies whether or not the current sample belong to the current label **i**?. The multi-label prediction for a new sample is determined by combining the classification results from all of these independent single-label classifiers.

Each of these classifiers will be built using attentive LSTM architecture as illustrated in Fig. 1. This deep neural network is usually very effective to encode sequences of words and is very powerful to learn on data which have long range dependencies. It considers each word in the posts with equal importance weight. The attention mechanism proposed to allow the model to pay attention to more important part of the students' posts. Therefore, this model can automatically concentrate on the important words that have greater impact on the final classification, to record the most important semantic information in each post. This model does not use any extra knowledge and outputs from NLP systems. The overall framework consists of four main layers as follows.

3.1   Word Embeddings Layer

Each students' post consists of n words, $s = \{w_1, w_2, ..., w_n\}$, where w_i is the ith word of the post. Each word in the posts will be converted into a vector x_i using word embedding. Word embedding is one of the most effective representation of post vocabulary nowadays. It has the capability of encoding the context of a word in a post, semantic as well as syntactic similarity, and the relation with other words, etc. In this paper, we use GloVe [10] which is an unsupervised learning algorithm for capturing representations for words in the vector form.

3.2   biLSTM Layer

Let $\mathbf{X} = (\mathbf{x}_1, \mathbf{x}_2, \ldots, \mathbf{x}_n)$ be a students' post consisting of the vector representations of n words in one post. At each location t, the outputs of RNNs express an intermediate representation based on $\mathbf{h}$ - a hidden state:

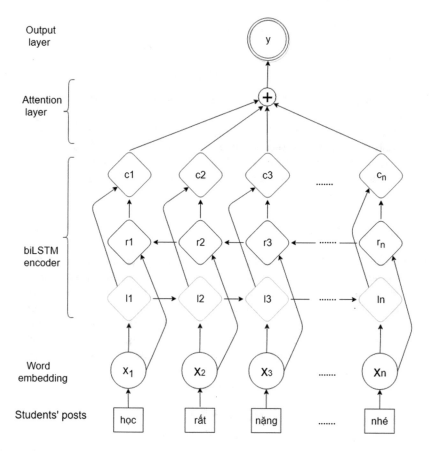

Fig. 1. An attention-based biLSTM for understanding students' learning experiences on social media.

$$\mathbf{y}_t = \sigma(\mathbf{W}_y \mathbf{h}_t + \mathbf{b}_y), \tag{1}$$

where $\mathbf{W}_y$ and $\mathbf{b}_y$ denote parameter matrix and vector. These are determined in the training process, σ denote the element-wise Softmax function. The hidden state $\mathbf{h}_t$ is updated using an activation function. It is a function of the previous hidden state $\mathbf{h}_{t-1}$ and the current input $\mathbf{x}_t$ as follows:

$$\mathbf{h}_t = f(\mathbf{h}_{t-1}, \mathbf{x}_t). \tag{2}$$

LSTM cells exploit a few gates to update the hidden state $\mathbf{h}_t$. These gates include an input gate $\mathbf{i}_t$, a forget gate $\mathbf{f}_t$, an output gate $\mathbf{o}_t$ and a memory cell $\mathbf{c}_t$. The update formula is given below:

$$\mathbf{i}_t = \sigma(\mathbf{W}_i \mathbf{x}_t + \mathbf{V}_i \mathbf{h}_{t-1} + \mathbf{b}_i), \tag{3}$$

$$\mathbf{f}_t = \sigma(\mathbf{W}_f \mathbf{x}_t + \mathbf{V}_f \mathbf{h}_{t-1} + \mathbf{b}_f), \tag{4}$$

$$o_t = \sigma(\mathbf{W}_o\mathbf{x}_t + \mathbf{V}_o\mathbf{h}_{t-1} + \mathbf{b}_o), \tag{5}$$

$$c_t = \mathbf{f}_t \odot \mathbf{c}_{t-1} + \mathbf{i}_t \odot tanh\ (\mathbf{W}_c\mathbf{x}_t + \mathbf{V}_c\mathbf{h}_{t-1} + \mathbf{b}_c), \tag{6}$$

$$\mathbf{h}_t = \mathbf{o}_t \odot tanh\ (\mathbf{c}_t), \tag{7}$$

where $\odot$ multiplication operator functions, $\mathbf{V}$ is a weight matrice, and $\mathbf{b}$ is vectors to be learned.

To improve the model performance, two LSTMs are trained on user utterances. The first on the utterance from left-to-right (l_i) and the second on a reversed copy of the utterance (r_i). The forward and backward outputs, l_i and r_i, should be combined into c_i by concatenation by default before being passed on to the next layer.

3.3   Attention Layer

Let $\mathbf{H}$ denote a matrix including output vectors $[h_1, h_2, ..., h_n]$ that biLSTMs layer produced, where n is the post length. You can just take the straight average these vectors and feed that to your classifier. But it is also true that not all of this information will be equally important. That is why we need attention to tell us which words are less important and which words are the most important. We will train a little neural network from $\mathbf{H}$ to vote on how important each word is. Let r be the representation of the post. r is created by a weighted sum of the output vectors as follows:

$$M = tanh(H) \tag{8}$$

$$\alpha = softmax(w^n M) \tag{9}$$

$$r = H\alpha^T \tag{10}$$

where w is a trained parameter vector and w^T is a transpose. A little alpha here tells you how important the cell is then you do the weighted sum and feed that into your classifier.

We get the last post-pair representation which will be used to classify as follows:

$$h^* = tanh(r) \tag{11}$$

3.4   Output Layer

This work exploits a softmax classifier[1] to guess the label y^* from a pre-defined set of classes Y for a student's post s. The model gets the hidden state h^* as input:

$$p(y|s) = softmax(W^{(s)}h^* + b^{(s)}) \tag{12}$$

$$y^* = argmax_y(p(y|s)) \tag{13}$$

[1] Instead of using this softmax function, you can also use the sigmoid function as an alternative. In fact, in the binary classification both sigmoid and softmax functions are the same where as in the multi-class classification softmax function is preferred.

4 Experiments

This section first presents about the dataset used to conduct experiments. Typical evaluation metrics are also described to estimate the effectiveness of the proposed method. Then, the detailed configuration to set up experiments is shown. Finally, this section expresses experimental results on this dataset.

4.1 Dataset

Data were collected from a forum of a famous university in Vietnam. The dataset contains 1834 posts relating to students' learning experiences of an information technology university. In this dataset, one post can fall into one or multiple categories. There are seven categories which are also the main problems/issues that students often meet in their studying process at the university. Figure 2 gives a description of the number of instances per labels in our dataset.

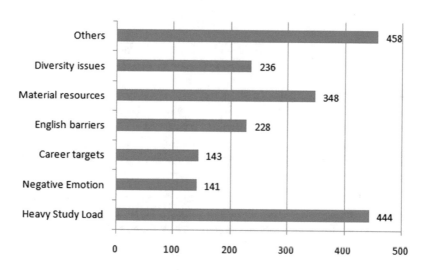

Fig. 2. Number of posts in each category of the dataset analyzed.

4.2 Evaluation Metrics

The evaluation metrics for the multi-class classification is slightly different with metrics for single-label task. In multi-label classification, a misclassification is not a hard wrong or right. A predicted set of labels which includes a subset of the gold classes should be considered better than a predicted set that does not contains any gold class. In this paper, we report both settings to evaluate the performance of the method.

In this situation, researchers [4] proposed two types of metrics which are example-based measures and label-based measures.

Example-Based Measures. These measures are calculated based on examples (in this case each post is considered as an example) and then averaged over all posts in the dataset. Suppose that we are classifying a certain post p, the gold (true) set of labels that p falls into is G, and the predicted set of labeled by the classifier is P, the example-based evaluation metrics are calculated as follows:

$$Acc = \frac{1}{M} \sum_{i=1}^{M} \frac{G_i \cap P_i}{G_i \cup P_i}$$

$$Prec = \frac{1}{M} \sum_{i=1}^{M} \frac{G_i \cap P_i}{P_i}$$

$$Rec = \frac{1}{M} \sum_{i=1}^{M} \frac{G_i \cap P_i}{G_i}$$

$$F1 = \frac{1}{M} \sum_{i=1}^{M} \frac{2 * Precision_i * Recall_i}{Precision_i + Recall_i}$$

here M is the number of posts in the corpus.

There are two more commonly used measures to estimate the effectiveness of multi-labeled classification which are micro-average F1 and macro-average F1. The former gives the same weight to each classification decision per post, while the latter gives the same weight to each label. They are variants of F1 used in different situation.

Label-Based Measures. These measures are measured on each label and then get averaged values over all labels in the dataset. Specifically, metrics of recall, precision, and F1 for each label l is calculated as follows:

$$F1 = \frac{2 * P * R}{P + R}$$

$$P = \frac{TP}{TP + FP}$$

$$R = \frac{TP}{TP + FN}$$

where TP is the number of posts that are correctly detected as the currently-considered label l. FP is the number of posts belonging to l but mis-identified to another label. FN is the number of posts of l but not recognized by the models.

Table 1. Experimental results of detecting students' learning experiences using example-based metrics

Methods	Accuracy	Precision	Recall	F1 micro	F1 macro
Decision Tree	0.565	0.548	0.571	0.583	0.558
Attentive LSTM	**0.612**	**0.587**	**0.629**	**0.635**	**0.597**

4.3 Experimental Setups

The model was implemented in *Python* programming language with several typical libraries such as *PyTorch, numpy, sklearn, utils*, etc. These libraries provide rich tools and options to support developments in NLP and many other research fields.

To create pre-trained embeddings of words, we gathered the raw data from Vietnamese newspapers ($\approx 7\,\text{GB}$ texts) to train the vector of word model using Glove[2]. The quantity of word embedding dimensions was fixed at 50.

For each label, we created a corresponding dataset which only focuses on the currently-considered label. On this dataset, we performed 5-fold cross-validation tests to evaluate the performance of the proposed attentive biLSTMs-based model on this dataset. The parameters were chosen by using the development set. We randomly select 10% of the training data as the development set. To detect students' learning experiences, we set the quantity of epochs equals 100, the batch size as 20, early stopping as True with 4-epoch patience, the rate of dropout at 0.5.

4.4 Experimental Results

In this paper, we compare the performance of the proposed model with the best results of previous work on this same dataset. The best performance of previous work is using Decision Tree method [14] in the same binary relevance setting. In that work, Tran et al. exploited C4.5 (J48). This algorithm is used to build a decision tree proposed by Ross Quinlan [11]. C4.5 begins with big sets of cases of known classes. These cases are represented by any mixture of properties both in nominal and numeric forms. The cases are carefully examined for patterns which allow the classes to be reliably discriminated. These patterns are then indicated as models that can be later used for classifying new unseen cases. The patterns emphasize on the ability of the models to be understandable as well as accurate. This C4.5 was ranked top #1 in the best 10 data mining algorithms published by Springer LNCS in 2008 [15]. Using this method, the baseline model achieved 58.3% in the micro-average F1 score, and 55.8% in the macro-average F1 score.

Table 1 showed experimental results of the baseline and the proposed method using example-based metrics. It should be noted that the higher the evaluation metrics, the better the performance of the models. As can be seen that the

---

[2] https://github.com/standfordnlp/GloVe.

attentive biLSTM model significantly boosted the performance of this task. It achieved the better results by around 4% on all metrics of accuracy, recall, precision, macro-average F1 and micro-average F1 scores. Specifically, the F1-micro score increased by 5.2% and the F1-macro score increased by 3.9%. This result suggested that the attention mechanism has significant effects on mining students' learning experiences in social media. In reality, it is quite effective in helping the model focus down on the words that are the most useful for classification of students' learning experiences.

Table 2. Experimental results of the attentive-based biLSTMs for detecting students' learning experiences using label-based metrics

	Study load	Negative emotion	Carrier targets	English barriers	Others	Material resources	Diversity issues
Precision	0.832	0.900	0.928	0.948	0.788	0.905	0.919
Recall	0.775	0.923	0.933	0.949	0.792	0.892	0.922
F1	0.788	0.910	0.921	0.944	0.776	0.895	0.914

Table 2 showed the performance of the attention-based biLSTMs method on each label using label-based metrics. We can see that the attentive biLSTM model yielded quite high scores. Most labels such as *Negative Emotion, English Barriers, Carrier Targets,* and *Diversity Issues* got more than 90% in the F1 score. *Material Resources* label got 89.5% in the F1 score. For the remaining two labels, *Heavy Study Load* and *Others,* the proposed method achieved around 78% in the F1 score. This result is quite promising due to the ambiguity problem in predicting these labels. Observing their samples in the dataset, we saw that these samples have a large overlap with the remaining labels. The model, therefore, is easy to make mistakes in prediction.

5   Conclusion

This paper presented a new approach to the task of determining students' learning experiences on social media. The previous systems still relied on traditional methods with manually-designed features. Building these features takes time and experts knowledge. One more challenge is that not all words in one post have the same important weight to the final prediction of the model. Therefore, this paper proposed an attention-based biLSTMs to solve these problems. This model utilizes neural attention mechanism with biLSTMs to automatically extract and capture the most critical semantic features in students' posts. We perform experiments on a Vietnamese benchmark dataset and experimental results express that the model achieves SOTA performance on this task for Vietnamese. The proposed method improves the performance by a large margin of 4% in terms of F1-micro score. It achieved 63.5% in the micro-average F1 score, and 59.7% in the macro-average F1 score.

This result is quite promising and could provide more complete and in-depth insights to help educational managers get necessary information in a timely fashion and make more informed decisions. In the future, we would like to exploit another deep neural network architecture to build a multi-label classifier in considering all the labels of each post in dependency when training the models. Another direction is to investigate more linguistic features to enrich the prediction models using external resources.

References

1. Aswini, M.S., Krishnamoorthy, I.: Social media mining to analyse students' learning experience. Int. J. Comput. Sci. Mob. Comput. **5**(2), 213–217 (2016)
2. Blessy, G.V.M., Prasanna, S.: Mining social networks for analyzing students learning experience and their problems. Int. J. Eng. Technol. (IJET) **8**(2), 1271–1274 (2016)
3. Chen, X., Vorvoreanu, M., Madhavan, K.: Mining social media data for understanding students' learning experiences. IEEE Trans. Learn. Technol. **7**(3), 246–259 (2014)
4. David, M.W.P.: Evaluation: from precision, recall and F-Factor to ROC, informedness, markedness & correlation. J. Mach. Learn. Technol. **2**(1), 37–63 (2011)
5. Gordon, J., Ludlum, J., Hoey, J.J.: Validating the NSSE against student outcomes: are they related? Res. High. Educ. **2008**(49), 19–39 (2008)
6. Jessiepriscilla, A., Kalaivani, V.: Analyzing social media data for understanding students learning experiences and predicting their psychological pressure. Int. J. Pure Appl. Math. **118**(7), 513–521 (2018)
7. Maruf, S., Martins, A.F.T., Haffari, G.: Selective attention for context-aware neural machine translation. In: Proceedings of NAACL-HLT 2019, 2 June–7 June 2019, Minneapolis, Minnesota, pp. 3092–3102 (2019)
8. Pande, A., Kinariwala, S.A.: Analysis of student learning experience by mining social media data. Int. J. Eng. Sci. Comput. **7**(5), 12215–12220 (2017)
9. Patil, S., Kulkarni, S.: Mining social media data for understanding students' learning experiences using memetic algorithm. Mater. Today **5**(1), pp. 693–699 (2018). Part 1
10. Pennington, J., Socher, R., Manning, C.D.: Proceedings of the 2014 Conference on Empirical Methods in Natural Language Processing (EMNLP), pp. 1532–1543 (2014)
11. Quinlan, J.R.: C4.5: Programs for Machine Learning. Morgan Kaufmann Publishers Inc., Burlington (1993). ISBN 1558602402. 1993
12. Sue, J.P., Linehan, C., Daley, L., Garbett, A., Lawson, S.: "I can't get no sleep": discussing #insomnia on Twitter. In: The Proceedings of the SIGCHI Conference on Human Factors in Computing Systems, USA, pp. 1501–1510 (2012). https://doi.org/10.1145/2207676.2208612
13. Takle, P.R., Gawai, N.: Interpreting students behavior using opinion mining. Int. J. Innov. Res. Comput. Commun. Eng. (An ISO 3297: 2007 Certified Organization) **3**(10), 9410–9419 (2015)
14. Tran, O.T., Thanh, N.V.: Understanding students' learning experiences through mining user-generated contents on social media. J. VNU Sci.: Policy Manag. Stud. **33**(2), 124–133 (2017)

15. Wu, X., Kumar, V., Ross Quinlan, J., et al.: Top 10 algorithms in data mining. Knowl. Inf. Syst. **14**(1), 1–37 (2008). https://doi.org/10.1007/s10115-007-0114-2
16. Yu, B.: The emotional world of health online communities. In: Proceedings of iConference 2011, pp. 806–807 (2011)
17. Zerihun, Z., Beishuizen, J., Van Os, W.: Student learning experience as indicator of teaching quality. Educ. Assess. Eval. Accountability. **24**(2), 99–111 (2012). https://doi.org/10.1007/s11092-011-9140-4
18. Zhao, Y., Ni, X., Ding, Y., Ke, Q.: Paragraph-level neural question generation with maxout pointer and gated self-attention networks. In: Proceedings of the 2018 Conference on Empirical Methods in Natural Language Processing, pp. 3901–3910 (2018)
19. Zhou, P., Shi, W., Tian, J., Qi, Z., Li, B., Hao, H., Xu, B.: Attention-based LSTM for aspect-level sentiment classification. In: Proceedings of the 54th Annual Meeting of the Association for Computational Linguistics, pp. 207–212 (2016)

Computing Residual Diffusivity by Adaptive Basis Learning via Super-Resolution Deep Neural Networks

Jiancheng Lyu$^{(\boxtimes)}$, Jack Xin, and Yifeng Yu

UC Irvine, Irvine, CA 92697, USA
{jianchel,jack.xin,yifengy}@uci.edu

Abstract. It is expensive to compute residual diffusivity in chaotic incompressible flows by solving advection-diffusion equation due to the formation of sharp internal layers in the advection dominated regime. Proper orthogonal decomposition (POD) is a classical method to construct a small number of adaptive orthogonal basis vectors for low cost computation based on snapshots of fully resolved solutions at a particular molecular diffusivity D_0^*. The quality of POD basis deteriorates if it is applied to $D_0 \ll D_0^*$. To improve POD, we adapt a super-resolution generative adversarial deep neural network (SRGAN) to train a nonlinear mapping based on snapshot data at two values of D_0^*. The mapping models the sharpening effect on internal layers as D_0 becomes smaller. We show through numerical experiments that after applying such a mapping to snapshots, the prediction accuracy of residual diffusivity improves considerably that of the standard POD.

Keywords: Advection dominated diffusion · Residual diffusivity · Adaptive basis learning · Super-resolution deep neural networks

1 Introduction

It has been a fundamental problem to characterize the large scale effective diffusion in fluid flows containing complex and turbulent streamlines [17]. In this paper, we consider the passive scalar model [11]:

$$T_t + (\boldsymbol{v} \cdot D)\, T = D_0 \, \Delta T, \tag{1}$$

where T is a scalar function, $D_0 > 0$ is a constant (the so called molecular diffusivity), $\boldsymbol{v}(\boldsymbol{x}, t)$ is a incompressible velocity field, D and Δ are the spatial gradient and Laplacian operators. In two dimension, the effective diffusivity tensor is given by [1]:

$$D_{ij}^E = D_0 \left(\delta_{ij} + \langle Dw_i \cdot Dw_j \rangle \right), \tag{2}$$

© Springer Nature Switzerland AG 2020
H. A. Le Thi et al. (Eds.): ICCSAMA 2019, AISC 1121, pp. 279–290, 2020.
https://doi.org/10.1007/978-3-030-38364-0_25

where $w = (w_1, w_2)$ is the unique mean zero space-time periodic vector solution of the *cell problem* [1]:

$$w_t + (\boldsymbol{v} \cdot Dw) - D_0 \Delta w = -\boldsymbol{v}, \tag{3}$$

and $\langle \cdot \rangle$ denotes space-time average over the periods. The term $\langle Dw_i \cdot Dw_j \rangle$ in (2) is a positive definite correction to $D_0 \delta_{ij}$.

Asymptotic behavior of D_{ij}^E can be solved when the flow is steady and periodic. For instance, the time independent cellular flow [4,5,13,19,20]

$$\boldsymbol{v} = (-H_y, H_x), \quad H = \sin x \sin y,$$

has been proved to generate effective diffusion following the square root law with dominated advection [5,6]:

$$D_{ij}^E = O(\sqrt{D_0}) \gg D_0, \quad D_0 \downarrow 0, \quad \forall i, j.$$

However, when the streamlines are time-dependent or fully **chaotic**, the enhancement can be quite different and difficult to solve analytically. A simple example is

$$\boldsymbol{v} = (\cos(y), \cos(x)) + \theta \cos(t) (\sin(y), \sin(x)), \quad \theta \in (0, 1], \tag{4}$$

where the first term is a steady cellular flow with a $\pi/4$ rotation and is perturbed by the a time-periodic flow. At $\theta = 1$, the flow (4) is fully chaotic [21] and was investigated in the Rayleigh-Bénard experiment [3]. Numerical simulations of the fully chaotic model [2,10,18] suggest that

$$D_{11}^E = O(1), \quad D_0 \downarrow 0, \tag{5}$$

hence the *residual diffusivity* phenomenon emerges.

As in (4)$\boldsymbol{v}$ is periodic in time, the solution of cell problem (3) subject to periodic boundary condition in space can be computed accurately by spectral method in Fourier basis. By expanding both w and $\boldsymbol{v}$ as Fourier series, (3) is equivalent to an ordinary differential equation (ODE) system. To solve the system numerically, one can approximate w using finitely many Fourier modes thereby the problem is reduced to solving for the periodic solution to a linear ODE system. The corresponding Poincaré map is constructed in [10] and the solution is found as the unique fixed point of it. The effective diffusivity D^E can be finally computed by (2). The drawback of the spectral approach is that the number of Fourier modes required by the truncated problem grows rapidly as $D_0 \downarrow 0$ due to the sharp gradient of the solution.

In [10], adaptive orthogonal basis vectors are constructed from snapshots of spectral solutions to handle the near singular solutions of (3) at small D_0. Particularly, at certain D_0 and θ sample, snapshots of the solution in one time period form a solution matrix and the adaptive basis consists of singular vectors for the top singular values of the matrix. Hence the linear ODE system with D_0 or θ that are close to the sampled values can be rewritten in terms of adaptive basis and solved similarly with Poincaré map. The number of adaptive basis functions is one or two hundred, far less than that of Fourier basis which is usually at least a few thousand. With the reduced adaptive basis, the relative error of computing residual diffusivity is no more than 6.5% with carefully selected samples.

The above procedure to generate adaptive basis vectors by taking snapshots of solutions and solving singular value decomposition (SVD) is the so called reduced order modeling [14, 16]. In fluid dynamics literature [7, 9], such technique is also referred to by proper orthogonal decomposition (POD). However, POD method relies heavily on collection of good snapshots data. Snapshots data with less representation capability would hardly recover the fully resolved solution. As shown in the test results of computing residual diffusivity, when snapshots are collected at certain D_0 and the solutions at a much smaller $D_1 \ll D_0$ are to be computed with reduced basis constructed at D_0, the errors can rapidly increase.

In this paper, we study a deep neural network (DNN) approach to alleviate the accuracy loss of POD and improve the error of reduced basis computation at D_1 based on prediction from snapshots at two values D_0^1 and D_0^2 (both above D_1). The idea is to train a mapping from snapshots (images) at D_0^1 to those at D_0^2 ($D_0^2 < D_0^1$). The mapping sharpens the images similar to what happens to solutions of (1) as D_0 becomes smaller. The mapping is applied to snapshots at D_0^2 for improving POD basis construction. As a proof of concept, we select a super-resolution DNN in the form of a generative adversarial network (SRGAN, [8]). We show that it serves our purpose well through numerical experiments where the mapping is constructed (trained) based on snapshots at D_0^1 and D_0^2, then tested (applied) at $D_1 < D_0^2$.

The paper is organized as follows. In Sect. 2, we describe the POD basis construction for (2), SRGAN architecture and its training objective. In Sect. 3, we present computational results on predicted residual diffusivity from POD basis with and without SRGAN. Concluding remarks are in Sect. 4.

2 Construction of Adaptive Basis via DNNs

2.1 Learning Thinner Structures

Consider spectral method for solving the cell problem below for D_{11}^E:

$$w_t + (\boldsymbol{v} \cdot \nabla) w - D_0 \, \Delta w = -v, \qquad (6)$$

where $\boldsymbol{v} = (v, \tilde{v})$, and

$$
\begin{aligned}
v(x, t) &= \cos(x_2) + \sin(x_2) \cos(t), \\
\tilde{v}(x, t) &= \cos(x_1) + \sin(x_1) \cos(t).
\end{aligned}
\qquad (7)
$$

Let w_k^N be the k-th mode of a $(2N+1)^2$ term Fourier approximation of w on the $[0, 2\pi]^2$ periodic domain [10]. Let v_k and $\tilde{v}_k$ be the k-th Fourier modes of v and $\tilde{v}$

respectively. In view of (7), v_k ($\tilde{v}_k$) equals zero unless $k = (0, \pm 1)$ ($k = (\pm 1, 0)$). The truncated ODE system on w_k^N is:

$$\frac{dw_k^N}{dt} + D_0 |k|^2 w_k^N + i \sum_{\|k-j\| \le N} \left[(k_1 - j_1) v_j(t) + (k_2 - j_2) \tilde{v}_j(t)\right] w_{k-j}^N = -v_k(t),$$

$$(8)$$

which reads in vector-matrix form: $d\mathbf{w}/dt = A(t)\mathbf{w} + \mathbf{v}(t)$. For a time discretization with N_t grid points on $[0, 2\pi]$, denote by $\{\hat{\mathbf{w}}_n^*\}_{n=0}^{N_t}$ a numerical time periodic solution to (8) for $D_0 = D_0^*$, a value where snapshots data are collected. Such a solution can be found by solving for the unique fixed point of the Poincaré map [10]. Define the solution matrix of size $(2N + 1)^2 \times N_t$:

$$W = \begin{bmatrix} \hat{\mathbf{w}}_0^* & \hat{\mathbf{w}}_1^* & \cdots & \hat{\mathbf{w}}_{N_t}^* \end{bmatrix}, \qquad (9)$$

and compute the SVD factorization $W = U \Sigma V^T$. Then one extracts columns $\mathbf{u}_j$ ($j = 1, \cdots, m$) of the matrix U corresponding to the largest $m \ll O(N^2)$ singular values, to form the adaptive orthogonal basis vectors and the matrix: $U_m = \begin{bmatrix} \mathbf{u}_1 & \mathbf{u}_2 & \cdots & \mathbf{u}_m \end{bmatrix}$. This is the end of basis training at a sampled value D_0^*. At $D_0 \ne D_0^*$, project $\mathbf{w}(t)$ to the span of column vectors of U_m or seek a vector of the form $U_m \mathbf{a}(t)$, where $\mathbf{a}(t) \in \mathbb{R}^m$ satisfies the ODE system in a much lower dimension (bar is complex conjugate, T is transpose):

$$\frac{d\mathbf{a}}{dt} = \bar{U}_m^T A(t) U_m \mathbf{a} + \bar{U}_m^T \mathbf{v}(t). \qquad (10)$$

Finding the time periodic solution to a time discrete version of (10) via Poincaré map to compute D^E by (2) in Fourier space, we completed the reduced order modeling.

The inverse Fourier transform of W gives the snapshot matrix in the physical domain:

$$S_p = \begin{bmatrix} \mathbf{I}_0^* & \mathbf{I}_1^* & \cdots & \mathbf{I}_{N_t}^* \end{bmatrix} \qquad (11)$$

where each column vector (snapshot) is an image after reshaping into a square matrix. The S_p is convenient for visualization and drawing a connection with image processing. Figs. 1 and 2 illustrate that internal layers in the physical domain snapshots (column vectors of S_p) emerge and get thinner as D_0 becomes smaller. For a better prediction of the nearly singular solutions at D_1 much smaller than D_0^*, it is helpful if the adaptive basis learned at D_0^* encodes certain thinner layered structures. Particularly, given the solution matrix W at D_0^* as (9), we look for a map $\mathcal{M}$ such that the physical domain snapshots (inverse Fourier transform) of $\mathcal{M}(W)$ have sharpened internal layers.

Suppose $D_0^1 > D_0^2$ are two values $\ge D_0^*$, and W^i is the Fourier domain solution matrix at D_0^i for $i = 1, 2$. Let $\mathcal{F}$ be column-wise Fourier transform on matrices, then columns of $\mathcal{F}^{-1}(W^i)$ are physical domain snapshots of W^i. When the W^i's are available, we use DNN to train a map $\mathcal{T}$ for the following regression problem:

$$\mathcal{T} : \mathcal{F}^{-1}(W^1) \to \mathcal{F}^{-1}(W^2). \qquad (12)$$

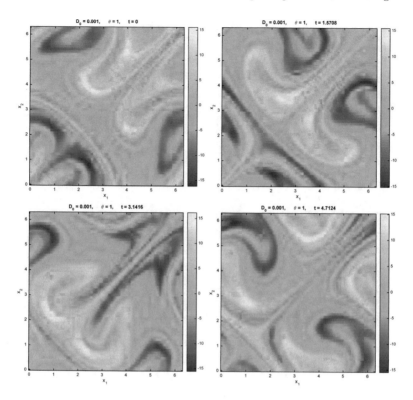

Fig. 1. Sampled snapshots of (6) at $D_0 = 10^{-3}$ with layered structures.

Since $D_0^2 < D_0^1$, $\mathcal{F}^{-1}\left(W^2\right)$ has thinner layered structures than $\mathcal{F}^{-1}\left(W^1\right)$. By solving the regression problem (12) via DNN, our goal is that the $\mathcal{T}\left(\mathcal{F}^{-1}\left(W^1\right)\right)$ inherits the image sharpening capability. Then $\mathcal{T}$ can be applied to solution matrix W^* at $D_0^* \leq D_0^2$ and $\mathcal{T}\left(\mathcal{F}^{-1}\left(W^*\right)\right)$ is expected to have thinner structures for better prediction of residual diffusivity. Finally, the adaptive basis with thinner structures will be obtained from SVD of

$$\mathcal{M}\left(W^*\right) = \mathcal{F}\left(\mathcal{T}\left(\mathcal{F}^{-1}\left(W^*\right)\right)\right).$$

2.2 Adversarial Network

We opt for the super-resolution generative adversarial network (SRGAN) [8] to train the map $\mathcal{T}$. As a generative adversarial network (GAN), SRGAN consists of a generator network G and a discriminator network D. The two networks compete in a way that D is trained to distinguish the real high-resolution (HR) images and those generated from low-resolution (LR) images, while G is trained to create fake HR images from LR images to fool D. We train the SRGAN with $\mathcal{F}^{-1}\left(W^1\right)$ as input data and $\mathcal{F}^{-1}\left(W^2\right)$ as target data so that the generator

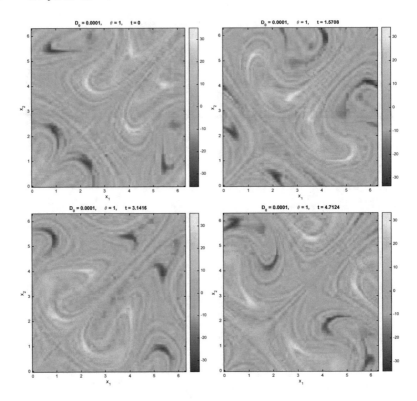

Fig. 2. Sampled snapshots of (6) at $D_0 = 10^{-4}$, formation of thinner layers.

G learn to generate thinner structures when it is fed with $\mathcal{F}^{-1}(W)$. In this approach, we realize $\mathcal{T}$ through a trained G.

The network architecture is shown in Fig. 3. The generator network G starts with a convolutional block with kernel size 9×9, followed by a few residual blocks. Here a convolutional block consists of a convolutional layer and a PReLU layer, a residual block is a convolutional block with kernel size 3×3 followed by a convolutional layer of the same kernel size and a shortcut from the input to output. There are two more convolutional layers with kernel size 3×3 and 9×9 after the residual blocks at the end of the network. The number of filters in all convolutional blocks are the same except for the last one. Note that we remove the two upscale layers in [8] since the snapshot sizes of $\mathcal{F}^{-1}(W^1)$ and $\mathcal{F}^{-1}(W^2)$ are the same.

The discriminator network D is defined by the architectural guidelines summarized in [12], see Fig. 3. It has eight convolutional blocks with PReLU layers replaced by LeakyReLU layers with slope parameter $\alpha = 0.2$. Moreover, there is a batch normalization layer before each LeakyReLU in the convolutional blocks. The kernel size is 3×3 in all convolutional blocks and the number of filters is doubled in the 3rd, 5th and 7th block. Those blocks are followed by a fully connected layer, a LeakyReLU layer and one more fully connected layer. Finally the feature map is fed in a sigmoid layer which gives the probability of real HR image and the reconstructed image.

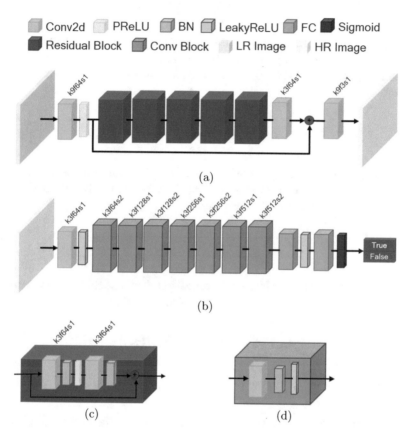

Fig. 3. Architecture of the generator and discriminator networks. (a) The generator network. (b) The discriminator network. (c) Residual block in the generator network. (d) Convolutional block in the discriminator network. We use a simple notation to indicate the Conv2d layer. For example, k9f64s1 indicates a convolutional layer with kernel size 9, number of filters 64 and stride 1.

As a binary classifier, the discriminator network is equipped with the cross entropy loss. Let us focus on the loss function of the generator network. Suppose $\mathcal{F}^{-1}\left(W^1\right)$ and $\mathcal{F}^{-1}\left(W^2\right)$ are real matrices of dimension $(2N+1)^2 \times N_t$, and the columns of $\mathcal{F}^{-1}\left(W^1\right)$ and $\mathcal{F}^{-1}\left(W^2\right)$ are x_i and y_i, $i = 1, 2, \ldots, N_t$. Following the formulation in [8], we define the loss function of the generator network as

$$l\left(G\right) = l_{MSE}\left(G\right) + 10^{-2}\, l_{VGG}\left(G\right) + 10^{-3}\, l_{Gen}\left(G\right). \tag{13}$$

In (13), l_{MSE} is the pixel-wise **MSE loss** defined as the sum of the squares of error at each pixel,

$$l_{MSE}(G) = \sum_{i=1}^{N_t} \|y_i - G(x_i)\|_2^2.$$

The l_{VGG} is the **VGG loss** based on layers of the pre-trained VGG-19 network [15]. Let ϕ be a feature map of VGG-19 and s_ϕ be its size, then the VGG loss is the average of squares of Euclidean distances between the feature representations of y_i and $G(x_i)$

$$l_{VGG}(G) = s_\phi^{-1} \sum_{i=1}^{N_t} \|\phi(y_i) - \phi(G(x_i))\|_2^2.$$

The generator network is expected to fool the discriminator network, so (13) contains l_{Gen} called **generative loss**. The l_{Gen} is defined based on the cross-entropy loss of the discriminator network

$$\sum_{i=1}^{N_t} \log[1 - D(G(x_i))], \tag{14}$$

where $D(G(x_i))$ means the binary classification result of the reconstructed HR image by the generator network G. In practice, we define

$$l_{Gen}(G) = \sum_{i=1}^{N_t} -\log D(G(x_i))$$

for better gradient behavior.

3   Experimental Results of Adaptive Basis from SRGAN

Let $D_0^1 = 10^{-2}$, $D_0^2 = 10^{-3}$. We solved for both W^1 and W^2 via spectral method with $N = 50$ and $N_t = 1500$, then train SRGAN with input data $\mathcal{F}^{-1}(W^1)$ and target data $\mathcal{F}^{-1}(W^2)$. The training of SRGAN includes two stages: (1) We train the generator G for 50 epochs to get a pretrained model; (2) We train the entire SRGAN for 200 epochs. Adam and SGD optimizers are applied to training of G and D respectively. We set batch size to be 32 and learning rate to be 10^{-4} for both optimizers. The training was carried out on a desktop with Nvidia graphics cards GTX 1080 Ti. We set $D_0^* = D_0^2$ in the following experiments.

Figure 4 shows two time slices of the input $\mathcal{F}^{-1}(W^2)$ (top) with $G(\mathcal{F}^{-1}(W^2))$ (bottom) at $D = D_0^2$. Columnwise, it can be seen that thinner layers are created by the network G. Due to identical dimension constraint of the input and output images, the up-scaling layers in SRGAN [8] have been removed. This adaptation lowers the fidelity of the generated images, as we see in each column of Fig. 4. However, the SRGAN generated snapshots are only used to construct reduced basis.

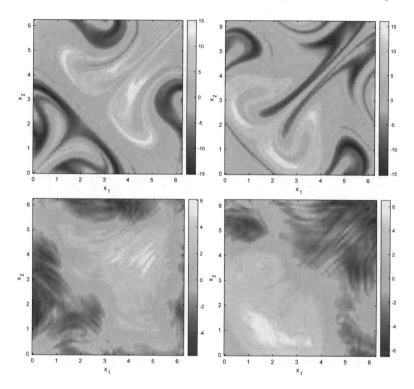

Fig. 4. Input (top) and SRGAN output (bottom) at $D_0^2 = 10^{-3}$.

The fact that sharp layers are generated by SRGAN training is more important for our task of computing residual diffusivity.

Set $D_0^1 = 10^{-2}$ and $D_0^2 = 10^{-3}$, the comparison of predictions of D_{11}^E by SVD and SRGAN assisted SVD is shown in Table 1. The number of adaptive basis is $m = 100$ for both methods.

Table 1. Comparison of $\hat{D}_{11,N}^{E,a}$ for flow (7) with $D_0^1 = 10^{-2}$, $D_0^2 = 10^{-3}$.

D_0		5×10^{-4}	4×10^{-4}	3×10^{-4}	2×10^{-4}	10^{-4}
$\hat{D}_{11,60}^{E}$		1.3847	1.3940	1.4105	1.4395	1.4951
$\hat{D}_{11,50}^{E,a}$	SVD	1.5258	1.5597	1.5969	1.6381	1.6854
	SRGAN	1.2429	1.2663	1.3056	1.3786	1.5293
Relative error	SVD	10.2%	11.9%	13.2%	13.8%	12.7%
	SRGAN	10.2%	**9.2%**	**7.4%**	**4.2%**	**2.3%**

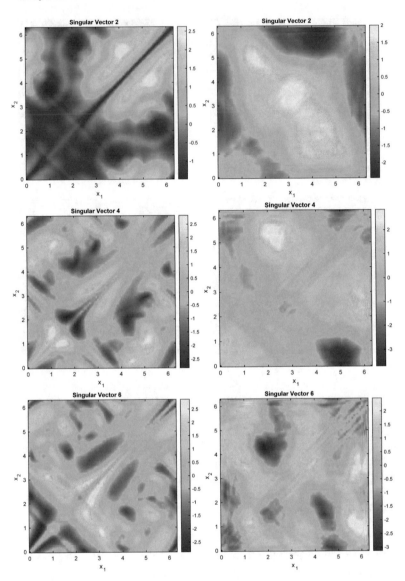

Fig. 5. Singular vectors of W^2 (left col.), $\mathcal{F}\left(G\left(\mathcal{F}^{-1}\left(W^2\right)\right)\right)$ (right col.).

When D_0^1 is closer to D_0^2, SRGAN assisted SVD may have even better predictions at smaller D_0. In Table 2, $D_0^1 = 5 \times 10^{-3}$, $D_0^2 = 10^{-3}$ and $N = 50$ and we predict the $\hat{D}_{11,60}^E$ at $D_0 = 3 \times 10^{-4}$, 2×10^{-4} and 10^{-4}. For $D_0^1 = 5 \times 10^{-3}$, $D_0^2 = 10^{-3}$, singular vectors of $\mathcal{F}\left(G\left(\mathcal{F}^{-1}\left(W^2\right)\right)\right)$ also have thinner structures than that of W^2, as shown in right column and left column of Fig. 5 respectively. Table 3 summarizes predictions for $D_0 = 2 \times 10^{-5}$ and 10^{-5} from $D_0^1 = 10^{-3}$, $D_0^2 = 10^{-4}$ and $N = 60$.

Table 2. Comparison of $\hat{D}_{11,N}^{E,a}$ for flow (7) with $D_0^1 = 5 \times 10^{-3}$, $D_0^2 = 10^{-3}$.

D_0		3×10^{-4}	2×10^{-4}	10^{-4}
$\hat{D}_{11,60}^{E}$		1.4105	1.4395	1.4951
$\hat{D}_{11,50}^{E,a}$	SVD	1.5969	1.6381	1.6854
	SRGAN	1.3111	1.3862	1.5015
Relative error	SVD	13.2%	13.8%	12.7%
	SRGAN	**7.0%**	**3.7%**	**0.4%**

Table 3. Comparison of $\hat{D}_{11,N}^{E,a}$ for flow (7) with $D_0^1 = 10^{-3}$, $D_0^2 = 10^{-4}$.

D_0		2×10^{-5}	10^{-5}
$\hat{D}_{11,60}^{E}$		1.6052	1.6301
$\hat{D}_{11,60}^{E,a}$	SVD	1.5107	1.5243
	SRGAN	1.6234	1.7120
Relative error	SVD	5.9%	6.5%
	SRGAN	**1.1%**	**5.0%**

4   Conclusions

Based on snapshots at two molecular diffusivity values, we trained an adapted super-resolution deep neural network (SRGAN) to model the internal layer sharpening effect of advection-diffusion equation as a nonlinear mapping. The mapping improves the quality of standard POD basis for low cost computation of residual diffusivity in chaotic flows. Though no other DNN model is known to assist POD in our setting, we shall explore how to improve the fidelity of the generated images in the current model.

Acknowledgements. The work was supported in part by NSF grants IIS-1632935, and DMS-1924548.

References

1. Bensoussan, A., Lions, J.-L., Papanicolaou, G.: Asymptotic Analysis for Periodic Structures. AMS Chelsea Publishing, Providence (2011)
2. Biferale, L., Cristini, A., Vergassola, M., Vulpiani, A.: Eddy diffusivities in scalar transport. Phys. Fluids **7**(11), 2725–2734 (1995)
3. Camassa, R., Wiggins, S.: Chaotic advection in a Rayleigh-Bénard flow. Phys. Rev. A **43**(2), 774–797 (1990)
4. Childress, S., Gilbert, A.: Stretch, twist, fold: the fast dynamo. Lecture Notes in Physics Monographs, No. 37. Springer (1995)
5. Fannjiang, A., Papanicolaou, G.: Convection enhanced diffusion for periodic flows. SIAM J. Appl. Math. **54**(2), 333–408 (1994)

290   J. Lyu et al.

6. Heinze, S.: Diffusion-advection in cellular flows with large Peclet numbers. Arch. Ration. Mech. Anal. **168**(4), 329–342 (2003)
7. Holmes, P., Lumley, J., Berkooz, G.: Turbulence, Coherent Structures, Dynamical Systems and Symmetry. Cambridge University Press, Cambridge (1998)
8. Ledig, C., Theis, L., Huszar, F., Cunningham, A., Acosta, A., Aitken, A., Tejani, A., Totz, J., Wang, Z., Shi, W.: Photo-realistic single image super-resolution using a generative adversarial network. In: CVPR, pp. 105–114 (2017)
9. Lumley, J.: The structures of inhomogeneous turbulent flows. In: Atmospheric Turbulence and Radio Wave Propagation, pp. 166–178 (1967)
10. Lyu, J., Xin, J., Yu, Y.: Computing residual diffusivity by adaptive basis learning via spectral method. Numer. Math.: Theory Methods Appl. **10**(2), 351–372 (2017)
11. Majda, A., Kramer, P.: Simplified models for turbulent diffusion: theory, numerical modelling, and physical phenomena. Phys. Rep. **314**, 237–574 (1999)
12. Nasrollahi, L., Metz, S., Chintala, S.: Unsupervised representation learning with deep convolutional generative adversarial networks. In: ICLR (2016)
13. Novikov, A., Ryzhik, L.: Boundary layers and KPP fronts in a cellular flow. Arch. Ration. Mech. Anal. **184**(1), 23–48 (2007)
14. Quarteroni, A., Rozza, G. (eds.): Reduced Order Methods for Modeling and Computational Reduction. MS&A, vol. 9. Springer (2014)
15. Simonyan, K., Zisserman, A.: Very deep convolutional networks for large-scale image recognition. In: ICLR (2015)
16. Sirovich, L.: Turbulence and the dynamics of coherent structures. Part I: coherent structures. Q. Appl. Math. **45**, 561–571 (1987)
17. Taylor, G.: Diffusion by continuous movements. Proc. London Math. Soc. **2**, 196–211 (1921)
18. Wang, Z., Xin, J., Zhang, Z.: Computing effective diffusivity of chaotic and stochastic flows using structure-preserving schemes. SIAM J. Numer. Anal. **56**(4), 2322–2344 (2018)
19. Xin, J., Yu, Y.: Sharp asymptotic growth laws of turbulent flame speeds in cellular flows by inviscid Hamilton-Jacobi models. Annales de l'Institut Henri Poincaré, Analyse Nonlineaire **30**(6), 1049–1068 (2013)
20. Xin, J., Yu, Y.: Front quenching in G-equation model induced by straining of cellular flow. Arch. Ration. Mech. Anal. **214**, 1–34 (2014)
21. Zu, P., Chen, L., Xin, J.: A computational study of residual KPP front speeds in time-periodic cellular flows in the small diffusion limit. Phys. D **311–312**, 37–44 (2015)

An Intensive Empirical Study of Machine Learning Algorithms for Predicting Vietnamese Stock Prices

Thanh-Phuong Nguyen[1], Tien-Duc Van[1], Nhat-Tan Le[2,3], Thanh-Tan Mai[4], and Khuong Nguyen-An[1(✉)]

[1] Faculty of Computer Science and Engineering,
University of Technology (HCMUT), VNU-HCM, Ho Chi Minh City, Vietnam
{1512591,1510824,nakhuong}@hcmut.edu.vn
[2] School of Engineering, Tan Tao University, Duc Hoa, Vietnam
tan.le@ttu.edu.vn
[3] Orient Commercial Joint Stock Bank (OCB), Ho Chi Minh City, Vietnam
tanln@ocb.com.vn
[4] Faculty of Mathematics and Statistics, Quy Nhon Univeristy, Quy Nhon, Vietnam
maithanhtan@qnu.edu.vn

Abstract. Predicting stock prices is a challenging task due to the highly stochastic nature of the financial market. Among many proposed quantitative approaches to tackle this problem, machine learning, in recent years, has become one of the most promising methods. However, machine learning is still new to a large part of Vietnamese investors community. This motivated us to take some first steps in using machine learning techniques on Vietnamese stock data, in particular top 20 listed stocks (according to market capitalization) of VN-Index in June 2019. The experimental results suggest that machine learning and hybrid methods give better performances in forecasting stock price fluctuation than ones achieved by traditional methods such as the Autoregressive Integrated Moving-average model. To realize our study, we implement a web-based tool and release its source code.

Keywords: Machine learning · ARIMA · ANN · Hybrid model · Vietnamese stock market

1 Introduction

Investing in stock exchange is one of the most risky and tough activities as it can yield in complete loss for investors. For the sake of reaping high profit, the investors should employ more and more efficient investment strategies. These strategies based on three different trading schools of thought: fundamental analysis, technical analysis, and quantitative technical analysis[1]. The fundamental

---

[1] https://muse.union.edu/2019capstone-hladikl/methods-of-stock-market-prediction-2/.

© Springer Nature Switzerland AG 2020
H. A. Le Thi et al. (Eds.): ICCSAMA 2019, AISC 1121, pp. 291–303, 2020.
https://doi.org/10.1007/978-3-030-38364-0_26

analysis involves the examination of the economic factors, such as balance sheets and income statements, that influence stock prices. While the technical analysis uses different types of indicators derived from the history of stock price and volumes to predict future values of stock price. Finally, the quantitative analysis[2], machine learning in particular – which we intend to use on our study, is a group of techniques that seek to understand stock market behaviors by using mathematical and statistical modeling, evaluation method, and the studies in the literature.

For the past decades, machine learning has been developed not only in theory, but also in industrial applications. In time series forecasting, some common machine learning methods that have been developed are Artificial Neural Networks (ANN) and Genetic Algorithms (GA), see [9]. Another form of ANN that is more appropriate for time series prediction is Recurrent Neural Network (RNN), see [12]. The use of hybrid models between to enhance accuracy prediction can be seen in [10, 13].

In time series forecasting, Autoregressive Integrated Moving-Average (ARIMA), a time series analysis method, has been applied widely for some sorts of time series data. Adebiyi et al. [1] studied the application of this method on stock price data. The study was conducted on public stock data obtained from the New York Stock Exchange (NYSE) and the Nigeria Stock Exchange (NSE). The authors used the Box-Jenkins approach in [3] to narrow down the collection of orders of ARIMA models. The results revealed that an ARIMA model will potentially yield better results in predicting stock prices on a short-term basis.

For investing tools, Kimoto et al. [8] presented a buying- and selling-time prediction system for stocks on the Tokyo Stock Exchange. The system operated in two stages combining Principal Component Analysis (PCA) and ANN. The study used PCA to select proper inputs from technical indicators to forecast the future values. The prediction system achieved highly accurate predictions and reaped excellent profit via the stock trading simulation.

In 2003, Zhang [14] proposed a hybrid method combining an ARIMA model and an ANN model. The author firstly assumed that the method is capable of taking advantages of the unique strength of ARIMA and ANN models in linear and non-linear modeling. In order to demonstrate the effectiveness of the hybrid method, this study conducted experiments on three well-known dataset including Wolf's sunspot data, the Canadian lynx data, and the British pound/US dollar exchange rate data. The experimental results showed that the combined method could be an effective way to improve the accuracy of prediction achieved by either of the models when used separately.

The work of Atilla et al. [2] provided a performance comparison of various combinations of ARIMA models, ANN models, Radial Basis Function Network (RBFN) models. The dataset consisting of the number of monthly tourist arrivals to Turkey between January 1984 and December 2003, was used for this study. The outcomes of this work indicated that the models with non-linear component gave a better forecasting performance.

---

[2] https://www.investopedia.com/terms/q/quantitativeanalysis.asp.

Regarding predicting Vietnamese stock behaviors, there have also been a few studies [5,11]. In [5], the author proposed a hybrid method combining Gaussian Process Regression (GPR) and Autoregressive Moving Average (ARMA). The result of experiments conducted on close price of VN-Index showed that each method when used separately yielded lower performance than when combined together. In [11], the authors employed an Firefly algorithm, in [7], in the process of training an Adaptive Neuro-Fuzzy Inference System (ANFIS). The study also conducted a performance comparison between this proposed system and ones that employ a hybrid algorithm, a back propagation algorithm and particle swarm optimization algorithm in training ANFIS on data of some stocks of Hanoi Stock Exchange (HNX). The experimental results indicated that the proposed system outperform the rest in terms of both performance and running time.

Our contributions are as follows. Firstly, each method mentioned above, which has its own advantages and disadvantages, is only applied on a particular dataset. Therefore, in this paper, we conduct a comprehensive comparison about the performance of three forecasting methods including ARIMA, ANN and the hybrid model of ARIMA and ANN on Vietnamese stock market, in particular, historical data of 20 stocks of VN-Index from January 2016 to June 2019.

Secondly, we aim at building a user-friendly tool for Vietnamese investors. Besides the main features that are to predict stock prices by three models used in this paper, we also append some needed features such as displaying real-time data, plotting some indicators (Moving average, Boillinger bands, etc.). Figure 4 shows the graphical user interface of our application. The details of this tool will be described in the last part of Subsect. 3.4.

The rest of our paper is organized as follows. In Sect. 2, we briefly recall the basic concepts about ARIMA, ANN and the hybrid model of ARIMA and ANN. In Sect. 3, we describe our data preparation, prediction procedure and the result of empirical experiments. Finally, Sect. 4 summaries the paper and envisions some future development directions.

2 Methodology

In this section, we briefly recall some of the fundamental concepts that will be used in subsequent sections.

2.1 Autoregressive Integrated Moving-Average

In an Autoregressive Integrated Moving-average model, future values are linearly dependent on several past values and random errors. Namely, this model includes three components which are Autoregressive (AR), Integrated (I) and Moving-average (MA). The underlying process generating the time series is written in the form of

$$x_t = \alpha_1 x_{t-1} + \alpha_2 x_{t-2} + \cdots + \alpha_p x_{t-p} + \epsilon_t + \beta_1 \epsilon_{t-1} + \beta_2 \epsilon_{t-2} + \cdots + \beta_q \epsilon_{t-q},$$

where x_t and ϵ_t are the actual value and error at the time t, respectively; $\{\alpha_i\}_{i=1...p}$ and $\{\beta_j\}_{j=1...q}$ are the coefficients (model parameters); p and q are integers and usually referred to as orders of the model (hyperparameters). The random errors, ϵ_t, are assumed to be independently and identically distributed with a zero mean value and a constant variance of σ^2. Furthermore, the I component is the number of times that the time series is taken differences in order to obtained a stationary time series. The order of this component is denoted by d.

Stationarity is a necessary condition of building an ARIMA model that is valid in terms of statistics [3]. A time series is called *stationary* when it's statistical characteristics, such as mean and autocorrelation structure, remain constant over time or at least in a long-term. Once the observed time series shows a trend and heteroscedasticity, differencing and `log` transformation are applied to the time series in order to remove the trend, and stabilize the variance before fitting the ARIMA model to the data.

When the ARIMA model is determined, which means that a set of orders is identified, the model parameters will be estimated. The estimation is conducted by minimizing an overall measure of errors. This can be done by iterative optimization procedures.

There are several sets of orders for ARIMA models that should be listed. The method used in order to specify a collection of sets of orders will be discuss in Sect. 3. Then, by validating the ARIMA model with respect to their sets of orders, the best model will be selected for forecasting purposes.

2.2 Artificial Neural Network

Multilayer Neural Network. Multilayer neural network is a form of an ANN containing perceptrons. These perceptrons are organized into many layers. A multilayer neural network architecture includes at least three layers: one input layer, one output layer, and one or more hidden layers. We will discuss the method we specify the architecture of a neural network in detail in Sect. 3.

Activation Function. An activation function is a non-linear function that helps the network approximate any non-linear functions. In our experiments, we use the `Tanh` function $f(x) = \frac{e^x - e^{-x}}{e^x + e^{-x}}$ as an activation function.

Backpropagation. Backpropagation is a method used in ANNs to calculate a gradient that is needed in the adjusting of the weights to be used in the network [4]. When we attempt to update the weights, we also minimize the error. To do this, one commonly used algorithm is *gradient descent*.

2.3 Hybrid Model of ARIMA and ANN

The hybrid approach of ARIMA and ANN combines all advantages of both ARIMA model and ANN model, producing better performance on some sort of time series data, see [14]. The main idea of this method comes from two perspectives. First, it is really difficult to determine whether a time series is generated by a linear or non-linear underlying structure in a process. Second, practical time series are rarely pure linear or non-linear. Therefore, it is assumed that every time series contains two components: a linear component and a non-linear component. Given a time series $\{x_t\}_{t=1..T}$ that is composed of linear autocorrelation structure and non-linear component

$$x_t = L_t + N_t + \epsilon_t,$$

where L_t represents the linear part, N_t represents the non-linear part and ϵ_t is a random error. First, the ARIMA model will be applied to capture the linear part, then the residuals from the linear model will contain only non-linear relationships. Let e_t denote the residuals at the time t, then

$$e_t = x_t - \hat{L}_t,$$

where $\hat{L}_t$ is the forecast value at the time t from the ARIMA model. Now, the residuals will be captured by the ANN model with n input,

$$e_t = f(e_{t-1}, e_{t-2}, ..., e_{t-n}) + \epsilon_t,$$

where f is the non-linear function that is determined by the ANN model. We should notice that "if the function f is not an appropriate one, the error term is not necessarily random. Therefore, the correct model identification is critical". Denote $\hat{N}_t$ the forecast value from f, then the predicted value is

$$\hat{x}_t = \hat{L}_t + \hat{N}_t.$$

The hybrid model exploits the unique features and combines the strength of both ARIMA and ANN models. However, this method is not universal in real-life problems [14].

3   Empirical Results

3.1   Data Collection

In order to conduct our experiments, stock price data was collected from the website investing.com[3] which is a global financial portal and internet brand owned by Fusion Media Limited. We collected data from this site because we found that AJAX requests for historical data of this website is easy to mimic.

To crawl data from Investing website, we used Selenium framework to get HTML contents. Then, we parsed HTML contents by using BeautifulSoup 4. The obtained dataset is organized as shown in Table 1. We collected historical stock price data of top 30 stocks of VN-Index from January, 2016 to June, 2019. Readers could find the public dataset on our GitHub[4]. However, in the scope of this paper, we only use the top 20 stocks in Table 4 to conduct experiments.

---

[3] https://www.investing.com/indices/vn-30-components.

[4] https://github.com/thanhphuong163/vn_stock_prediction/tree/master/Data.

Table 1. An example of historical data in database.

date	name	ticket	index
22-05-2019	Cotec Construction JSC	CTD	VN 30 (VNI30)
22-05-2019	Mobile World Investment Corp	MWG	VN 30 (VNI30)
22-05-2019	Phu Nhuan Jewelry JSC	PNJ	VN 30 (VNI30)
22-05-2019	No Va Land Investment Group	NVL	VN 30 (VNI30)

close	open	high	low	volume	change_per
115300	115100	116000	115100	379500	-0.35
87000	87000	87983	86705	6731400	0.34
107400	106400	109000	106400	6358500	0.94
57700	59000	59200	57700	3901800	-2.2

3.2   Data Preprocessing

To feed data to machine learning models, such as an ANN, it is necessary to transform time series datasets into supervised learning datasets.

The key function for transformating a time series dataset into a supervised learning dataset used in this paper is the `shift()` function of Pandas framework. This function can create copies of a time series which are pushed forward (`nan` values will be put to the front) and pulled backward (`nan` values will be put to the end). Copies of the time series which are pushed forward will be columns of lag observations. These columns will be considered as input patterns. Meanwhile, copies that are pulled backward will be columns of forecast observations, also known as output patterns.

Table 2 illustrates an example of transformation from a time series dataset into a supervised learning dataset. We have a time series on the left table, then we re-frame it into supervised learning data on the right table. The columns **lag_1** and **lag_2** are copies of the origin time series which were pushed forward. They represent the price in the past. The column **forecast_0** is the original time series and represents the price at present. The columns **lag_1**, **lag_2** and **forecast_0** are considered as input features of a machine learning model. Finally, the column **forecast_1** is the forecast observations or output.

3.3   Evaluation Method

Mean Absolute Percentage Error. Mean absolute percentage error (MAPE) is a measure of prediction accuracy of forecasting models. It is usually used as a loss function for regression problems in machine learning. This measure is computed by the following formula

Table 2. Converting time series data into unsupervised learning data.

Date	Value	Date	lag_2	lag_1	forecast_0	forecast_1
2019-06-14	104300.0	2019-06-14	nan	nan	104300.0	106300.0
2019-06-13	106300.0	2019-06-13	nan	104300.0	106300.0	106800.0
2019-06-12	106800.0	2019-06-12	104300.0	106300.0	106800.0	106500.0
2019-06-11	106500.0	2019-06-11	106300.0	106800.0	106500.0	107000.0
2019-06-10	107000.0	2019-06-10	106800.0	106500.0	107000.0	107000.0
2019-06-07	107000.0	2019-06-07	106500.0	107000.0	107000.0	108000.0
2019-06-06	108000.0	2019-06-06	107000.0	107000.0	108000.0	nan

$$\text{MAPE} = \frac{100}{n} \sum_{t=1}^{n} \left| \frac{\hat{y}_t - y_t}{y_t} \right|,$$

where y_t is the actual price and $\hat{y}_t$ is the predicted price. The error is calculated by taking the mean of the absolute differences of $\hat{y}_t$ and y_t divided by y_t again; multiplying 100 to make it a percentage error.

3.4   Time Series Cross-Validation Procedure

Time series cross-validation is a more sophisticated version of cross-validation [6]. To conduct this type of cross-validation, the data is organized into a series of training/test set splits as depicted in Fig. 1. For each training/test set split, we fit model with a collection of hyperparameters into a training set. The fitting models will be used to predict the price of the next day corresponding to the price in the test set. Once we obtain the predictions, we compute the error of each prediction and choose the least value among these errors. The same procedure is applied to the rest of training/test set splits. The total error of model is the mean of these least errors over the series training/test set splits.

In the rest of this subsection, we describe how we specify a collection of hyperparameters for each model used in our experiments subsequently.

Validation of Autoregressive Integrated Moving-Average. We use two techniques including the trial-and-error technique and the grid-search technique to determine the model which yields the best performance. An ARIMA model is specified by a set of hyperparameters (p, d, q), where p is the order of the AR component; d is the order of the I component; and q is the order of the MA component. By using the trial-and-error technique, we observed that the orders of the AR model and the MA model in ARIMA model usually fall into range $[0, 3]$ and range $[0, 6]$, respectively. Furthermore, the bigger order is, the longer training time is.

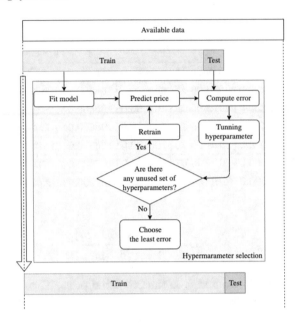

Fig. 1. The illustration of time series cross-validation.

Validation of Artificial Neural Network. With an ANN model, we use the trial-and-error approach to determine the proper architecture for fitting the model. After many trials, we find that ANN models including two hidden layers produced better results than others, and the number of units in each layer was approximately 10 units. Another observation is that the models with one to three input units yield better results than the ones using a larger number of input units. From these observations, we employed the grid-search technique for seeking the best model with the number of units in the first hidden layer from three to seven and the number of units in the second layer from one to three. An ANN model is specified by a set of hyperparameters (i, k, ℓ), where i, k, ℓ are the number of units in the input, the first hidden and the second hidden layer; and the output layer only consist of one unit.

Validation and Implementation of Hybrid Model of ARIMA and ANN. Firstly, we fit ARIMA models to the data and selected the best model fitting to the linear component of the time series. Next step, we compute the residuals (non-linear component of the times series) on training set to be fitted by ANN models. To find the best ANN model, we still employ the grid-search technique with the same setup as in the second experiment (the one that used an ANN model). To the best of our knowledge, this model has no public source code or built-in model in any frameworks. Therefore, we implement this model ourselves based on description of Zhang [14], the implementation[5] is conducted following the flow described in the diagram in Fig. 2.

3.5 Results and Analysis

To illustrate the numerical results in this paper, we will use the close prices of Phu Nhuan Jewelry JSC. Figure 3 shows the differences between the predicted prices and the actual prices of three models.

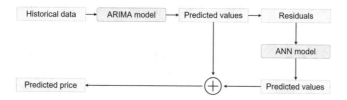

Fig. 2. Diagram of the hybrid ARIMA+ANN model.

- The red solid line is the actual prices,
- The blue dash-dot line is the one-step predicted prices of the ARIMA model,
- The yellow dashed line is the one-step predicted prices of the ANN model,
- The green dotted line is the one-step predicted prices of the hybrid model of ARIMA and ANN.

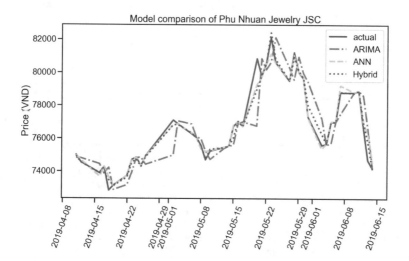

Fig. 3. Model comparison of three models on Phu Nhuan Jewelry JSC.

---

[5] The private repository of the tool is located at https://github.com/thanhphuong163/ vn_stock_predictiontree/master/Models. Access to this repository is granted upon requests.

In the figure, the predicted values of the ANN model and the hybrid model seem to be closer to the actual values than the ARIMA model. Indeed, Table 3 tells us that the hybrid model is the best fitting model.

Table 3 shows the MAPE of each model applied on the top 20 stocks of VN-Index. The MAPEs which are bold denote that they are the best performance of three models for each stock. ANN models outperform on nine stocks, and the hybrid models give better performance on 11 stocks. This result suggests that ANN and the hybrid models give the best performance on the majority of stocks. This implies that machine learning and hybrid methods produce more accurate predictions on Vietnamese stock data compared to ARIMA - a traditional time series analysis method.

Furthermore, below each MAPE is the range of absolute percentage errors of each model. This range includes minimum and maximum errors over the process of time series cross-validation. Ranges of the ANN and the hybrid models seem to be subsets of ranges of ARIMA models. This result implies that ANN and hybrid models give more stable performance than ARIMA models.

The most clearly related works to our study are the ones by Atilla et al. [2], Huynh Quyet Thang [5] and Nguyen Nhu Hien et al. [11]. Atilla et al. used the same models to our study. However, these models were applied on a different dataset. In [5,11], the authors studied on Vietnamese stock market, but they examined their models on VN-Index and some stocks of HNX. Moreover, the authors also used Root Mean Squared Error (RMSE) while we used MAPE as an evaluation measure. Therefore, we made no comparison as well as conclusion between our study and the above-mentioned papers. However, since MAPE is a relative value (percentage), we can compare results of different stocks without noticing what the unit is.

Table 3. MAPE of 20 listed stocks of VN-Index (%).

Ticket	ARIMA	ANN	Hybrid	Ticket	ARIMA	ANN	Hybrid
CTD	0.5857	0.1604	**0.1198**	MWG	0.6174	0.1425	**0.0629**
	[0.0, 3.6516]	[0.0016, 2.0693]	[0.0031, 0.8525]		[0.0, 2.3114]	[0.0016, 0.9554]	[0.0008, 0.3441]
DHG	0.7067	0.0957	**0.0852**	NVL	0.452	**0.1137**	0.1379
	[0.0, 2.2226]	[0.0034, 0.4585]	[0.0004, 0.7587]		[0.0, 1.9389]	[0.0017, 1.0121]	[0.0015, 1.1799]
ROS	1.1189	0.2524	**0.1841**	DPM	0.8825	**0.1359**	0.1433
	[0.0, 5.8533]	[0.0008, 1.9342]	[0.0034, 1.5147]		[0.0, 5.3179]	[0.0006, 0.9245]	[0.0034, 0.5912]
FPT	0.7411	0.0961	**0.0731**	PNJ	0.8613	0.1917	**0.1578**
	[0.0, 2.2657]	[0.0013, 0.6104]	[0.0015, 0.4405]		[0.0, 4.9514]	[0.0074, 1.423]	[0.0017, 1.5642]
GMD	0.4653	**0.087**	0.1001	REE	0.5697	0.0972	**0.0921**
	[0.0, 3.252]	[0.0011, 0.689]	[0.0028, 0.5355]		[0.0, 3.0789]	[0.0003, 0.9905]	[0.0022,0.4002]
CII	0.479	**0.0666**	0.0841	STB	0.5098	0.0835	**0.0785**
	[0.0, 2.2832]	[0.0022, 0.2758]	[0.0013, 1.0391]		[0.0, 3.2084]	[0.0002, 0.5221]	[0.0016, 0.6935]
HDB	0.2801	**0.0559**	0.0649	SAB	0.8686	0.1832	**0.1246**
	[0.0, 1.2336]	[0.0, 0.2153]	[0.0003, 0.3026]		[0.0, 3.6254]	[0.0041, 1.3722]	[0.0015, 0.7896]
HPG	0.6342	**0.1022**	0.1228	SSI	0.5672	0.1304	**0.0903**
	[0.0, 4.4476]	[0.0006, 0.6058]	[0.0019, 0.4752]		[0.0, 2.591]	[0.004, 0.8443]	[0.0015, 0.6173]
MSN	0.6406	**0.0831**	0.1126	TCB	0.6748	**0.0644**	0.2471
	[0.0, 3.4088]	[0.0016, 0.6518]	[0.0003, 0.6386]		[0.0, 4.7885]	[0.0005, 0.2711]	[0.0024, 0.8618]
MBB	0.5	0.1313	**0.107**	SBT	0.5261	**0.0863**	0.105
	[0.0, 2.4699]	[0.0048, 0.7402]	[0.0, 1.0252]		[0.0, 6.1373]	[0.0002, 0.6595]	[0.0016, 0.8092]

4   Conclusion

Stock price prediction is an active research area over the last few decades. The accuracy of forecasting performance plays an important role for in any investor's decision. In Vietnamese stock market, where most investors still use traditional methods, even their intuition, in stock trading, our study has proved the effectiveness of machine learning techniques in stock price prediction. By experiments, we have shown that machine learning and hybrid methods such as ARIMA+ANN outperformed traditional methods such as ARIMA on Vietnamese stock market. This means that there are many Vietnamese stocks having both linear and non-linear underlying structures in their historical data. Moreover, machine learning also gives more stable forecasting results than traditional time series analysis methods.

Nonetheless, the dynamic of stock price in market is not only influenced by its own past, but also by a variety of factors such as news, financial indicators, political conditions, human psychology, etc. Machine learning is an approach which can combine all possible factors to improve the performance of stock price prediction. Therefore, using market factors as input features for machine learning models might be a promising research direction.

Acknowledgements. The authors would like to thank Nguyen Van Thanh and Nguyen Huynh Huy for their comments helping to improve the manuscript of this work significantly.

A   Appendix

A.1   List of Top 20 Stocks of VN-Index

Table 4. List of top 20 stocks of VN-Index.

Symbol	Name of company
CTD	Cotec Construction JSC
DHG	DHG Pharmaceutical JSC
ROS	Faros Construction Corp
FPT	FPT Corp
GMD	Gemadept Corp
CII	Ho Chi Minh City Infrastructure Investment JSC
HDB	Ho Chi Minh City Development Joint Stock Commercial Bank
HPG	Hoa Phat Group JSC
MSN	Masan Group Corp
MBB	Military Commercial Joint Stock Bank

(*continued*)

Table 4. (*continued*)

Symbol	Name of company
MWG	Mobile World Investment Corp
NVL	No Va Land Investment Group Corp
DPM	Petro Vietnam Fertilizer and Chemicals Corp
PNJ	Phu Nhuan Jewelry JSC
REE	Refrigeration Electrical Engineering Corp
STB	Sai Gon Thuong Tin Commercial Joint Stock Bank
SAB	Saigon Beer Alcohol Beverage Corp
SSI	Saigon Securities Incorporation
TCB	Vietnam Technological and Commercial Joint Stock Bank
SBT	Thanh Thanh Cong Tay Ninh JSC

A.2   Application Interface

Figure 4 shows the interface of the tool mentioned in Footnote 5.

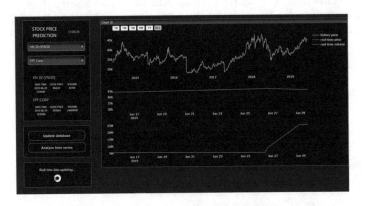

Fig. 4. Graphical user interface of the application.

References

1. Adebiyi, A., et al.: Stock price prediction using the ARIMA model. In: UKSim-AMSS 16th International Conference on Computer Modelling and Simulation (2014)
2. Aslanargun, A., et al.: Comparison of ARIMA, neural networks and hybrid models in time series: tourist arrival forecasting. J. Stat. Comput. Simul. **77**(1), 29–53 (2007)
3. Dimitrios Asteriou, S.G.H.: Applied Econometrics, 2nd edn. Palgrave Macmillan (2011)

4. Goodfellow, I.J., et al.: Deep Learning. MIT Press, Cambridge (2016)
5. Huynh Quyet, T.: Vietnam stock index trend prediction using Gaussian process regression and autoregressive moving average model. Research and Development on Information and Communication Technology (2018)
6. Hyndman, R., et al.: Forecasting: principles and practice
7. Johari, N., et al.: Firefly algorithm for optimization problem. Appl. Mech. Mater. **421** (2013)
8. Kimoto, T., et al.: Stock market prediction system with modular neural networks. In: IJCNN (1990)
9. Kolhe, M., et al.: GA-ANN for short-term wind energy prediction. In: 2011 Asia-Pacific Power and Energy Engineering Conference, pp. 1–6 (2011)
10. Luxhøj, J.T., et al.: A hybrid econometric—neural network modeling approach for sales forecasting. Int. J. Prod. Econ. **43**(2), 175–192 (1996)
11. Nhu, H., et al.: Prediction of stock price using an adaptive Neuro-Fuzzy inference system trained by Firefly algorithm. In: 2013 International Computer Science and Engineering Conference, ICSEC 2013, pp. 302–307 (2013)
12. Saad, E.W., et al.: Comparative study of stock trend prediction using time delay, recurrent and probabilistic neural networks. IEEE Trans. Neural Netw. **9**(6), 1456–1470 (1998)
13. Wedding, D.K., et al.: Time series forecasting by combining RBF networks, certainty factors, and the Box-Jenkins model. Neurocomputing **10**(2), 149–168 (1996). Financial Applications, Part I
14. Zhang, P.: Time series forecasting using a hybrid ARIMA and neural network model. Neurocomputing **50**, 159–175 (2003)

Improvement of Production Layout in the Furniture Industry in Indonesia with the Concept of Group Technology

Orchida Dianita$^{(\boxtimes)}$, Thomas Djorgie, and M. K. Herliansyah

Universitas Gadjah Mada, Yogyakarta 55281, Indonesia
orchida.dianita@ugm.ac.id

Abstract. The furniture industry in Indonesia is multiplying, with average report rate was reached US$ 1,627 billion in 2017. According to the massive potential of the furniture industry in Indonesia, a method to increase its productivity and efficiency is indispensable. Group technology in cell manufacturing as an approach to achieve its previous objective is used to improve the production layout. Rank Order Clustering (ROC) and Hollier method are used to classifying machines and process in the production layouts. The Discrete Event Simulation (DES) as a validation tool proven that the layout improvement is successfully reducing the total travel time 24.6% and total travel distance up to 16.1%.

Keywords: Group Technology · Rank Order Clustering · Furniture industry

1 Introduction

The furniture industry in Indonesia plays an important role and become one of the leading export contributors. In 2016, the furniture industry in Indonesia reached up to US$ 1.6 billion in export. The number then increased in 2017, which it reaches up to US$ 1.627 billion. Based on its sector clustering, the furniture industry could be classified as a craft sector. The craft sector ranked as the 2nd of the highest export contributor in 2015 and 2016 up to 39.01% from total. However, the problems found in the furniture production floor become an obstacle that should not be neglected.

Some of the problems found in the furniture production floor such as low coordination between the workstations, the non-updated information of finished parts and work in progress part current status, duplication and unnecessary amount of some part in the production line, irregular and inefficient flow and transport part. The problems in the production floor could be simplified by minimizing the flow of part transport time, grouping part into family based on the process similarity, and forming the Cell Manufacturing (CM) [1]. CM is denoted as part of Group Technology (GT), which aims to increase the productivity and efficiency in regards to flow time and production cost [2]. GT is suitable for job shop production system, namely high variation products to produced in relatively small number [3].

A grouping technique to build the CM layout can be done with several methods such as heuristics and clustering approach [4–6]. CM implementations carry some benefits, such as reducing the non-value added time and increase group efficacy [7, 8].

H. A. Le Thi et al. (Eds.): ICCSAMA 2019, AISC 1121, pp. 304–310, 2020.
https://doi.org/10.1007/978-3-030-38364-0_27

2   Methods

Clustering Method. To form the CM, a Rank Order Clustering (ROC) methods applied. Firstly, four selected furniture products are taken as a sample model named SN-5, SN-6, SN-7, and SN-8. In general, the Operation Process Chart (OPC) to make the furniture products consist of six stages, including crosscutting, spindling, mortis, pinning, drilling, and profiling. Forty-three different parts are used to form the furniture products, clustered by considering the Production Flow Analysis (PFA) which hence, forms 14 part groups. The details of 14 grouping parts are shown in Table 1.

Table 1. Part clustering.

Cluster	Parts name
A	(GN, GDB/SB, SD, SDB, SRP) SN-6; (SLTB, GDB) SN-7; (SS) SN-5
B	(TGN) SN-6; (TGN) SN-7; (TGN) SN-8
C	(SDB, SDD) SN-8
D	(GN/SB) SN-5
E	(PLD) SN-5
F	(GN) SN-7
G	(FSDR, FDS) SN-6; (SS) SN-7; (FSDR, SS) SN-8
H	(SLTB, SRP, SD) SN-5; (SRP) SN-7; (SLTB, GBD, SRP) SN-8
I	(KB) SN-5; (KB, KD) SN-6; (KB, KD) SN-7; (KB, KD) SN-8
J	(GN, SB) SN-8; (TGN) SN-5
K	(KD) SN-5
L	(SB,SDB/SDD) SN-7
M	(FSDR) SN-7
N	(GBD) SN-5

Secondly, ROC used to form the CM group by using the previous 14 parts group. Three times iteration generates new parts/machines matrix shown in Table 2. The grey area indicated the CM clustering group. CM 1 consist of I, A, K, F, and D. Subsequently, M, E, C, G, N, L, B, J, and H included in CM 2.

From Table 2, two CM groups formed. However, fifteen exceptional elements found which is not included in CM clustering groups. Hence, the production process of 15 exceptional elements conducted outside the CM clustering group.

The Machine Sequencing. As the CM groups identified, the sequence of the machine is determined to minimize the Backtracking Move (BTM). Hollier methods are used to specify the flow parts efficiency in both CM 1 and CM 2. The flow part within machines in CM 1 is shown in Table 3 followed with flow diagrams in Fig. 1. In addition, Table 4 presented the CM 2 within machines flow parts, and Fig. 2 illustrated its diagram flows.

Table 2. Iteration matrix of parts/machines.

Parts/Machines Group	Cross cut 1	Pro fil 1	Spin-dle 1	Mor-tis	Bor 2	Ten-on 1	Ten-on 2	Pro-fil 2	Bor 1	Spin dle 2	Cross cut 2
I	1	1	1	1	1						
A	1	1	1			1					
K	1	1		1	1	1					
F	1		1	1	1			1			
D	1			1				1			
M	1						1		1	1	
E		1	1					1			1
C		1			1		1				1
G		1						1	1		
N		1									1
L			1	1		1		1			1
B				1				1	1	1	
J						1		1	1	1	1
H							1	1			1

Table 3. Cell manufacturing 1 within machines flow part.

From/To	Crosscut 1	Profil 1	Spindle 1	Mortis	Bor 2	Tenon 1	From Sums
Crosscut 1			25800	300	5200	10000	**41300**
Profil 1							**0**
Spindle 1	13200			1000	2800	22000	**39000**
Mortis		9000			2300		**11300**
Bor 2		2300		8000			**10300**
Tenon 1		30000		2000			**32000**
To sums	**13200**	**41300**	**25800**	**11300**	**10300**	**32000**	

Layout Improvement. Following the CM grouping, the improvement of production layout is made. The production layout improvement is built attributed to part/machine and flow diagrams. The illustration of previous production layout is shown in Fig. 3 followed with the improved layout in Fig. 4.

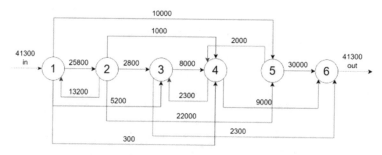

Fig. 1. Cell manufacturing 1 flow diagrams.

Table 4. Cell manufacturing 2 within machines flow part.

From/To	Tenon 2	Profil 2	Bor 1	Spindle 2	Crosscut 2	From Sums
Tenon 2		21000	3400			**24400**
Profil 2						**0**
Bor 1	1000	10500		2700		**14200**
Spindle 2	900	2100	13400		1300	**17700**
Crosscut 2	23400			2900		**26300**
To sums	**25300**	**33600**	**16800**	**5600**	**1300**	

Simulation Model. Simulation model represent the real system which built to testing the problem solutions. In this study, simulation model is conducted to verify the efficiency results of the production layout improvement. Company policy, production layout, production resources, and duration in each production process are considered when simulate the model. Utilize the Flexsim software, the simulation model of total transport time and total travel distance between previous layout and improved layout is gathered.

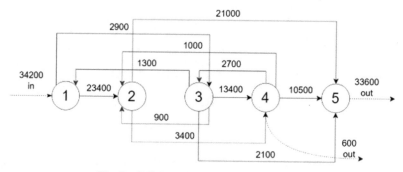

Fig. 2. Cell manufacturing 2 flow diagrams.

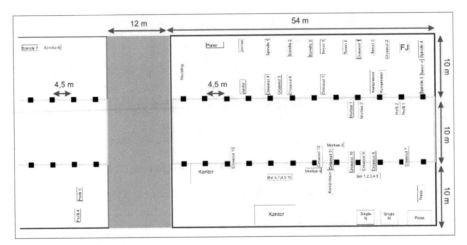

Fig. 3. Production layout before improvement.

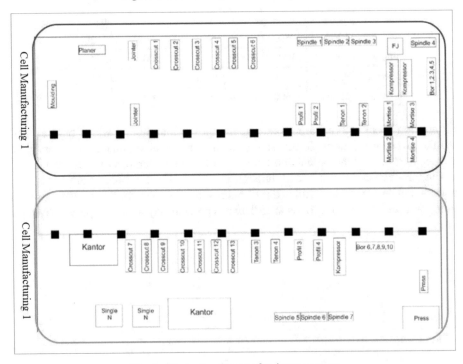

Fig. 4. Production layout after improvement.

3   Results

The results of simulation model between improved layout and non-improved layout shows significant contribution regarding the travel distance, time efficiency and cost savings. Table 5 shows the comparison of actual layout and improved layout in terms of total travel distance and total travel time.

Table 5. Comparison of actual layout and improvement layout.

Model	Total travel distance (km)	Total transport time (working days)
Actual layout	4317,301	138,50
Improvement layout	3621,93	104,48

Information from Table 5 indicated that the improved layouts could reduce the total travel distance 16.1% (695.37 km). Meanwhile, the total transport time is reduced up to 24.6% (34.02 working days) after the improvement.

Additionally, the furniture company set the minimum payment of transporter person is 27.500,00 (Indonesian Rupiah) per day. As the improvement layout could reducing up to 34.02 working days, thus the company management could save 7.649.600,00 (Indonesian Rupiah).

4 Discussions and Conclusions

The implementation of GT in the furniture industry is proven could benefits the company, indicated in time efficiency and cost savings after the layout improvement. The clustering methods such as ROC helps the formation of CM and mapping the flow diagrams in each cell. Additionally, this study is adopting the similar furniture products. Further investigation in different parts, and product families or various industries to explore the GT implementation is needed. Furthermore, various grouping techniques and algorithm to build the CM groups is fascinating to explore.

References

1. Brennan, R.W.: Modeling and Analysis of Manufacturing Systems. By R. G. ASKIN and C. R. STANDRIDGE (John Wiley and Sons, 1993) ISBN 0-417-51418-7 [Pp.461] £16 50 (softback). Market: undergraduate/practising engineer. Int. J. Comput. Integr. Manuf. 8, 155–156 (2007). https://doi.org/10.1080/09511929508944639
2. Taylor, P., Lee, K., Ahn, K.: GT efficacy : a performance measure for cell formation with sequence data 37–41 (2013). https://doi.org/10.1080/00207543.2013.794317
3. Vakharia, A.J., Wemmerlov, U.: Designing a cellular manufacturing system: a materials flow approach based on operation sequences. IIE Trans. (Inst. Ind. Eng. 22, 84–97 (1990). https://doi.org/10.1080/07408179008964161
4. Harhalakis, G., Nagi, R., Proth, J.M.: An efficient heuristic in manufacturing cell formation for group technology applications. Int. J. Prod. Res. 28, 185–198 (1990). https://doi.org/10.1080/00207549008942692
5. Lee, C.S., Hwang, H.: A hierarchical divisive clustering method for machine-component grouping problems. Eng. Optim. 17, 65–78 (1991). https://doi.org/10.1080/03052159108941061
6. Suzić, N., Stevanov, B., Ćosić, I., Anišić, Z., Sremčev, N.: Customizing products through application of group technology: a case study of furniture manufacturing. Stroj. Vestnik/J. Mech. Eng. 58, 724–731 (2012). https://doi.org/10.5545/sv-jme.2012.708

7. Pattanaik, L.N., Sharma, B.P.: Implementing lean manufacturing with cellular layout: a case study. Int. J. Adv. Manuf. Technol. **42**, 772–779 (2009). https://doi.org/10.1007/s00170-008-1629-8
8. Satheeshkumar, V.: Evaluation of cell formation algorithms and implementation of MOD-SLC algorithm as an effective cellular manufacturing system in a manufacturing industry. Int. J. Curr. Eng. Technol. **2**, 183–190 (2014). https://doi.org/10.14741/ijcet/spl.2.2014.33

Reinforcement Learning in Stock Trading

Quang-Vinh Dang[(✉)] [ID]

Industrial University of Ho Chi Minh city, Ho Chi Minh City, Vietnam
dangquangvinh@iuh.edu.vn

Abstract. Using machine learning techniques in financial markets, particularly in stock trading, attracts a lot of attention from both academia and practitioners in recent years. Researchers have studied different supervised and unsupervised learning techniques to either predict stock price movement or make decisions in the market.

In this paper we study the usage of reinforcement learning techniques in stock trading. We evaluate the approach on real-world stock dataset. We compare the deep reinforcement learning approach with state-of-the-art supervised deep learning prediction in real-world data. Given the nature of the market where the true parameters will never be revealed, we believe that the reinforcement learning has a lot of potential in decision-making for stock trading.

Keywords: Reinforcement learning · Machine learning · Stock trading

1 Introduction

Searching for an effective model to predict the prices of the financial markets is an active research topic today [13] despite the fact that many research studies have been published for a long time [3,11]. In the midst of financial markets prediction, stock price prediction is considered as one of the most difficult tasks [44]. Among the state-of-the art techniques, machine learning techniques are the most widely chosen techniques in recent years, given the rapid development of the machine learning community. The other reason is that the traditional statistical learning algorithms can not cope with the non-stationary and non-linearity of the stock markets [15].

In general, there exists two main approaches to analyze and predict stock price which are technical analysis [23] and fundamental analysis [39]. The technical analysis looks into the past data of the market only to predict the future. On the other hand, the fundamental analysis takes into account other information such the economic status, news, financial reports, meeting notes of the discussion between CEOs, etc.

The technical analysis relies on the efficient market hypothesis (EMH) [25]. The EMH states that all the fluctuation in the market will be reflected very quickly in the price of stocks. In practice, the price can be updated in the magnitude of milliseconds [8], leading to a very high volatility of the stocks. In recent

H. A. Le Thi et al. (Eds.): ICCSAMA 2019, AISC 1121, pp. 311–322, 2020.
https://doi.org/10.1007/978-3-030-38364-0_28

years the technical analysis attracts a lot of attention due to a simple fact that we have enough information just by looking to the historical stock market, which is public and well-organized, compared to the fundamental analysis where we need to analyze unstructured dataset.

Compared to the supervised learning techniques and at a certain level, unsupervised learning algorithms, are widely used in stock price prediction, to the best of our knowledge the reinforcement learning for stock price prediction has not yet received enough support as it should be. The main issue of supervised learning algorithms is that they are not adequate to deal with time-delayed reward [18,22]. In other words, supervised learning algorithms focus only on the accuracy of the prediction *at the moment* without considering the delayed penalty or reward. Furthermore, most supervised machine learning algorithms can only provide *action recommendation* on particular stocks[1], using reinforcement learning can lead us directly to the *decision making* step, i.e. to decide how to buy, hold or sell any stock.

In the present paper we study the usage of reinforcement learning in stock trading. We review some related works in Sect. 2. We present our approach in Sect. 4. We describe and discuss the experimental results in Sect. 5. We conclude the paper and draw some future research directions in Sect. 6.

2 Literature Review

There are two main applications of using machine learning in the stock markets: stock price prediction and stock trading.

Stock price prediction can be divided into two applications: price regression or stock trend prediction. In the first application, the researchers aim to predict exactly the numerical price, usually based on day-wise price [15] or closed price of a stock. In the second approach, the researchers usually aim to predict the turning point of a stock price, i.e. when the stock price change the moving direction from up to down or vice versa [44]. Traditionally time-series forecasting techniques such as ARIMA and its variant [26,43] are adapted from the econometric literature. However, these methods cannot cope with non-stationary and non-linearity nature of the stock market [2].

It is claimed that the stock price reflects the belief or opinions of the market on the stock rather than the value of the stock itself [7]. Several research studies propose to analyzing the social opinions to predict the stock price. In the research study of [33], the authors used Google Trends, i.e. to analyze the Google query volumes for a particular keyword or search term, they can measure the attention level of the public to a stock. The research is based on one idea that a decision making process will start by information collection [37].

Over centuries, the researchers and practitioners have developed many technical indicators to predict the stock price [29]. Technical indicators are defined

---

[1] According to NASDAQ standard, recommendation from analysts can be Strong Buy, Buy, Hold, Underperform or Sell. Reference: https://www.nasdaq.com/quotes/analyst-recommendations.aspx. Accessed on 07-September-2019.

as a set of tool that allow us to predict the future stock market by solely looking to the historical market data [31]. Originally the technical analysis are not
highly supported in academia [27] even though it is very common in practice [35].
Nevertheless, with the development of the machine learning community, technical analysis gains attention of researchers in recent years. [32] derived what
the authors called "Trend Deterministic Data Preparation" from ten technical
indicators then combined them with several machine learning techniques such
as Support Vector Machine (SVM) or Random-Forest for the stock price movement prediction. The "Trend Deterministic Data Preparation" are simply the
indication from the technical indicators that the price will go up or down, so the
approach of [32] can be considered as ensemble learning from local experts [17].

The authors of [44] inherited the idea of Japanese candle stick in stock analysis [30] to develop a status box method combined with probabilistic SVM to
predict the stock movement. We visualized a Japanese candlestick in Fig. 1. The
status box developed by [44] is presented in Fig. 2. The main idea is instead of
focusing on only one time period as the traditional Japanese candlestick method,
the status box focus on a wider range of time, that allow us to overcome the
small fluctuation in the price.

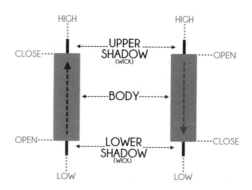

Fig. 1. Japanese candle stick using in stock analysis. A candlestick shows us the highest
and lowest price of a stock in a period of time, as well as opening and closing price of
this stock.

It is quite straightforward to combine technical analysis and fundamental
analysis into a single predictive model. The authors of [42] combined news analysis with technical analysis for stock movement prediction and claimed that the
combination yields a better predictive performance than any single source.

Deep learning, both supervised and unsupervised techniques, have been utilized for stock market prediction. One of very first research work in this segment
belongs to the work of [40] published in 1996 to use recurrent neural networks
(RNN) in ARIMA-based features. Many other feature extraction methods based
on supervised or unsupervised learning have been developed since then [21].

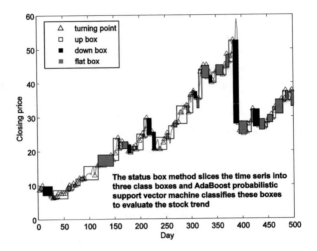

Fig. 2. A status box based on [44].

The authors of [4] used three different unsupervised feature extraction methods: principal component analysis (PCA), auto-encoder and restricted Boltzmann machine (RBM) for the auto-regressive (AR) model. In the same direction, the authors of [24] designed a multi-filters neural network to automatically extract features from stock price time-series. The authors combine both convolutional and recurrent filters to one network for the feature extraction task. In [45] the authors used Empirical Mode Decomposition [34] with neural networks for the feature extraction.

3   Our Contribution

Compared to supervised/unsupervised learning approaches, the difference of our approach is that we generate a trading strategy rather than only stock price prediction as in existing research studies [10, 19]. Stock price prediction definitely is a very important task, but eventually we need to build a strategy to decide what to buy and what to sell in the market that requires a further research step. In this study, we employ a simple baseline greedy strategy given the prediction of a supervised learning algorithm (RNN-LSTM) but definitely studying a strategy based on the prediction of another algorithm is not a trivial task.

Several research works in using reinforcement learning have been presented [1, 6, 22] in literature. However, the work of [22] is presented in 2001 using only TD(0) algorithm, while the works of [6] or [1] used external information such as news for the trading task. In our approach we do not use any external information but only the historical data of stock prices for the trading.

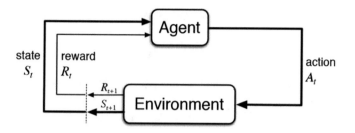

Fig. 3. The interaction between agent and environment in reinforcement learning.

4   Reinforcement Learning

Reinforcement learning [38] is visualized in Fig. 3. Different from supervised learning techniques that can learn the entire dataset in one scan, the reinforcement learning agent learns by interacting repeatedly with the environment. We can imagine the agent as a stock trader and the environment as the stock market [22]. At a time step t, the agent performs an action A_t and receives a reward $R_{t+1} = R(S_t, A_t)$. The environment then move to the new state $S_{t+1} = \delta(S_t, A_t)$. The agent needs to learn a policy $\pi : S \to A$, i.e. learn to react to the environment in which it can maximize the total reward as:

$$V^\pi(S_t) = \sum_{k=0}^{\infty} \gamma^k R_{t+k+1} \tag{1}$$

Here, the coefficient γ represents the decay factor, usually consider as interest rate in finance, reflects the "one dollar today is better than one dollar tomorrow" statement. It means any trading strategy should beat the risk-free interest rate because otherwise a reasonable investor should not invest to this strategy at all - she should invest money to the risk-free rate such as buying T-bonds or opening a saving account. The latter option will give her a better profit and lower risk. However in high-frequency trading and short period of time we can set γ close to 1. The optimal policy is notated as π^*.

In this paper we employ Deep Q-learning [28] by approximate the optimal policy function by a deep neural network. The term "Deep" here refers to Deep Convolutional Neural Networks (CNNs) [36]. Here we parameterize the Q-function by a parameter set θ.

In our settings, the actions are similar to other stock trading studies. The possible actions include buy, hold or sell. We defined the rewards as the profit (positive, neutral or negative) after each action.

The loss function is:

$$L(\theta) = \frac{1}{N} \sum_{i \in N} (Q_\theta(S_i, A_i) - Q'_\theta(S_i, A_i))^2 \tag{2}$$

with

$$Q'_\theta = R(S_t, A_t) + \gamma max_{A'_i}(S'_i, A'_i) \tag{3}$$

The network is updated by a normal gradient descent:

$$\theta \leftarrow \theta - \alpha \frac{\partial L}{\partial \theta} \quad (4)$$

In deep Q-network, the gradient of the loss function is calculated as:

$$\nabla_{\theta_i} L(\theta_i) = E_{S,A \sim P(.),S' \sim \epsilon}[(R_{t+1} + \gamma max_{A'} Q(S_{t+1},A') - Q(S,A,\theta_i))\nabla_{\theta_i} Q(S,A,\theta_i)] \quad (5)$$

5   Experiments

5.1   Datasets

We use the daily stock price of more than 7,000 US-based stocks collected up to 10-November-2017[2]. For each stock, we always use the period of time from 01-January-2017 until 10-November-2017 for testing, and the data from 01-January-2015 until 31-December-2016 as the training set. Hence, there are 504 samples for training and 218 samples for testing. The sample size is so small compared to well-known supervised learning problem such as ImageNet [20] that contains one million labelled images, but as we will present in Sect. 5.2, we still can generate positive profit strategies.

The stock price of Google is displayed in Fig. 4.

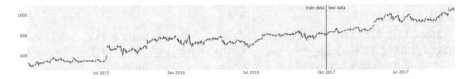

Fig. 4. The stock price of Google from 01-Jan-2015 to 10-November-2017

5.2   Experimental Results

We evaluate three variants of Deep Q-learning which is vanilla Deep Q-learning [28], double DQN [12] and dueling double DQN [41].

We used RNN-LSTM model with a greedy strategy approach as the baseline model. According to a recent time-series prediction[3] organized in July 2019, the RNN-LSTM model achieved the highest performance in forecasting a financial time-series. The greedy strategy means that we buy every-time we predict the stock will go up and sell if we predict the stock will go down.

We visualize the performance of four models vanilla DQN, double DQN, dueling DQN and the baseline LSTM model on the Google stock (code: GOOGL) in Figs. 5, 6 and 7 respectively. The profit of each model are described in the

---

[2] https://www.kaggle.com/borismarjanovic/price-volume-data-for-all-us-stocks-etfs.
[3] https://www.isods.org/.

Fig. 5. Performance of vanilla DQN on Google stock. The profit on the test period is −838.

Fig. 6. Performance of Double DQN on Google stock. The profit on the test period is 1430.

Fig. 7. Performance of Dueling Double DQN on Google stock. The profit on the test period is 141.

figure. In each figure, we visualize the profit of each model in general, the profit of each model against the stock price and the profit of each model against the standard deviation of the stock price.

Generally speaking, the Deep Q-network yields the highest average profit compared to the Double Deep Q-Network and the Dueling Deep Q-Network. The result is consistent to other studies [14]. However, as expected, Deep Q-Network yields a higher volatility compared to two other methods. We display the distribution of profit generated by each model in Fig. 8. It is clear that Double Q-Network seeks for profit, hence sometimes it generates the negative profit. We note that the LSTM model combined with the greedy algorithm is much more stable than other models because we buy and sell immediately when we detect any signal of changing the price.

As described above, we visualize the profit against mean of stock price and standard deviation of stock price in the testing period in Figs. 9 and 10. The main idea is to see if a model behave differently or not in different segments of stock price. The general trends for all models are the expected profit range are higher given the higher stock price and stock volatility. The results are consistency with the principle of no arbitrage in finance [9] which basically stated that we cannot expect a higher profit without facing a higher risk.

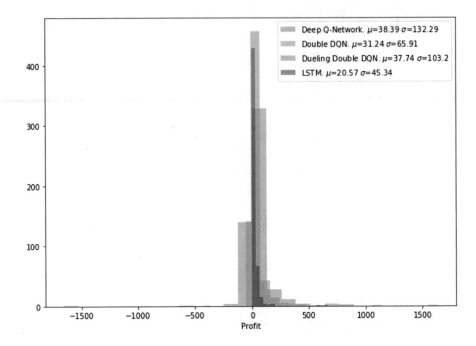

Fig. 8. Distribution of profit of three models

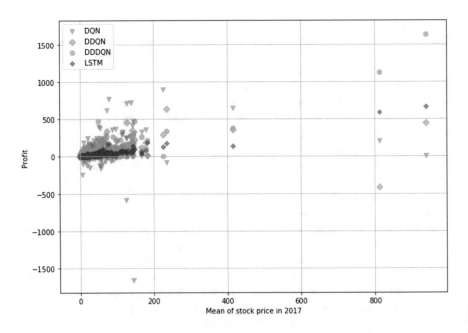

Fig. 9. The profit against mean of stock price in the testing period

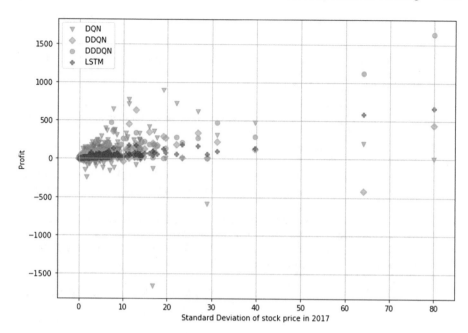

Fig. 10. The profit against standard deviation of stock price in the testing period

6   Conclusions

In this paper we study the usage of Deep Q-Network for stock trading. We evaluated the performance of Deep Q-Network in large-scale real-world datasets. Deep Q-Network allow us to trade the stock directly without taking further optimization step like other supervised learning methods. Using only few hundreds samples, reinforcement learning algorithms variants based on Q-learning can generate the strategies that on average earning a positive profit.

In the future, we plan to incorporate multiple stock trading, i.e. portfolio management strategies, into the study. Furthermore, we will introduce different constraints into the model, for instance the maximum loss one can resist while using a model. Another approach is to integrate simulated behavior of users in non-cooperative or cooperative markets [5, 16].

Acknowledgment. We would like to thank the anonymous reviewer for valuable comments.

References

1. Azhikodan, A.R., Bhat, A.G., Jadhav, M.V.: Stock trading bot using deep reinforcement learning. In: Innovations in Computer Science and Engineering, pp. 41–49. Springer, Singapore (2019)
2. Bisoi, R., Dash, P.K.: A hybrid evolutionary dynamic neural network for stock market trend analysis and prediction using unscented Kalman filter. Appl. Soft Comput. **19**, 41–56 (2014)
3. Bradley, D.A.: Stock Market Prediction: The Planetary Barometer and how to Use it. Llewellyn Publications, Woodbury (1948)
4. Chong, E., Han, C., Park, F.C.: Deep learning networks for stock market analysis and prediction: methodology, data representations, and case studies. Expert Syst. Appl. **83**, 187–205 (2017)
5. Dang, Q., Ignat, C.: Computational trust model for repeated trust games. In: Trustcom/BigDataSE/ISPA, pp. 34–41. IEEE (2016)
6. Deng, Y., Bao, F., Kong, Y., Ren, Z., Dai, Q.: Deep direct reinforcement learning for financial signal representation and trading. IEEE Trans. Neural Netw. Learn. Syst. **28**(3), 653–664 (2017)
7. Elton, E.J., Gruber, M.J., Brown, S.J., Goetzmann, W.N.: Modern Portfolio Theory and Investment Analysis, 9th edn. Wiley, Hoboken (2014)
8. Florescu, I., Mariani, M.C., Stanley, H.E., Viens, F.G.: Handbook of High-frequency Trading and Modeling in Finance, vol. 9. Wiley, Hoboken (2016)
9. Föllmer, H., Schied, A.: Stochastic Finance: An Introduction in Discrete Time, 4th edn. Walter de Gruyter, Berlin (2016)
10. Göçken, M., Özçalici, M., Boru, A., Dosdogru, A.T.: Stock price prediction using hybrid soft computing models incorporating parameter tuning and input variable selection. Neural Comput. Appl. **31**(2), 577–592 (2019)
11. Granger, C.W.J., Morgenstern, O.: Predictability of Stock Market Prices. Heath Lexington Books, Lexington (1970)
12. van Hasselt, H., Guez, A., Silver, D.: Deep reinforcement learning with double q-learning. In: AAAI, pp. 2094–2100. AAAI Press (2016)
13. Henrique, B.M., Sobreiro, V.A., Kimura, H.: Literature review: machine learning techniques applied to financial market prediction. Expert Syst. Appl. **124**, 226–251 (2019)
14. Hester, T., Vecerík, M., Pietquin, O., Lanctot, M., Schaul, T., Piot, B., Horgan, D., Quan, J., Sendonaris, A., Osband, I., Dulac-Arnold, G., Agapiou, J., Leibo, J.Z., Gruslys, A.: Deep q-learning from demonstrations. In: AAAI, pp. 3223–3230. AAAI Press (2018)
15. Hiransha, M., Gopalakrishnan, E.A., Menon, V.K., Soman, K.: NSE stock market prediction using deep-learning models. Procedia Comput. Sci. **132**, 1351–1362 (2018)
16. Ignat, C.L., Dang, Q.V., Shalin, V.L.: The influence of trust score on cooperative behavior. ACM Trans. Internet Technol. (TOIT) **19**(4), 46 (2019)
17. Jacobs, R.A., Jordan, M.I., Nowlan, S.J., Hinton, G.E., et al.: Adaptive mixtures of local experts. Neural Comput. **3**(1), 79–87 (1991)
18. Jangmin, O., Lee, J., Lee, J.W., Zhang, B.T.: Adaptive stock trading with dynamic asset allocation using reinforcement learning. Inf. Sci. **176**(15), 2121–2147 (2006)
19. Jiang, X., Pan, S., Jiang, J., Long, G.: Cross-domain deep learning approach for multiple financial market prediction. In: IJCNN, pp. 1–8. IEEE (2018)

20. Krizhevsky, A., Sutskever, I., Hinton, G.E.: Imagenet classification with deep convolutional neural networks. In: NIPS, pp. 1106–1114 (2012)
21. Längkvist, M., Karlsson, L., Loutfi, A.: A review of unsupervised feature learning and deep learning for time-series modeling. Pattern Recogn. Lett. **42**, 11–24 (2014)
22. Lee, J.W.: Stock price prediction using reinforcement learning. In: ISIE 2001. 2001 IEEE International Symposium on Industrial Electronics Proceedings (Cat. No. 01TH8570), vol. 1, pp. 690–695. IEEE (2001)
23. Lo, A.W., Mamaysky, H., Wang, J.: Foundations of technical analysis: computational algorithms, statistical inference, and empirical implementation. J. Financ. **55**(4), 1705–1765 (2000)
24. Long, W., Lu, Z., Cui, L.: Deep learning-based feature engineering for stock price movement prediction. Knowl. Based Syst. **164**, 163–173 (2019)
25. Malkiel, B.G., Fama, E.F.: Efficient capital markets: a review of theory and empirical work. J. Financ. **25**(2), 383–417 (1970)
26. Menon, V.K., Vasireddy, N.C., Jami, S.A., Pedamallu, V.T.N., Sureshkumar, V., Soman, K.: Bulk price forecasting using spark over NSE data set. In: International Conference on Data Mining and Big Data, pp. 137–146. Springer, Cham (2016)
27. Mitra, S.K.: How rewarding is technical analysis in the indian stock market? Quant. Financ. **11**(2), 287–297 (2011)
28. Mnih, V., Kavukcuoglu, K., Silver, D., Graves, A., Antonoglou, I., Wierstra, D., Riedmiller, M.A.: Playing Atari with deep reinforcement learning. CoRR **abs/1312.5602** (2013)
29. Nazário, R.T.F., e Silva, J.L., Sobreiro, V.A., Kimura, H.: A literature review of technical analysis on stock markets. Q. Rev. Econ. Financ. **66**, 115–126 (2017)
30. Nison, S.: Japanese Candlestick Charting Techniques: A Contemporary Guide to the Ancient Investment Techniques of the Far East. Penguin, New York (2001)
31. Park, C.H., Irwin, S.H.: What do we know about the profitability of technical analysis? J. Econ. Surv. **21**(4), 786–826 (2007)
32. Patel, J., Shah, S., Thakkar, P., Kotecha, K.: Predicting stock and stock price index movement using trend deterministic data preparation and machine learning techniques. Expert Syst. Appl. **42**(1), 259–268 (2015)
33. Preis, T., Moat, H.S., Stanley, H.E.: Quantifying trading behavior in financial markets using Google trends. Sci. Rep. **3**, 1684 (2013)
34. Rilling, G., Flandrin, P., Goncalves, P., et al.: On empirical mode decomposition and its algorithms. In: IEEE-EURASIP Workshop on Nonlinear Signal and Image Processing, vol. 3, pp. 8–11. NSIP 2003, Grado (I) (2003)
35. Schulmeister, S.: Profitability of technical stock trading: has it moved from daily to intraday data? Rev. Financ. Econ. **18**(4), 190–201 (2009)
36. Sewak, M.: Deep Reinforcement Learning - Frontiers of Artificial Intelligence. Springer, Singapore (2019)
37. Simon, H.A.: A behavioral model of rational choice. Q. J. Econ. **69**(1), 99–118 (1955)
38. Sutton, R.S., Barto, A.G.: Reinforcement Learning: An Introduction. MIT Press, Cambridge (2018)
39. Thomsett, M.C.: Getting Started in Fundamental Analysis. Wiley, Hoboken (2006)
40. Wang, J., Leu, J.: Stock market trend prediction using ARIMA-based neural networks. In: ICNN, pp. 2160–2165. IEEE (1996)
41. Wang, Z., Schaul, T., Hessel, M., van Hasselt, H., Lanctot, M., de Freitas, N.: Dueling network architectures for deep reinforcement learning. In: ICML. JMLR Workshop and Conference Proceedings, vol. 48, pp. 1995–2003. JMLR.org (2016)

42. Zhai, Y.Z., Hsu, A.L., Halgamuge, S.K.: Combining news and technical indicators in daily stock price trends prediction. In: ISNN (3). Lecture Notes in Computer Science, vol. 4493, pp. 1087–1096. Springer, Heidelberg (2007)

43. Zhang, G.P.: Time series forecasting using a hybrid ARIMA and neural network model. Neurocomputing **50**, 159–175 (2003)

44. Zhang, X., Li, A., Pan, R.: Stock trend prediction based on a new status box method and adaboost probabilistic support vector machine. Appl. Soft Comput. **49**, 385–398 (2016)

45. Zhou, F., Zhou, H., Yang, Z., Yang, L.: EMD2FNN: a strategy combining empirical mode decomposition and factorization machine based neural network for stock market trend prediction. Expert Syst. Appl. **115**, 136–151 (2019)

A Survey on Forecasting Models for Preventing Terrorism

Botambu Collins[1], Dinh Tuyen Hoang[1,3], HyoJeon Yoon[1], Ngoc Thanh Nguyen[2], and Dosam Hwang[1(✉)]

[1] Department of Computer Engineering, Yeungnam University, Gyeongsan, South Korea
botambucollins@gmail.com, hoangdinhtuyen@gmail.com,
hjyoon314@ynu.ac.kr, dosamhwang@gmail.com
[2] Faculty of Computer Science and Management,
Wroclaw University of Science and Technology, Wroclaw, Poland
Ngoc-Thanh.Nguyen@pwr.edu.pl
[3] Faculty of Engineering and Information Technology, Quang Binh University, Dong Hoi,
Vietnam

Abstract. The security and welfare of any country are very crucial with states investing heavily to protect their territorial integrity from external aggression. However, the increase in the act of terrorism has given birth to a new form of security challenges. Terrorism has caused untold suffering and damages to civilian lives and properties and hence, finding a lasting solution to terrorism becomes inevitable. Until recently, the discourse on the nature and means of combating terrorism have largely been debated by politicians or statesmen. This study attempts an appraisal of machine learning survey on models in preventing terrorism via forecasting. Since the act of terrorism is dynamic in nature, i.e. their strategies change as counterterrorism methods are improved thereby requiring a more sophisticated way of predicting their moves. Some models discussed in this study include the Hawkes process, STONE, SNA, TGPM, and Dynamic Bayesian Network (DBN) which are all geared towards predicting the likelihood of a terrorist attack.

Keywords: Terrorism · Counterterrorism · Forecasting models · Counterterrorism models

1 Introduction and Background

Nowadays, Terrorism is a very crucial issue affecting our contemporary world. Terrorism derived its meaning from the word "terror" dating back to the French revolution around the 1790s, pertaining to the French government use of force to intimidate and coerce the populace to submission [4, 10]. Until recently, the debate about terrorism and means of combating it has always been viewed as a political science discourse. Thus, the effort towards combating terrorism has been in the hands of politicians. Contrary to this, is the fact that the weapons used to carry out an act of terrorism are manufacture by engineers and hence, there is a need for an engineering model in handling terrorism. Before we proceed, it is crucial for us to understand the meaning of terrorism.

© Springer Nature Switzerland AG 2020
H. A. Le Thi et al. (Eds.): ICCSAMA 2019, AISC 1121, pp. 323–334, 2020.
https://doi.org/10.1007/978-3-030-38364-0_29

The discourse on the nature and meaning of terrorism has been a conflicting one since there has not been an agreement on the definition of terrorism, [18]. The FBI defined terrorism as "the unlawful use of force and violence against persons or property to intimidate or coerce a government, the civilian population, or any segment thereof, in furtherance of political or social objectives". UN Security Council Resolution [18] surmises a widely definition of terrorism which states that any "criminal acts, including against civilians, committed with the intent to cause death or serious bodily injury, or taking of hostages, with the purpose to provoke a state of terror in the general public or in a group of persons or particular persons, intimidate a population or compel a government or an international organization to do or to abstain from doing any act" Tavares [16] surmises short term goals of terrorism which include;

(1) Gaining publicity and media attention [4, 16] (2) destabilizing political community and (3) waging havoc to the economies. The definition of terrorism holds divergence views, and this has been the major problem in countering terrorism since one state sees a group and designate it as a terrorist while another state simply has a contrary view. "one-man terrorist is another man freedom fighter" [4, 10, 18]. Although the British and her allied designated the Irish Republican Army (IRA) as a terrorist group, they saw themselves as freedom fighters fighting for a just cause and they had sympathizers and support, the problem of who a terrorist is, rely on the state's interest [10].

2   Rationale for Terrorism

Many arguments have been advanced by scholars of peace and conflict studies, political science and sociology as to why people engaged in act of terrorism [1, 4, 16]. Understanding why people join terrorism or become a terrorist is a necessary step toward combating terrorism, the following are some of the causes of terrorism.

2.1 Economic and Social Deprivation

Some scholars have argued that poverty and economic deprivation is one of the fundamental causes of terrorism and its related activities [12]. When people are poor, they may engage themselves in an unscrupulous act just to make end means. ISIS which is now believed to be one of the largest terrorist organization on Earth recruited Europeans and other members online by promising huge wages as well as other benefits such as offering fighters free women. As a result, their numbers quickly multiplied within a short-term period. [12] contends that persistent poverty and oppression can lead to hopelessness and despair and thus giving birth to terror as it becomes easy for a terrorist organization to recruit such poverty enshrined individuals. Arguing on the fact of poverty, it is argued that not all terrorist group emerges as a result of poverty and not every poverty enshrines individual to join or is motivated to become a terrorist. Poverty cannot be the bases for someone to commit an act of terror, for instance, Osama Bin Laden, one of the leaders of the Al-Qaeda terrorist network was from a wealthy family yet he became a leader of a terrorist group.

2.2 State Weakness (Failed State) (Syria, Iraq, Libya, CAR, Somali)

Failed states or state failure are those states that are unable to deliver political, social, economic goods to its citizens as a result of political upheavals [13, 19]. Such a state is characterized by insecurity and chaos. When a state fails or collapses, there is high crime wave, and the proliferation of weapons becomes imminent. The Fund for Peace alludes that "state is failing when its government is losing physical control of its territory or lacks a monopoly on the legitimate use of force". Failed state or state weakness has often been the cause of terrorism, most failed states become safe havens for terrorist organization [17]. For instance, ISIS took advantage of Libya as a failed State and expanded its network in North Africa, same is true in Syria and Iraq where ISIS or "Daesh" gained enough land and resource control. It is therefore important to take into account state weakness or failed state when finding a suitable means of combating terrorism.

2.3 External Caused Based on Invasion (Iraq, Afghanistan, Libya)

Although the USA and her allies had often played a pivotal role in fighting terrorism, they also hold a greater share of causing terrorism through promoting radicalization as well as causing state failure. State failure has been used as a means for imperialism and invasion by stronger states [17]. Most of the countries in which the US and her allied invaded in the name of "getting rid of dictator" has often cause more harm than good, the US and British invasion of Iraq and Afghanistan in the name of getting rid of Saddam Hussein and the so-called Weapon of Mass Destruction (WMD) of which no such weapons were found in Iraq. The main motive behind the Invasion of both Iraq and Afghanistan stemmed from the anger of the September 11 attacked and thus such invasion had massive support. The White House posits that "The events of 9/11, taught us that weak states, like Afghanistan, can pose a great danger to our national interests as a strong state" [1]. After the invasion of Iraq, a stable state, it later became a failed state with supporters of Hussein joining rebels faction, with abject poverty and misery, Iraq became a breeding ground for most terrorist groups, same as in Afghanistan, war has far-reaching consequences such as destruction, poverty which come after the war has ended and sometimes famine. As already discussed above, when people are poor they become a source of employment by terrorist organization and this was true in Iraq and Afghanistan, also the case of Libya, one of the wealthy state in the world at the time suffered invasion from NATO and became a failed state and today it is a safe haven for ISIS and other terrorist groups.

2.4 Religious Fundamentalism and Extremism

This is view as a type of terrorism in itself and also as cause of terrorism, most of the renowned terrorist organization today are link to religious movement especially Islam, not to say that Islam is a religion of terror but group such as Al Shabaab, Al Nusra Front, the AlQIM, Boko Haram, Al Qaeda usually glorify Allah (God) in their action. ISIS' main goal is to install a global caliphate and make Islam the sole religion on Earth, while Boko Haram main motive is to eradicate Western values and makes Nigeria an Islamic republic and also seek to install a caliphate in the whole of Africa. Almost all Islamic link terrorism group use the slogan "Allahu Akbar" meaning God is the greatest. Religious fanatics and extremist often believe that they are certain rules in which believers must obey, and they often see non-believers as a threat or as enemies, and hence they need to be eradicated. In most cases, the perpetrator of these acts is made to believe that such action is ordained by God and there is no other God except the God in which their belief system is inclined with. There is constant propaganda that one is guaranteed of a good life in heaven, he will be welcome by 72 virgins after a suicide act has been committed and this made suicide bombers even more happy to carry out such an act.

3   Types of Terrorism

They exist a plethora of terrorism types such as State terrorism, Nuclear terrorism, Bioterrorism, Narcoterrorism, Ecoterrorism with each aim at achieving maximum damage to its victims. Some counterterrorism strategies are such that one must understand the type and nature of terrorism before deriving a model at combating it. We will discuss the relevant type of terrorism such as Nuclear terrorism, cyberterrorism.

3.1   Nuclear Terrorism

The aftermath of the Atomic bomb that meltdown the city of Hiroshima and Nagasaki are still fresh in our mind, with the devastating effect still present till date one will wonder if a terrorist could lay hand on such dangerous weapon. The non-proliferation treated geared toward preventing states from acquiring nuclear weapons was successful, in part, because they addressed the motives of non-nuclear states aspiring to acquire nuclear weapon such as Iran. But presently, it is more complex because the effort is now directed towards non-state actors which are usually invisible in the case of a terrorist organization. Nuclear terrorism refers to diverse means in which terrorist can utilize nuclear material to achieve their objectives, this includes; the bombing and destruction of nuclear facility to wreak havoc and cause economic damage to the community, the purchase, production, and utilization of nuclear material to coerce, maimed and instill fear into a civilian population. The action of *Aum Shinrikyo* in Japan where a religious cult terrorist group attacked the Tokyo subway with sarin gas shows that terrorism has no limit [5]. So far, two terrorist groups had struggled to acquired nuclear weapons, al-Qaeda and Aum Shinrikyo [5, 8] both groups struggled multiple time to buy nuclear weapons from Russian unsuccessful, Aum Shinrikyo in an effort even recruited scientist with a plan of developing nuclear weapon and even bought a ranch in Australia to enrich Uranium [5].

3.2   Cyber Terrorism

The rapid advancement of information and communication system has paved a new wave of technological terrorism also refer to as cyberterrorism, the psychological fear of cyber terrorism is more than the actual action, the media, politician, military and security experts make cyberterrorism sound more frightening than conventional terrorism [21]. Imagine if someone could hack the nuclear defense system and push the red button? What about the traffic system, the banks, and the dams? these are the frightening stories about cyberterrorism in which the media and politicians expressed to the public. Weimann [21] contends that there is some sort of financial benefit to both politicians and military personnel's when such fear is expressed to the masses thereby promoting massive funding to protect the cyberspace and consequently an increase to defense budget even without any real threat [21]. While they have been more contested meaning to cyberterrorism owing to the fact that no real such case of cyberterrorism has happened, [15, 21]. Denning [6] surmises a broadly acceptable definition of cyberterrorism and thus alludes that "the convergence of cyberspace and terrorism where unlawful attacks and threats of attack against computers, networks, and the information stored therein are carried out to intimidate or coerce a government or its people in furtherance of political or social objectives, and should result in violence against persons or property, or at least cause enough harm to generate fear. Attacks that lead to death or bodily injury, explosions, plane crashes, water contamination, or severe economic loss would be examples. Serious attacks against critical infrastructures could be acts of cyberterrorism, depending on their impact." A terrorist can use the computer as a weapon such as stealing the identity of people or users, spreading computer viruses and malware, hacking, destroying or manipulating important data [7].

4 Means of Combating Terrorism

Counterterrorism is the means employs by states or security expert to prevent terrorism. Usually, the best way to counterterrorism is predicting the likely attack of a terrorist and this is where machines learning plays a vital role. In the past countering terrorism has been the affair of politicians with the decision to deploy the military and security experts, but because the action of a given terrorist is unknown. It is necessary to start by classifying the attacked type, weapons type, target time based on the previous history. The prevention of terrorism in the past has been based on the cause and nature of terrorism, for instance, in a situation where a failing state or failed state is the cause of terrorism like Libya, Somali, Iraq already discussed above, such a state is given assistance both in term of security and financial aid in order to strengthen the institution of the state. Also, the spillover effect is also another issue, where a particular country is having conflict, the probability P that the conflict will move to a given neighbor T is $T = 1/n$ and therefore such country is required to strengthen its security and prevent the flow of combatant from country X to Y. However, all these strategies worked to a certain degree but terrorism has dramatically increase thereby making mankind to look for a new approach in preventing terrorism. These new approaches include the use of machine learning models to predict and prevent a future terrorist attack, these approaches will be vividly explained below.

4.1 Counterterrorism Model, Machine Learning Approach

In this section, we review some of the approaches that had been used to predict or forecast future terrorist attack (Table 1). The use of computer technology in predicting the likelihood of a terrorist attack is a new trend in research.

Table 1. Showing various model of forecasting terrorist attack.

No.	Model	Proposed model	Uses/action	Authors
1	SNA	STONE	Identify and predicting the successor of a removed terrorist from Network (Al-Qaeda, Hamas, Hezbollah, Lashkar-e-Taiba)	Subrahmanian (2016)
		CT-SNAIR	Tracking and detecting terrorist scenarios and target	Weinstein et al. (2009)
		SMD	Had great success rate to predict if one is terrorist or not.	Coffman and Marcus (2014)
2	DBN	TAPM	Detect the rate of attack transport areas	Manoj (2009)
3	TGPM		Predict the group responsible for an attack.	Sachan and Roy (2012)
4	MMURT		Predict the method of radicalization and terrorism	Chuang and Maria (2016)
5	Hawkes process		Predict the likely attack of terrorist e.g. IRA	Hawkes (1971), Tench (2018)

The nature of terrorist attacks is such that one must always be ready because a terrorist act in a surprising way no one least expected and at such, predicting their actions is a better way of countering terrorism.

4.2 Social Network Analysis Model

The advancement of the internet has ushered in a new form of communication, unlike the tradition media, the social media is becoming increasingly popular with terrorist using the social media in several ways to achieve their objectives, the radicalization, and recruitment of ISIS members were mostly done via the social media and hence the social network analysis model has become a useful tool in tracking and identifying terrorism [20]. The Social Network Analysis SNA is a method used in appraising human social interaction and such an approach has been used to investigate patterns and community structures [19]. The simulated datasets modeled (SDM) which was championed by [19] with a two-class classification for terrorist and no-terrorist and got 93% prediction rate. This model is useful to identify whether someone is a terrorist or not.

Weinstein and colleagues [22] used simulation of terrorist attacks based on real information about past attacks and proposed a model known as Counter-Terror Social Network Analysis and Intent Recognition (CT-SNAIR) which was used in tracking and detecting a terrorist. The use of intent recognition tool was vital in detecting a potential terrorist target. They made use of real-time scenarios herein referred to as forensic scenarios from the September 2004 bombing of the Australian embassy in Indonesia as well as non-real scenarios which were developed in a prior project for research in social network analysis. The utilization and application of a new Terror Attack Description Language (TADL) and with the use of the Hidden Markov Model (HMM), the criteria for modeling and simulation of terrorist attacks was developed.

Shaping Terrorist Organization Network Efficacy (STONE)

Subrahmanian [20] used social network analysis and proposed a model known as Shaping Terrorist Organization Network Efficacy (STONE) which has three key objectives:

- Quantifying Terror Network Lethality
- Predicting Successors of a Removed Terrorist
- Identifying Who to Remove

This method takes into cognizance the network structure of a terrorist organization as well as the vertices herein which represent operational roles, it is assumed that every individual in a network has a role to play irrespectively of the hierarchical structure. The word used in the study such as "removed" or "delete" referred to the targeted head of network or most important person in the network. The killing or neutralization of a known terrorist such as Osama Bin Laden is considered to have been "removed" from the Al-Qaeda organizational network. The network here refers to the organizational network rather than a social network, STONE is a framework to identify a set of k nodes whose removal will optimally destabilize a terrorist network. The STONE approach was tested on four terrorists' networks namely Al-Qaeda, Hamas, Hezbollah, Lashkar-e-Taiba and got 80% success rate of prediction.

Predicting Successors of a Removed Terrorist

Once a terrorist has been removed from a network, he will be replaced and predicting his replacement and deleting him from the network is a fundamental step in destroying the network. In order to predict who will replace a removed terrorist, the author made 4 assumptions as seen below.

Assumption I. It is assumed that the replacement vertex u is not too far away from v in ON

Assumption II. Assume that the probability that v is replaced by a vertex u depends on u's rank in the network.

Assumption III. We assume that individuals with a higher weight than v does not desire v's position.

Assumption IV. It is believed that ON will rebuild itself to be maximally lethal thus when v is deleted from the network, u is selected in a way that maximizes lethality".

For instance, it was already forecasted that should Osama Bin Laden been neutralized or removed, he will be replaced by his right-hand man Ayman al-Zawahiri and hence removing him too become inevitable. One of the most fundamental aspects of this algorithm is the *Person Successor Problem (PSP)*: which set of k ($k > 0$) terrorists should be removed in order to minimize the lethality of the terrorist network.

Suppose ON = (V, E, wt, $\wp$)is an organizational network, r ∈ V is a vertex, and B is a logical condition on VP. The *person successor problem* is the problem of finding a vertex v ∈ V such that the replacing person must meet the conditions of the "removed".

The logical condition *B* preventing who can replace a removed terrorist *r* is simple and understandable. For instance, if *r* is an IT engineer (i.e. $\wp(r,$ IT engineer) = 1), then replacing him with a *cook(chef)* would not yield fruit. Therefore, the professional involves are required to outline that replacement node must have certain features of an IT engineer which may be difficult finding another IT engineer and hence, leading to network destabilization. Furthermore, the strength of this model lies in the fact that it is not based on removing organizational head such as leader but identifying who plays a pivotal role in the organizational network in a way that removing such an individual will create network malfunction and difficult to find a replacement. In terrorist a network, bomb-makers, computer experts and machines operators often play key roles than those just giving orders, those giving command can easily be replaced if removed such as the replacement Osama Bin Laden.

4.3   Dynamic Bayesian Network

According to Hudson and colleagues [11] Bayesian network is a useful tool for antiterrorism risk assessment. In a profiler installation security system planner, Bayesian network was used, and it gives the user more accurate prediction in managing attacks.

Manoj [14] develops a terrorist attack prediction model TAPM using dynamic Bayesian networks (DBNs). His study was motivated based on the September 11 attack where terrorist hijacked four US airplanes, the method is aimed at predicting future attack in transport facilities such as airports, metro and subway systems, bridges, and tunnels. This Dynamic Bayesian Network model is based on the premise of uncertainty and being dynamic and ready at all time for uncertainty. Manoj [14] explains that in a situation where a passenger at a given airport is suspected of being a terrorist and intelligence are gathered about the particular passenger and maybe possible detention and interrogation, and after investigation for about two to three days, they discovered such person was not a terrorist, the security system may be relax, and it will restart the process again and give another prediction taking into cognizance the previous failure. An example is the multiple threats to the New York subway system which indicate that DBN may be capable of updating the threat level over time if new information is available.

4.4   Terrorist Group Prediction Model (TGPM)

Terrorist Group Prediction Model (TGPM) or Predicting the Perpetrator Model (PPM) is one of the best models in predicting the likelihood of which group is responsible for an attack (Fig. 1). When we predict which group is responsible for an attack, counterattack measures can be developed based on the group's history. Terrorist behavior is dynamic based on counterterrorism response, attackers copy tactics from other terrorist group based on success rate, if a terrorist group T successfully carry out an attack in target x, the weapons type and approach used will likely be employed by another terrorist group and thus, this approach study the behavior of terrorist group and predict which group is responsible for an attack. Terrorist Group Prediction Model (TGPM) championed by

Abishek Sachan and Devshri Roy [2] uses similar features of past terrorist attacks and forecast the perpetrator or the group responsible. They employ certain concepts such as Group Detection Model (GDM) Crime Prediction Model and Offender Group Detection Model (OGDM). The study focuses on India as the case study and using a dataset from the Global Terrorism Database (GTD) which includes the history of a terrorist attack in India from 1998 to 2008. This method makes use of input data such as attack type, location of the attack, suicide attack, weapon type, target type, etc., which were used to calculate the group weighted point. The percentage of attacks by each group is calculated. Clusters are created based on the percentage of attacks of each group and the parameter weights. Association of the input data with the formed clusters is calculated and the highest value is chosen. The group name matching to the cluster with the highest association value is predicted to be the group responsible for the attack. The results show 80.41% accuracy rate. The main limitation to this model is the fact that it is a reactive or responsive model rather than proactive, predicting which group is responsible for an action which has already taken place rather than preventing such action yields very little fruit. Couple to that, today's terrorism has gained enormous media attention such that a group will quickly claim responsibility for an action is carried out without necessarily waiting for a predictor. This, therefore, makes such a model redundant and not suitable. In addition, since this model relies heavily on data from historical evidence, it becomes more difficult to predict and identify smaller groups with little history should they carry out an attack using similar techniques with patterns that match that of a larger known terrorist organization. Lastly, this model is not a dynamic model and terrorist actions are evolutional and dynamic in nature. For instance, the predictor was using weapon type and attack type as a feature and the terrorist group manufacture a new weapon and attack strategy which is not found in their history, the model will not be accurate. Nonetheless, the TGPM remains a good source of knowledge in the research for countering terrorism.

4.5 Mathematical Models for Understanding Radicalization and Terrorism MMURT

In this section, Chuang and Maria [3] proposed Mathematical Models for understanding Radicalization and Terrorism MMURT. The increase in terrorism is as a result of radicalization and therefore understanding how and why people become radicalized is a necessary step in preventing terrorism. This approach is based on the notion that everyone has an opinion and the belief of an individual in a given society can change or be influenced via peer to peer contact or through the media [3]. Radicalization usually takes place through networks of peers and can be enhanced through the internet or other technological means mostly in the form of web-based recruitment material or chat rooms, ISIS are zealous for this style and often use internet-savvy group, that uses social media to recruit western foreign fighters in Syria and Iraq [3].

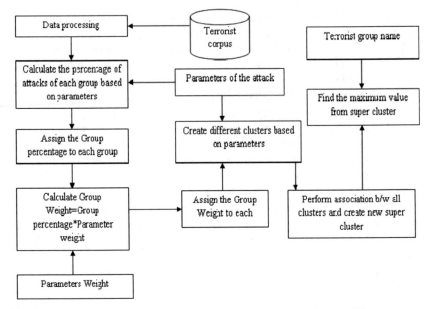

Fig. 1. Terrorist Group prediction Model (TGPM) Source; [2]

4.6 Hawkes Process Modeling

The Hawkes process is useful in predicting event occurrences within a short-term. Developed in 1971, the Hawkes self-exciting point process model explained that event in itself is self-exciting and independent from each other in such a way that a successful occurrence of an event will trigger another occurrence, i.e. a given event raises the chances of another event in the future [14]. Although it was developed to predict the possible occurrence of an Earthquake, Tench (2018) used it as a machine learning approach to countering terrorism by predicting terrorist acts such as the case of the Irish Republican Army (IRA). Terrorist actions are evolutional and similar in cases where they are successful, just like the Hawkes process model, when a terrorist successfully carries out an attack, there is excitement emanating from the success and such excitement increases the chances of further attack within a short-term period.

5 Conclusions and Future Works

Determining the likelihood that a terrorist may carry out an attack is a crucial step in counterterrorism strategy. States are putting countless efforts in securing their territorial boundaries from external aggression (aggression from another state). But, there is a new form of security challenges emanating from terrorism that global efforts at combating it are becoming a potent challenge. Our study is therefore important for not only in academia but for security experts and policymakers to draw ways of combating and preventing terrorism. Most information about terrorist activities is gotten from the media and terrorism database site such as the Global Terrorism Database. So far, no survey was found on models predicting terrorism and it makes it more difficult to compare previous studies. furthermore, another challenge is the fact that most of the models proposed did not focus on a specific case study and hence study is needed to compare a case by a case study of several terrorist groups. One study such as the Hawke process was applied to IRA terrorist group and was successful, but this model did not take into cognizance the goals of such terrorist group and hence, in future work, one needs to take into consideration the types of a terrorist group and their goals. The IRA, for instance, has as main objective to gain an autonomous state of Ireland and their action became dormant when Ireland became an independent state, while other groups such as ISIS has a wider goal such as installing a global caliphate and making the world an Islamic one, finding measures of tackling them is even more difficult.

Acknowledgment. This research was supported by the Basic Science Research Program through the ational Research Foundation of Korea (NRF) funded by the Ministry of Science, ICT & Future Planning (2017R1A2B4009410), and the National Research Foundation of Korea (NRF) grant funded by the BK21PLUS Program (22A20130012009).

References

1. Bassil, Y.: The 2003 Iraq war: operations, causes, and consequences. OSR J. Human. Soc. Sci. (JHSS) (2012). ISSN 2279-0837
2. Bunn, M.G., Malin, M.B., Roth, N.J., Tobey, W.H.: Preventing Nuclear Terrorism: Continuous Improvement or Dangerous Decline? pp. 12–31. Havard University Press (2016)
3. Chuang, Y.l., D'Orsogna, M.R.: Mathematical models of radicalization and terrorism. arXiv preprint arXiv:1903.08485 (2019)
4. Crenshaw, M.: The causes of terrorism. Comparative Politics **13**, 379–399 (1981)
5. Daly, S., Parachini, J., Rosenau, W.: Aum shinrikyo, al qaeda, and the kinshasa reactor: implications of three case studies for combating nuclear terrorism. Technical report, Rand Corp Santa Monica CA (2005)
6. Denning, D.E.: Cyberterrorism: Testimony before the special oversight panel on terrorism committee on armed services. Focus on Terrorism 9 (2000)
7. Gordon, S., Ma, Q.: Convergence of Virus Writers and Hackers: Fact or Fantasy?. Symantec Security White paper, Cupertine (2003)
8. Gross, M.L., Canetti, D., Vashdi, D.R.: Cyberterrorism: its effects on psychological well-being, public confidence and political attitudes. J. Cybersecurity **3**, 49–58 (2017)
9. Hawkes, A.G.: Spectra of some self-exciting and mutually exciting point processes. Biometrika **58**(1), 83–90 (1971)
10. Hoffman, B.: Inside Terrorism, pp. 266–302. Columbia University Press, New York (2006)
11. Hudson, L.D., Ware, B.S., Laskey, K.B., Mahoney, S.M.: An application of Bayesian networks to antiterrorism risk management for military planners. Rapport technique (2005)
12. Humenberger, M.: Dynamics and root causes of recent major terrorist attacks in Europe. Seminar on Current Issues of European and Transatlantic Security (2017)
13. Iqbal, Z., Starr, H.: Bad neighbors: failed states and their consequences. Conflict Manag. Peace Sci. **25**(4), 315–331 (2008)
14. Jha, M.K.: Dynamic bayesian network for predicting the likelihood of a terrorist attack at critical transportation infrastructure facilities. J. Infrastructure Syst. **15**(1), 31–39 (2009)
15. Klein, J.J.: Deterring and dissuading cyberterrorism. J. Strategic Secur. **8**(4), 23–38 (2015)

16. Krieger, T., Meierrieks, D.: What causes terrorism? Public Choice **147**, 3–27 (2011)
17. Le Gouriellec, S.: The strategic threat of weak states: when reality calls theory into question. Semantic scholar (2015)
18. Odhiambo, E.O.: Religious fundamentalism and terrorism. J. Global Peace Conflict **2**(1), 187–205 (2014)
19. Rotberg, I.: When States Fail: Causes and Consequences. Princeton University Press, Princeton (2010)
20. Subrahmanian, V., et al.: Computational analysis of terrorist groups: Lashkar-e-Taiba. Springer, Heidelberg (2012)
21. Weimann, G.: Cyberterrorism: how real is the threat? [electronic version], vol. 119. Special Report (United States Institute of Peace) (2004)
22. Weinstein, C., et al.: Modeling and detection techniques for counter-terror social network analysis and intent recognition. IEEE (2009)

On the Design of a Privacy Preserving Collaborative Platform for Cybersecurity

Thanh-Hai Nguyen$^{(\boxtimes)}$, Vincent Herbert, and Sergiu Carpov

CEA LIST, Design Architectures and Embedded Software Division,
Point Courrier 172, 91120 Palaiseau Cedex, France
{thanhhai.nguyen,vincent.herbert,sergiu.carpov}@cea.fr

Abstract. Nowadays, cyber-attacks are targeting mobile devices, bank accounts, connected vehicles and cyber-physical systems. These attacks are becoming more complex and are raising safety problems when targeting physical environment. An efficient way to protect against these attacks is making several security actors collaborate in defining appropriate countermeasures. However, in practice, security actors refrain from collaborating to avoid sharing their proprietary security processes. These processes represent a critical knowledge as they reflect these actors brand images. In this work, we investigate the use of homomorphic encryption to define a privacy preserving framework for sharing processes between different cybersecurity actors and for providing confidential data analysis. We describe a high level design for a secure cloud platform managing encrypted data. The data analysis algorithms provided by the cloud platform are designed with our open source tool Cingulata, which enables designers to implement any data analysis function, compile it and run it on homomorphically encrypted data.

Keywords: Homomorphic encryption · Private data sharing agreements · European data economy · Secure data exchange · Platform security · GDPR

1 Introduction

A survey by the European Union Agency for Network and Information Security (ENISA) [14] affirms that three-quarters of the businesses faced cybersecurity issues during the last decade. The majority of respondents believed that their organisations had been victims of targeted attacks, and almost a third of them reported a significant business impact. In addition, the advent of Internet of Things (IoT) raised the number of connected devices and so increased the attack surfaces of many IT systems. The increased number of attacks resulted in a great demand of security protection mechanisms. These mechanisms in turn require

This work is sponsored by the H2020 European Research Council funding (Grant agreement ID: 700294, and ID: 727528) for the C3ISP and KONFIDO projects (http://www.c3isp.eu/, https://konfido-project.eu/).

H. A. Le Thi et al. (Eds.): ICCSAMA 2019, AISC 1121, pp. 335–345, 2020.
https://doi.org/10.1007/978-3-030-38364-0_30

collecting data, either for incident reporting, for attack prediction or for system protection improvement.

Current IT services generate an increasing number of data regarding several events. Events come from various origins such as: system logs for access control, hardware traps raised by hypervisors, applications audit trails that identify business transactions. The diverse nature of these events with the 5 V[1] storage challenges of big data, make it difficult to perform either real-time detection of exceptional events-aimed to promptly take corrective actions, or deep and narrow historical data analysis aimed at identifying warning signs of security threats.

In the work with C3ISP [4] and KONFIDO [20] projects, we specify a platform to create an efficient and flexible framework for secure data analytics where data access and data analysis operations are regulated by multi-stakeholders data-sharing agreements.

The rest of this paper is structured as follows. In Sect. 2, we evaluate the benefits of developing a flexible framework, which allows confidential and collaborative information sharing and analysis among relevant security stakeholders. Based on the architecture of the C3ISP framework, we further propose a solution for privacy aware cloud platform design with a support for homomorphic encryption in Sect. 3. We evaluate this platform via an industrial use-case detailed in Sect. 4. Finally, the Sect. 5 concludes the paper by presenting the limitations of this work and its future enhancements.

2   Information Sharing and Privacy-Aware Design Principles

Critical infrastructures like energy, water and nuclear ones receive increased attention globally due to its importance to our society or national defense. Fortunately, thanks to increased awareness and attention by government agencies and company management in these domain, the security of these systems is improving. However, substantial challenges remain a bottleneck to information sharing between these companies, the government and regulation authorities due to the critical nature of data and sovereignty issues.

In this section, we summarize some of the actions taken to date - with emphasis on the European Union experience - to encourage and facilitate improved information sharing.

2.1   Benefits of Information Sharing in Mitigating Cyber and Physical Threats

A simple way to look at the information sharing is as data flows between government, private business and citizens. Essentially optimal information sharing allows an open flow of ideas and concerns from the government to the businesses

---

[1] 5 V big data challenges: Data Volume, Velocity, Variety, Veracity and Value.

and citizens, and equivalent flows of information from the businesses/private citizens back to the government. And, of course, the flow of information between the citizens and the businesses is also desired. Unfortunately, the open and unimpeded flow of information between these different institutions and individuals can be slowed-or even blocked due to a variety of challenges such as politics, intellectual property protection, legal concerns, sovereignty and ultimately a lack of trust.

There is a significant advantage on sharing information in terms of increasing the capability of attacks detection and prevention. In the past years a number of security attacks with very serious consequences have been performed. Noticeable examples are the Flame and Conficker malware [8] which caused a loss pf several millions of dollars. These attacks have been successfully tackled thanks to a collaboration among security and business companies. These companies shared relevant information, whose collaborative analysis has been vital in detecting the features to prevent upcoming attacks of this kind. Indeed, as noticed also by the activities of the WG2, of the Network and Information Security (NIS) EU, on Information Sharing [19], there are several benefits in information sharing for cybersecurity (including incident notification) as well as several barriers to be removed. Finally, the report on Cybersecurity Policies and Critical Infrastructure Protection [18] confirms that sharing information on security threats between security actors is the best way to reinforcement against cyberattacks.

Some benefits of information sharing for cybersecurity are:

- Faster attacks prediction and then reaction
- Collaborative threats analysis
- Increasing knowledge on attacks thwarting

However, there are also several barriers to sharing cybersecurity data. Some of these barriers are related to the provision of proper access control mechanisms to cybersecurity data [19]. Indeed, organizations fear the risk related to their brand images and the loss of their reputation if the shared information reveals publicly their cybersecurity incidents There are also other concerns related to the compliance with legislation such as the GDPR with restricts the access to personal information.

Some barriers of cybersecurity information sharing are:

- Lack of trust on entities sharing data
- Lack of such systems as commodity
- Lack of control on shared data

In our works, we aim at developing a technological and procedural framework to unleash the power of information sharing for collaborative analytics for cyber protection in a confidential manner. The framework allows data producers/consumers to easily express their preferences on how to share their data, which operations can be performed in such data and with whom the resulting data can be shared etc. This entails a framework that combines several technologies for expressing and enforcing data sharing agreements as well as technologies to perform data analytics operations, in a way compliant to these agreements. Among these technologies, we can mention data-centric policy enforcement mechanisms and data analysis operations directly performed on encrypted data provided by multiple producers/consumers or prosumer.

2.2 Privacy-Aware Platform Design

The privacy-aware platform design consists of defining the component patterns and criteria in the respectful ways to approach privacy and data collection. A right way of exploring data will allow detect unauthorized data requests, malicious third-party tracking and offboarding experience. One of interesting book to read and apply in practices is the one of Fair Information Practices proposed by the Department of Health, Education, and Welfare (HEW) in a 1973 study entitled Records, Computers, and the Rights of Citizen [16]. The resulting recommendations are the following:

1. Provide full description of data collection and exploitation purpose: The full description on data usage is considered as a critical element of public data collection, and was a sharing consent form of the Fair Information Practices. This consent form allows to prosumer know in the transparent way the usage on their data and the constraint according to the privacy policy under which it was collected like: the duration of data exploitation and purpose, the measures in terms of storage for data protection, the analysis model based on which customer data would be tread. From the point of view of legal implications, privacy-related architecture design need to clarify and figure out the questions of legal capacity, adequate understanding in a simple way for architecture design in order to highlight any residual and legal risks [10] during architecture audit. Another purpose for the sharing consent form is a pedagogical purpose, it implies user awareness about the presence of data collection and data usage.

2. A minimal and sufficient data collection for a specific analysis model: "Personal data" could be revealed the behavior, thoughts, and/or preferences of an individual, it is very important for product development, but very sensible for the development and testing phase. Any data that are used in these phases need to be anonymized or homomorphically encrypted before.

3. A minimal identification of data with individuals: in the use case of telco, the software of 360-degree customer management sometimes requires collecting customer information like user equipment, user position (GPS) and user behavior when surfing the web. In this critical point, when defining a plan of privacy - aware platform design, an independent and isolated authentication, identification with authorization authority would be planned to implement and it could be necessary and sufficient for future data processing systems.

4. Minimize and secure data retention: Data usage need to be directly related to a feature of an application. If data is restrained and stored, then it must be protected in an auditable secure way. The minimum security requirement is to store data with encryption technology in a manner that data disclosure is difficult or impossible without decryption process. In addition, the proof of

data nonreusability is always required, which ensures to customers inability to reuse of stored data.

3 Privacy Preserving Computation on Cloud Platforms

In this section, we describe the basic component of a secure cloud platform with distributed constraints.

3.1 Secure Content-Based Routing

Secure Content-Based Routing (CBR) is a known event - driven architecture which supports to the communication of machine processors or micro-services in distributed cloud platform. This technical design described with high level in [13,18] allows routing messages based on their content rather than their destination and simplifying with reducing linkage between micro-services or diverse functionalities in cloud application. The first principle of this technique is to favor data production/consumption decoupling: synchronisation decoupling with maximizing asynchronous and anonymous communication; space decoupling: unknown data producers and data consumers; time decoupling: production and consumption at different times. The second one is to allow scalability in terms of message volume per minute, data volume per second and finally connection volume (producers and consumers) at a given instant. In the context of cloud application development, this kind of technique is known as distributed event-based systems, distributed publish-subscribe systems, distributed messaging service, message-oriented middleware, active databases.

To secure data communication between producer (publisher) and consumer (subscriber), a communication mechanism described in [15] can be improved by adding an extra security layer to the process. This mechanism ensures and reduces the threat to the data confidentiality and integrity. Moreover, using SSL connection with strong authentication may offer a secure protocol for exchanging confidential data content like cryptographic keys between both ends of the communication chain. As publications and subscriptions are encrypted and signed, it becomes easier to detect a malicious attack or data corrupted. Figure 1 presents a secure event driven architecture

3.2 Secure Data Access Control

The standard security requirements of confidentiality, integrity and availability (CIA-triad) are at the core of our data-centric protection vision. In addition to these traditional security requirements, also non-repudiation, authentication, authorisation, and accountability are among the top priorities of private data sharing.

1. Confidentiality: the property of protecting the secrecy of data and only disclose it to authorised parties under the policies specified in their Data Sharing

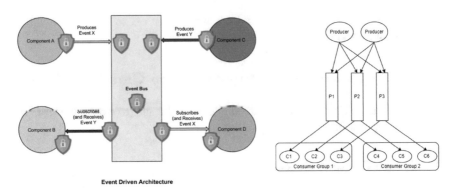

Fig. 1. Multiple produces send messages to a partitioned log, consumed by multiple consumer groups.

Agreements (DSAs). Specifically, encryption has been traditionally used for preserving data confidentiality in transit and at rest, and just lately homomorphic encryption has started emerging as a novel approach for ensuring data confidentiality in use;

2. Integrity: the property of preserving the data from fraudulent modification;
3. Availability: the capability to have the data available when they are requested (this is important for a quick reaction to attacks).
4. Non-repudiation: the fact that one party cannot deny of having submitted data, can help as a deterrent countermeasure to limit a malicious party to submit bad data;
5. Authentication and authorisation: both the data sharing and the consumption of analytics results have to be protected by access control mechanisms and conditions. DSAs regulate how access and usage control protect the cyber threat information;
6. Accountability: especially to address compliance mandatory requirements or to help internal investigations, assess the correctness of system processing, etc., our platform has to be able to trace and identify the right entities or people that participate in the DSA-regulated federation and be able to understand that the policies stated have been correctly and effectively enforced.

3.3   Homomorphic Encryption-Based Cloud Analysis Platform

Homomorphic encryption (HE) is an encryption method which allows to perform computation on encrypted data without decrypting it. Such schemes are known to be very useful to construct privacy preserving protocols even in its classical version. As example, homomorphic encryption has been used as a key-tool in the popularisation of electronic-based voting scheme. Another application of homomorphic encryption is Private Information Retrieval, which is a communication efficient interactive protocol which allows a user to retrieve an item in a database without revealing which item he is looking for. This paradigm has found a number of applications in numerous contexts: private searching, keyword search, private storage, anonymous authentication, etc. Another very popular scenario which makes the benefits of homomorphic encryption is cloud computing: a user relies on some computing resources from a cloud provider to perform expensive computation on sensitive data. These scenarios have in common that Fully HE (FHE) encryption is used as a method which allows the scrambling of data in order to protect their confidentiality via the execution of algorithm on encrypted data. In real world applications using FHE encryption, one or several entities interact with the cloud. To preserve privacy of each user, the data are sent encrypted over the cloud. The service provider processes the received data homomorphically and sends the encrypted result to an end user (owning the FHE parameters and, hence its secret key). The latter one decrypts the result using its own decryption key. Here, the service provider can compute almost any functions over the encrypted data and acts transparently with respect to each entity using only public information and encrypted data (Fig. 2).

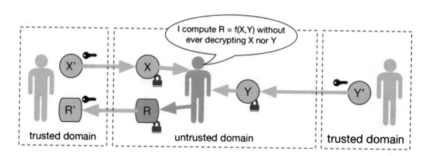

Fig. 2. Homomorphic encryption allows computation on encrypted data without decrypting it in untrust environnement.

The underlying mathematical objects used to conceive fully homomorphic encryption schemes are euclidean lattices. The security of almost all known FHE construction relies on the problem of finding short vector or basis in a high dimensional lattice. Gentry's solution relies on ideal lattices over algebraic number fields. In 2012, Brakerski, Gentry and Vaikuntanathan (BGV) [2] improved this scheme without using bootstrapping; they proposed a generalized construction secure under the popular Learning With Errors assumption and its ring variant. Then Brakerski [1] proposed a new scale invariant scheme that does not require modulus switching. In 2012, Fan and Vercauteren (FV) [6] proposed a ring variant scheme and improved its efficiency. The so-called BGV and FV cryptosystems which are already implemented in version Cingulata 1.0 [17]. The 3rd generation of FHE with fast bootstrapping techniques called TFHE - Fast Fully Homomorphic Encryption over the Torus based on [5,9] is released in the version Cingulata 2.0 since June 2019.

Our Cingulata open source [3] offers a compiler chain with high-level language development targeting HE execution based on manipulating Boolean circuits.

That is a directed graph $G = (V, A)$ which vertices are either inputs, outputs or operators (XOR, AND) and which arcs corresponds to data transfers. The following constraints are imposed to a compiler targeting HE execution [7,12]:

- No ifs (unless regularized by conditional assignment).
- No data dependant loop termination (need upper bounds).
- Array dereferencing/assignment in $O(n)$ (vs $O(1)$).
- Algorithms always realize (at least) their worst-case complexity!

In terms of technology design, this compilation chain is composed of 3 layers: a front-end, a middle-end and a back-end. The front-end transforms code written in C++ into its Boolean circuit representation. The middle-end layer optimizes the Boolean circuit produced by the front-end. The back-end homomorphically executes the Boolean circuit over encrypted data. Two HE libraries are supported by our Cingulata compiler: (i) an in-house implementation of [6] and (ii) the publicly available TFHE library.

A simple "hello world" example written using Cingulata is:

```
CiInt a{CiInt::u8};          // create an unsigned 8-bit variable
CiInt b{CiInt::u8v(42)};     // use helper function to create
                             // an unsigned 8-bit
CiInt c{-1, 16, false};      // or manually specify value,
                             // size and signedness

a.read("a");                 // read variable a and b
b.read("b");

c = a + b;
```

Using the FV cryptosystem this program is homomorphically executed in under 5 s and using TFHE in under 1 s.

4 Use Case: A Privacy Preserving Detection of Brute Force Attacks

This analytics works on connection request logs and identifies whether the destination addresses belong to malicious hosts. Connection logs are directly taken from a router that use the Netflow V9 protocol and pushes the information to a client software that collects the logs. This service is run with a combination of a Data Manipulation Operations component before linking to the homomorphic encryption operation. We use CEF [11] format to crawl log data, the format CEF is an extensible, text-based format designed to support multiple device types by offering the most relevant information. Message syntax's are reduced to work with ESM normalization. Specifically, CEF defines a syntax for log records comprised of a standard header and a variable extension, formatted as key-value pairs.

The following example illustrates a CEF message using Syslog connection:

```
20 01:25:12 host
CEF:0|Router_Vendor|Router_CED|1.0|100| Connection
Detected|5|src=192.168.1.31 spt=22126 dst=
214.141.161.177 dpt=24920 proto=TCP
end=1505462161000 dtz=EuropeBerlin
```

For a high performance with trade-off between the security levels and keep FHE analysis in practical usage, we propose four configurations dubbed:

- BLACK_LIST_FULL
- BLACK_LIST_HIGH
- BLACK_LIST_MEDIUM
- BLACK_LIST_LOW

They permit to establish a trade-off between security and efficiency by encrypting respectively the last $i = 4, 3, 2, 1$ byte(s) of each IPv4. Let us take an example with an address IPv4, it is with this form a.b.c.d. We use $[x]_{fhe}$ fhe to design the data x encrypted with FHE.

- BLACK_LIST_FULL : $[a]_{fhe}.[b]_{fhe}.[c]_{fhe}.[d]_{fhe}$
- BLACK_LIST_HIGH : $a.[b]_{fhe}.[c]_{fhe}.[d]_{fhe}$
- BLACK_LIST_MEDIUM : $a.b.[c]_{fhe}.[d]_{fhe}$
- BLACK_LIST_LOW : $a.b.c.[d]_{fhe}$

We have used a compute server $W3520$ @2.67 GHz, 128 GB Ram with 48 cores CPU and consider a result of 32 bit, if the result is positive then the IP is belong to Blacklist, otherwise it is not. In terms of performance of analysing a list size of 320 IPs, running only with 8 CPU, it means that we compute 1 CPU for 40 IPs, i.e processing sequentially with 4 slots of 10 IPs. Running fully with 48 cores and using parallel processing, we can verify up to 1920 IPs ($6 \times 320IPs$) per request with the same length of time.

FV. With option FV library in Cingulata, the runtime is around 56 s and 115 s with respectively BLACK_LIST_LOW and BLACK_LIST_MEDIUM configurations whereas it takes 170 s and 415 s with BLACK_LIST_MEDIUM and BLACK_LIST_FULL.

TFHE. With option TFHE library in Cingulata, using BLACK_LIST_FULL and BLACK_LIST_HIGH configuration, it consumes respectively 17 s and 13 s whereas with BLACK_LIST_MEDIUM we obtain the result after 8 s and BLACK_LIST_LOW we only wait after 4 s.

	FV*		TFHE*	
	320 IPs[a]	1920 IPs[b]	320 IPs[a]	1920 IPs[b]
BLACK_LIST_LOW	56 s	56 s	4 s	4 s
BLACK_LIST_MEDIUM	115 s	115 s	8 s	8 s
BLACK_LIST_HIGH	170 s	170 s	13 s	13 s
BLACK_LIST_FULL	415 s	415 s	17 s	17 s

*: a computed server $W3520$ @2.67 GHz, 128 GB Ram with 48 cores CPU.
[a]: with the same compute server but with 8 CPU.
[b]: with the same compute server but with 48 CPU.

5   Conclusion

In this paper, we have shown the benefits of information sharing in mitigating cyber and physical threats, we have discussed privacy-aware design principles for user awareness in sharing data consent, and implicating the clarification for design architects to be aware of data usage and data collection as well as auditability of privacy - related design with regards to legal consequences. We then concluded by describing a high level design for two critical components: secure content - based routing and secure data access control which are integrated in our cloud - based analysis platform with homomorphic encryption technology. We presented the Cingulata compilation tool-chain for fully homomorphic encryption technology and concluded by the use case detection of brute force attacks. We show the Cingulata performance with TFHE and FV libraries, 3rd generation of FHE technology which was already integrated in our Cingulata library and available in open source github. In comparison with our solution dedicated to enterprise partner which currently is in development, this open source compiler still shows some constraints of usage, integration and implementation for industrial usages. However, the compiler design allowing developers working with high-level languages and the HE performance based on TFHE library are impersonated. For the future work activities, we will improve the performance of Cingulata open source and also define optimisation works for Boolean circuits with optimizing in memory consumption and execution times, we also prepare some works to adapt Cingulata with Artificial Intelligence technology.

References

1. Brakerski, Z.: Fully homomorphic encryption without modulus switching from classical GapSVP. In: Advances in Cryptology - CRYPTO 2012. Lecture Notes in Computer Science, pp. 868–886 (2012)
2. Brakerski, Z., Gentry, C., Vaikuntanathan, V.: (Leveled) fully homomorphic encryption without bootstrapping. In: Proceedings of the 3rd Innovations in Theoretical Computer Science Conference on - ITCS 2012 (2012)
3. Carpov, S., Dubrulle, P., Sirdey, R.: Armadillo: a compilation chain for privacy preserving applications. In: Proceedings of the 3rd International Workshop on Security in Cloud Computing, pp. 13–19 (2015)

4. Chadwick, D.W., Fan, W., Constantino, G., Lemos, R.D., Cerbo, F.D., Herwono, I., Mori, P., Sajjad, A., Wang, X.S., Manea, M.: A cloud-edge based data security architecture for sharing and analyzing cyber threat information. Future Gener. Comput. Syst. **102**, 710–722 (2019)
5. Chillotti, I., Gama, N., Georgieva, M., Izabachène, M.: Faster fully homomorphic encryption: bootstrapping in less than 0.1 seconds. In: ASIACRYPT 2016, Part I, pp. 3–33. Springer (2016)
6. Fan, J., Vercauteren, F.: Somewhat practical fully homomorphic encryption. Cryptology ePrint Archive Report 2012/144 (2012). http://eprint.iacr.org/
7. Fau, S., Sirdey, R., Fontaine, C., Melchor, C.A., Gogniat, G.: Towards practical program execution over fully homomorphic encryption schemes. In: 3PGCIC, pp. 284–290 (2013)
8. Fung, C.J., Boutaba, B.: Design and management of collaborative intrusion detection networks. In: IFIP/IEEE International Symposium on Integrated Network Management (2013)
9. Chillotti, I., Gama, N., Georgieva, M., Izabachène, M.: Improving TFHE: faster packed homomorphic operations and efficient circuit bootstrapping. In: Proceedings of ASICACRYPT 2017. LNCS, vol. 10624, pp. 377–408. Springer (2017)
10. Lidz, C.W.: Informed Consent: A Study of Decision Making in Psychiatry. Guilford, New York (1984)
11. McAfee: Common Event Format - McAfee, September 2018. https://kc.mcafee.com/resources/sites/MCAFEE/content
12. Melchor, C.A., Fau, S., Fontaine, C., Gogniat, G., Sirdey, R.: Recent advances in homomorphic encryption: a possible future for signal processing in the encrypted domain. IEEE Sig. Process. Mag. **30**(2), 108–117 (2013)
13. Nabeel, M., Shang, N., Bertino, E.: Efficient privacy preserving content based publish subscribe systems. In: Proceedings of the 17th ACM Symposium on Access Control Models and Technologies (SACMAT 2012). ACM, New York (2012)
14. Trimintzios, P., Gavrila, R.: On national and international cyber security exercises: survey, analysis and recommendations. ENISA European Union Agency for Cybersecurity (2012). http://www.enisa.europa.eu/activities/Resilience-and-CIIP/cyber-crisis-cooperation/cce/cyber-exercises/exercise-survey2012
15. Pires, R., Pasin, M., Felber, P., Fetzer, C.: Secure content-based routing using Intel software guard extensions. In: Proceedings of the 17th International Middleware Conference (Middleware 2016). ACM, New York (2016)
16. Richardson, E.L., Weinberger, C.W.: Records, computers, and the rights of citizens. U.S. Department of Health, Education, and Welfare (1973)
17. C.L. team: Homomorphic encryption technology. github (2018)
18. UNISS: Cybersecurity policies and critical infrastructure protection (2018). https://uniss.org/cyber-security-policies-and-critical-infrastructure-protection/
19. NISP WG2: NISP WG2 plenary report information sharing and incident notification. Network and Information Security Group (2013)
20. eHealth Acceptance Factors and KONFIDO Adaptation Strategy (2018). https://konfido-project.eu

Secure and Robust Watermarking Scheme in Frequency Domain Using Chaotic Logistic Map Encoding

Phuoc-Hung Vo[1,2](✉) 🆔, Thai-Son Nguyen[1], Van-Thanh Huynh[1], Thanh-C Vo[1], and Thanh-Nghi Do[2]

[1] School of Engineering and Technology, Tra Vinh University, Tra Vinh, Vietnam
{hungvo,thaison,hvthanh,vothanhc}@tvu.edu.vn
[2] College of Information Technology, Can Tho University, Can Tho, Vietnam
dtnghi@cit.ctu.edu.vn

Abstract. The paper proposes a secure and robust stereo image watermarking scheme based on chaotic logistic map encoding and frequency transformations. The chaotic logistic map encoding is employed to confuse both original stereo images and watermarked images. Since the high sensitivity to initial conditions introduced in the chaotic logistic map, a huge key space is provided for encoding images. Thus, the proposed system has a strong secure capability to resist brute-force and statistical attacks, in which the evaluation goes through various analysis methods such as sensitive key analysis, correlative adjacent pixels analysis. Moreover, to enhance the robustness of the watermarked image, the discrete cosine transform (DCT) and the singular value decomposition (SVD) are exploited for this purpose. Based upon the properties of the SVD and utilizing the advantages of the DCT, a watermarked image is embedded in the singular value. Performance evaluations show that the proposed watermarking scheme for stereo images is highly secure as well as strongly robust against different kinds of attacks.

Keywords: Data hiding · Watermarking · DCT-SVD · Chaos · Logistic map

1 Introduction

The explosion of computer networks and information technology in recent years have led to the revolution of the communication field. The transmission of information among activities has become more prominent. Every minute, a huge amount of digital information is exchanged among different users through the Internet. However, digital information is totally different from conventional information. The ability to copy and alter information and unauthorized distribution has changed dramatically. Thus, there is a need to have a secure method for sharing information. Depending on the application, different security method is considered. Digital watermarking is one of the most common techniques that has been applied in copyright protection and authentication for over the last two decades [1–4].

© Springer Nature Switzerland AG 2020
H. A. Le Thi et al. (Eds.): ICCSAMA 2019, AISC 1121, pp. 346–357, 2020.
https://doi.org/10.1007/978-3-030-38364-0_31

In a digital image watermarking technique, a watermark is inserted into the cover image to protect it from unlawful practice. According to the watermark inserting process, digital image watermarking techniques can be grouped into spatial domain techniques or frequency domain techniques. In the spatial domain algorithms, the watermark is directly embedded in the cover image by changing pixel values [5–7]. The advantages of spatial domain methods lie in easy implementation and low computational complexity. But these methods are not enough robust to resist against image processing attacks. In contrast, in the frequency domain approaches, the watermark is embedded in coefficients in which pixel values are transformed into frequency coefficients. Many watermarking schemes in the frequency domain have been proposed [8–13]. For example, a water-mark is hidden inside the coefficients of the frequency domain after transforming. In 2010, Lu et al. [8], proposed a robust watermarking scheme in which a watermark is inserted into the frequency coefficients after the discrete Fourier transform. In 2015, AL-Nabhani et al. [10], developed discrete wavelet transform (DWT) to embed a binary watermark image in selected coefficient blocks. Besides, there are many hybrid trans-form domain watermarking techniques such as [9], to increase the embedding capacity. Authors successfully combined integer wavelet transform (IWT) with Discrete Gould transform (DGT) for medical images. In 2018, Kang et al. proposed a novel hybrid of discrete cosine transform (DCT) and singular value decomposition (SVD) in DWT domain [11] that offers the high robustness and quality. In general, frequency domain watermarking can withstand image processing attacks better in comparison with spatial domain watermarking.

Drawing on the state-of-the-art analysis of methods, watermarking algorithms were designed for mono-images. Recently, the advance of new technologies, many three-dimensional (3D) videos/images have been generated. Those images are taken by two cameras horizontally aligned and separated at a scalable distance similar to the distance between our eyes [14–16]. The stereoscopic 3D display that can be achieved by projecting on the left and right views to the specially designed screen is called a stereo image. By providing information about the 3D structure of scenes, stereo images are being used in various applications such as computer vision, medical surgery and autonomous navigation [17]. Of course, not except for digital information, unauthorized users have become possible to easily alter, copy and distribute stereo images over open networks without the permission of the original authors [18]. Ensuring the precision and integrity of stereo images is important. Digital watermarking technology is currently considered as an effective solution for authentication and copyright protection.

There are several previous works in the field of stereo image digital watermarking [16, 19, 20]. In [19], the authors proposed stereo image coding based on the DCT and SVD. Disparity extracted from the stereo pair is used as watermark and is embedded in the left views of the stereo image. The extracted watermark is normally good in visuality, but the scheme is weak against some attacks such as JPEG compression, cropping, filtering. Zhou et al. [16] propose a watermarking technique using hierarchical tamper detection strategy and stereoscopic matching in 3-D multimedia which can perform well for tamper detection and self-recovery in stereo images. In 2017, we devoted a robust hybrid watermarking scheme for stereo images copyright protection in transform domain [20]. Since the left and right views of a stereo image are not independent of each other, the method matched a block in one view to the corresponding block in the other view in DCT domain. Then watermark is embedded in similar block pairs based on SVD which can meet a robust watermarking, compared to other algorithms.

Most of existing stereo image watermarking schemes concentrate on the visual quality and signal processing attacks. These approaches are just considering around payload and robustness. Therefore, if eavesdroppers know watermarking, they can evacuate. Once the information is revealed, such schemes are no longer secured [3]. To overcome this problem, a secure and robust watermarking scheme is proposed in the present paper, in which the original image and the watermark image are encrypted using the chaotic logistic map. In addition, to enhance the robustness of the scheme, a cipher watermark image is embedded in the cipher original image in the DCT and SVD transform.

The rest of this paper is structured as follows. We, firstly, introduce the basic background and mathematical theory in Sect. 2. Next, the proposed scheme is presented in Sect. 3. Then, performance evaluations of the proposed system are exhibited in Sect. 4. Finally, a conclusion is drawn in Sect. 5.

2   Preliminaries

In this section, we discuss step by step the procedure of chaotic logistic map which is used to encrypt a plain image in the proposed scheme. Subsequently, we outline the discrete cosine transform as well as the singular value decomposition.

2.1   Chaotic Logistic Map

A chaotic logistic map is a mathematical technique that deals with a nonlinear dynamic system. It is a kind of one-dimensional chaotic mapping, which is widely used for protecting image content in recent years [4, 21]. A logistic map function is defined as the following Eq. (8).

$$x_i = \mu x_{i-1}(1 - x_{i-1}) \text{ with } i \text{ running from } 1 \tag{1}$$

where $3.57 < \mu < 4$ and $0 < x_0 < 1$ are the parameters and the initial values. When the initial conditions are met, the generated sequence element $x_1, x_2, \ldots, x_n$ are neither periodic nor convergent. In addition, the function is very sensitive to initial conditions. In other words, even if the difference in initial values is so small, the iteration results are widely diverging.

2.2   Discrete Cosine Transform (DCT) and Quantized DCT (QDCT)

DCT is one of the techniques for transformations of the signal presentation from the spatial domain to the frequency domain. In the spatial domain, a digital image is considered as a two- dimensional (2D) matrix, in which the gray-scale values of pixels are ranged from 0 to 255. Two-dimensional DCT (2D-DCT) function is adopted to convert images [22]. The 2D DCT function is defined as follows:

$$F(u, v) = c(u)c(v)\frac{2}{n}\sum_{x=0}^{N-1}\sum_{y=0}^{N-1} f(x, y)\cos\left(\frac{(2x + 1)u\pi}{2N}\right)\cos\left(\frac{(2y + 1)v\pi}{2N}\right) \tag{2}$$

where N refers to the image block size, $c(u) = c(v) = \begin{cases} \frac{1}{\sqrt{2}}, if\ u = 0, v = 0; \\ 1, otherwise. \end{cases}$ and $F(u, v)$ is the DCT coefficient and $f(x, y)$ is the pixel value.

The inverse transform (IDCT) function is operated as follows:

$$f(x, y) = \frac{2}{N} \sum_{u=0}^{N-1} \sum_{v=0}^{N-1} c(u)c(v)F(u, v)\cos\left(\frac{(2x + 1)u\pi}{2N}\right)\cos\left(\frac{(2y + 1)}{2N}v\pi\right)$$

(3)

Quantization DCT is the step where we discard information which is not visually significant. The quantization of every coefficient in the 8×8 block is divided by a corresponding quantization value. The quantized DCT (QDCT) coefficient is calculated as the following Eq. (4):

$$F(u, v)_Q = round\left(\frac{F(u, v)}{Q(u, v)}\right)$$

(4)

where $Q(.)$ is a quantization table.

2.3 Singular Value Decomposition (SVD)

SVD is a linear algebraic algorithm. A matrix A is decomposed as a product of column orthogonal matrix U, diagonal matrix S with r (rank of A matrix) nonzero elements called singular values of A matrix, and transpose orthogonal matrix V, that can be described as:

$$A = U \cdot S \cdot V^T$$

(5)

where $U \cdot U^T = I; V \cdot V^T = I.$

SVD is a useful tool to extract geometric features from an image. Thus, the basic idea behind the SVD technique in the watermarking scheme is to find singular values of the image and embed the watermark in it. As doing such method, watermarking scheme withstands geometric attacks.

3 Proposed Scheme

3.1 Watermarking Embedding

Watermarking embedding process is classified into three stages. Stereo image encoding is presented in the first stage. Next, block matching is implemented in the second stage. Finally, watermarking embedding is implanted in the third stage.

Stage 1: Stereo Image Encoding. A cryptosystem normally includes two mutually independent stages: confusion and diffusion. In this paper, we only address the confusion, in which all the image pixels are scrambled without changing their value.

Assuming that, P_L and P_R are the left and the right plain-images of stereo pair sized of m-by-n. each image is reshaped to a 1-by-$m \times n$. To break the relationship between adjacent pixels, a sequence random number is generated by using chaotic logistic map given by the Eq. (8) with two initial conditions $x_0 = 0.1$ and $\mu = 3.75$. The sequence random number $(x_1, x_2, \ldots, x_{m \times n})$ is a useful indicator of image pixel permutations. The plain-image P_L and P_R are scrambled with the sequence random numbers to obtain the cipher-image S_L and S_R, respectively.

Stage 2: Block Matching. Since the left and right views of the stereo image are considered as similar. To improve the performance of a stereo image watermarking scheme, inter-correlation between two views is matching. The process is presented in the following steps:

Step 1: Decompose S_L, and S_R (of size $m \times n$) into non-overlapping blocks $B_{i,j}^L$ and $B_{i,j}^R$ sized of 8×8 pixels, with $i \in [0, (m/8)-1]$ and $j \in [0, (n/8)-1]$.

Step 2: Transform image blocks using 2D-DCT to obtain $DB_{i,j}^L$ and $DB_{i,j}^R$. After forward DCT transformation, the element in the uppermost left is the DC coefficient and the rest elements are AC coefficients

Step 3: Quantize DCT blocks to acquire QDB_i^L and QDB_i^R

Step 4: Calculate the difference value from the left and right blocks based on the searching area. Figure 1 displays the searching area. Which is introduced in [23] and the formulation is defined as follows:

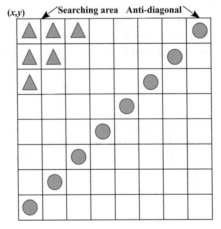

Fig. 1. Searching area and Anti-diagonal area in block (Triangle shape: Searching area, Circle shape: Anti-diagonal area)

$$Dif_{i+u,j+v} = \sum_{x,y=0}^{x+y<3} \left[QDB_{i,j}^L(x, y) - QDB_{i+u,j+v}^R(x, y) \right]^2 \quad (6)$$

where $u, v \in (-k, k)$, $1 \leq k \leq 5$ indicates neighbor blocks; $QDB(.)$ quantized-DCT coefficients. Two blocks are considered as the most similar block pair when the smallest different value is obtained, i.e. $\min(Dif_{i+u,j+v}) \leq T$ is satisfied, where T denotes as similar threshold.

Stage 3: Watermark Embedding. To improve the security of the proposed system, the binary watermark image is scrambled using a chaotic logistic map before inserting to cover image. Moreover, to enhance the robustness of the scheme, SVD is executed on matrix A which is formed by two anti-diagonals of pair matched blocks. The singular values are modified to embed the scrambled watermark image W, which is implemented as follows:

$$S_w = S + \sigma \cdot W \tag{7}$$

where, σ is the robustness factor of the system. Then SVD is performed on S_w to obtain S_1 such that

$$SVD(s_W) = U_1 \cdot S_1 \cdot V_1 \tag{8}$$

Next, the matrix A' is accomplished using reverse transformation on matrices U, S_1, and V according to the following equation

$$A' = U \cdot S_1 \cdot V^T \tag{9}$$

Then, coefficients of A' have been distributed back to matched block pair and then apply IDCT.

Finally, watermarked stereo image is formed in which perfect combination of IDCT blocks and decoding are implemented.

3.2 Watermarking Extracting

Once the embedding process is accomplished the watermarked image together with the original image and keys, watermark image will be extracted. The process includes two stages described as follows:

Stage 1: This stage will be implemented similarly to the first and second stages of the watermarking embedding process. Result of the stage is that the matrix A' is achieved.

Stage 2: The watermark extraction process is realized by following steps:

Step 1: Apply $SVD(A')$ to decompose it into 3 matrices U', S'_w and V':

$$SVD(A') = U' \cdot S'_w \cdot V' \tag{10}$$

Step 2: Calculate S'_1 based on U_1, V_1 and S'_w according to the following equation

$$S'_1 = U_1 \cdot S'_w \cdot V_1^T \tag{11}$$

Step 3: Extract watermark W' as follows

$$W' = \frac{S'_1 - S}{\sigma} \tag{12}$$

4 Performance Evaluations of the Proposed System

In most watermarking algorithms, researchers normally consider two kinds of attacks on watermarking. The first one is signal processing such as filtering, additive noise and compression and the other is using geometric conversions such as cropping, rotating, or some other similar methods. However, an efficient watermarking is not only tolerant of the signal processing and geometric attacks but also resistant to statistical and brute-force attacks. Our watermarking system is an acceptable trade-off for the mentioned attacks. In this section, we discuss the security analysis and the robustness of the proposed scheme. The analysis includes key sensitive analysis and adjacent pixels correlation analysis.

4.1 Sensitive Key Analysis

It is known that chaotic logistic map is highly sensitive to every initial condition even with minute changes for an image encryption proposal, which guarantees to withstand a brute-force attack [24]. To evaluate the key sensitivity, we, firstly, scramble the plain image with two initial values x_0 and μ of Eq. (8) as 0.7589 and 3.67, respectively. Subsequently, we change the initial value x_0 by adding 10^{-10} to value x_0. After that, we rearrange the cipher image. Figure 2 displays the results of the key sensitivity.

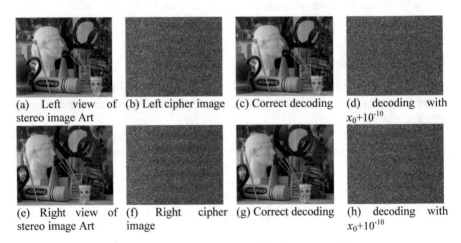

(a) Left view of stereo image Art   (b) Left cipher image   (c) Correct decoding   (d) decoding with x_0+10^{-10}

(e) Right view of stereo image Art   (f) Right cipher image   (g) Correct decoding   (h) decoding with x_0+10^{-10}

Fig. 2. The results of the key sensitivity analysis: (a)

4.2 Adjacent Pixels Correlation Analysis

It is known that each pixel in an image is usually highly correlated with its adjacent pixels, which can be exploited by hackers to find out the relationship between the plain image and the cipher image. Thus, in order to test the security of the system, pixels correlation analysis between the plain image and the cipher image are conducted. We randomly select 2048 pairs of adjacent pixels in a horizontal, vertical and diagonal direction from the plain image as well as the cipher image. The correlation of adjacent pixels can be visually analyzed by plotting. Figure 3 displays the correlation distributions of the plain stereo image 'Art' with the red component before and after encoding.

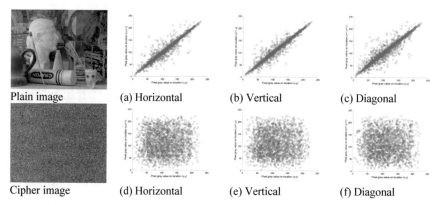

| Plain image | (a) Horizontal | (b) Vertical | (c) Diagonal |
| Cipher image | (d) Horizontal | (e) Vertical | (f) Diagonal |

Fig. 3. Correlation visualization for 2048 randomly selected adjacent in plain and cipher images

Moreover, the correlation function is used to calculate the correlation of the adjacent pixels [25], given by:

$$r_{xy} = \frac{cov(x, y)}{\sqrt{D(x)}\sqrt{D(y)}} \tag{13}$$

where:

$$cov(x, y) = \frac{1}{N}\sum_{i=1}^{N}(x_i - E(x)) \times (y_i - E(y)) \tag{14}$$

$$D(x) = \frac{1}{N}\sum_{i=1}^{N}(x_i - E(x))^2 \tag{15}$$

$$D(y) = \frac{1}{N}\sum_{i=1}^{N}(y_i - E(y))^2 \tag{16}$$

$$E(x) = \frac{1}{N}\sum_{i=1}^{N}x_i \tag{17}$$

$$E(y) = \frac{1}{N}\sum_{i=1}^{N}y_i \tag{18}$$

with x_i, y_i are pixel values of two adjacent pixels.

The Fig. 3 and Table 1 show that the correlation coefficients of the adjacent pixels in the plain and cipher image are far part. In other words, the system successfully transforms high correlation coefficients in the plain image to very low correlation coefficients in the cipher image.

Table 1. Correlation coefficients of adjacent pixels

Art stereo image	Plain image			Cipher image		
	Horizontal	Vertical	Diagonal	Horizontal	Vertical	Diagonal
Left view						
Red component	0.9694	0.9648	0.9461	0.0298	0.0467	0.0361
Green component	0.9769	0.9834	0.9635	0.0088	−0.0034	0.0461
Blue component	0.9738	0.9827	0.9613	0.0244	−0.0430	−0.0069
Right view						
Red component	0.9746	0.9727	0.9533	0.0397	0.0480	0.0173
Green component	0.9704	0.9811	0.9589	0.0111	−0.0164	−0.0168
Blue component	0.9741	0.9834	0.9619	0.0245	0.0129	−0.0119

(a) logo watermark   (b) Left view of Laundry   (c) Left view of Art

Fig. 4. Logo watermark and left view of test stereo images

4.3 Imperceptibility and Robustness Evaluation of the Watermarking System

To evaluate the watermark imperceptibility and robustness of the system, a logo water-mark sized of 128 × 128 and stereo image dataset from Middlebury [26]. Figure 4 displays a logo watermark and some of test stereo images.

Table 2 gives the values of peak signal-to-noise (PSNR) and structural similarity (SSIM) index for test images. The PSNR (dB) of the watermarked stereo images is about 43 for W. Zhou et al.'s scheme. While the PSNR of the watermarked stereo images of the proposed scheme is larger than 56 (dB), which is satisfied with watermark transparency. Additionally, watermark transparency is also evaluated by the SSIM index [27]. Table 2. shows that SSIM of the proposed method is completely satisfied for watermark transparency.

Bit correlation error (BCR) is normally used to evaluate the robustness of the water-marking method [20]. Table 3 displays the PSNR of the watermarked image after attack and BCR of the restored watermark. Even PSNR of the watermarked image is so slow, the restored watermarks are up to 86%.

Table 2. PSNR and SSIM of watermarked images

Stereo images	Zhou et al.'s scheme [16]		Proposed scheme	
	Left view	Right view	Left view	Right view
Laundry	42.80/0.978	42.79/0.977	56.99/0.999	59.39/0.999
Art	43.20/0.987	43.10/0.980	57.24/0.999	59.85/0.999

Table 3. Watermark robustness to common images processing operations

Attacks	PSNR	BCR
Salt & Pepper 0.05	18.35	0.88
Cropping 6.25%	20.06	0.86
Gaussian 0.05	19.14	0.88

5 Conclusion

Thus, paper presents a secure and robust watermarking scheme which enhances the security of the system by protecting the watermark image and embedding positions. The performance evaluations demonstrate that the proposed scheme meets a good imperceptibility of each view in comparison with the similar state-of-the-art algorithm. The PSNR, SSIM metrics verified the better performance of the proposed watermarking scheme.

Acknowledgements. The research is funded by Tra Vinh University, Viet Nam.

References

1. Kadhim, I.J., Premaratne, P., Vial, P.J., Halloran, B.: Comprehensive survey of image steganography: techniques, Evaluations, and trends in future research. Neurocomputing **335**, 299–326 (2019). https://doi.org/10.1016/j.neucom.2018.06.075
2. Ali, M., Ahn, C.W., Pant, M.: A robust image watermarking technique using SVD and differential evolution in DCT domain. Optik **125**, 428–434 (2014). https://doi.org/10.1016/j.ijleo.2013.06.082
3. Yadav, G.S., Ojha, A.: Chaotic system-based secure data hiding scheme with high embedding capacity. Comput. Electr. Eng. **69**, 447–460 (2018). https://doi.org/10.1016/j.compeleceng.2018.02.022
4. Safi, H.W., Maghari, A.Y.: Image encryption using double chaotic logistic map. In: 2017 International Conference on Promising Electronic Technologies (ICPET), pp. 66–70 (2017). https://doi.org/10.1109/ICPET.2017.18
5. Abraham, J., Paul, V.: An imperceptible spatial domain color image watermarking scheme. J. King Saud Univ. Comput. Inf. Sci. **31**, 125–133 (2019). https://doi.org/10.1016/j.jksuci.2016.12.004

6. Mathur, S., Dhingra, A., Prabukumar, M., Agilandeeswari, L., Muralibabu, K.: An efficient spatial domain based image watermarking using shell based pixel selection. In: 2016 International Conference on Advances in Computing, Communications and Informatics (ICACCI), pp. 2696–2702 (2016). https://doi.org/10.1109/ICACCI.2016.7732468
7. Parekh, M., Bidani, S., Santhi, V.: Spatial Domain Blind Watermarking for Digital Images. In: Pattnaik, P.K., Rautaray, S.S., Das, H., and Nayak, J. (eds.) Progress in Computing, Analytics and Networking. pp. 519–527. Springer Singapore (2018)
8. Lu, W., Lu, H., Chung, F.-L.: Feature based robust watermarking using image normalization. Comput. Electr. Eng. **36**, 2–18 (2010). https://doi.org/10.1016/j.compeleceng.2009.04.002
9. Selvam, P., Balachandran, S., Pitchai Iyer, S., Jayabal, R.: Hybrid transform based reversible watermarking technique for medical images in telemedicine applications. Optik **145**, 655–671 (2017). https://doi.org/10.1016/j.ijleo.2017.07.060
10. AL-Nabhani, Y., Jalab, H.A., Wahid, A., Noor, R.M.: Robust watermarking algorithm for digital images using discrete wavelet and probabilistic neural network. J. King Saud Univ. Comput. Inf. Sci. **27**, 393–401 (2015). https://doi.org/10.1016/j.jksuci.2015.02.002
11. Kang, X., Zhao, F., Lin, G., Chen, Y.: A novel hybrid of DCT and SVD in DWT domain for robust and invisible blind image watermarking with optimal embedding strength. Multimed Tools Appl. **77**, 13197–13224 (2018). https://doi.org/10.1007/s11042-017-4941-1
12. Shehab, A., Elhoseny, M., Muhammad, K., Sangaiah, A.K., Yang, P., Huang, H., Hou, G.: Secure and robust fragile watermarking scheme for medical images. IEEE Access. **6**, 10269–10278 (2018). https://doi.org/10.1109/ACCESS.2018.2799240
13. Vo, P.-H., Nguyen, T.-S., Huynh, V.-T., Do, T.-N.: A novel reversible data hiding scheme with two-dimensional histogram shifting mechanism. Multimedia Tools Appl. **77**, 28777–28797 (2018). https://doi.org/10.1007/s11042-018-5991-8
14. Troccoli, A., Kang, S.B., Seitz, S.: Multi-view multi-exposure stereo. In: Third International Symposium on 3D Data Processing, Visualization, and Transmission (3DPVT 2006), pp. 861–868 (2006). https://doi.org/10.1109/3DPVT.2006.98
15. Orozco, R.R., Loscos, C., Martin, I., Artusi, A.: Chapter 4 - multiview HDR video sequence generation. In: Dufaux, F., Le Callet, P., Mantiuk, R.K., Mrak, M. (eds.) High Dynamic Range Video, pp. 121–138. Academic Press (2016). https://doi.org/10.1016/B978-0-08-100412-8.00004-8
16. Zhou, W., Jiang, G., Luo, T., Yu, M., Shao, F., Peng, Z.: Stereoscopic image tamper detection and self-recovery using hierarchical detection and stereoscopic matching. JEI **23**, 023022 (2014). https://doi.org/10.1117/1.JEI.23.2.023022
17. Scharstein, D.: View Synthesis Using Stereo Vision (1997)
18. Zhou, W., Yu, L., Wang, Z., Wu, M., Luo, T., Sun, L.: Binocular visual characteristics based fragile watermarking scheme for tamper detection in stereoscopic images. AEU – Int. J. Electron. Commun. **70**, 77–84 (2016). https://doi.org/10.1016/j.aeue.2015.10.006
19. Rawat, S., Gupta, G., Balasubramanian, R., Rawat, M.S.: Digital watermarking based stereo image coding. In: Ranka, S., Banerjee, A., Biswas, K.K., Dua, S., Mishra, P., Moona, R., Poon, S.-H., Wang, C.-L. (eds.) Contemporary Computing, pp. 435–445. Springer, Heidelberg (2010)
20. Vo, P., Nguyen, T., Huynh, V., Do, T.: A robust hybrid watermarking scheme based on DCT and SVD for copyright protection of stereo images. In: 2017 4th NAFOSTED Conference on Information and Computer Science, pp. 331–335 (2017). https://doi.org/10.1109/NAFOSTED.2017.8108087
21. Younus, Z.S., Hussain, M.K.: Image steganography using exploiting modification direction for compressed encrypted data. J. King Saud Univ. Comput. Inf. Sci. (2019). https://doi.org/10.1016/j.jksuci.2019.04.008
22. Ghanbari, M.: Standard Codecs: Image Compression to Advanced Video Coding. Institution Electrical Engineers (2003)

23. Yang, W.-C., Chen, L.-H.: Reversible DCT-based data hiding in stereo images. Multimed Tools Appl. **74**, 7181–7193 (2015). https://doi.org/10.1007/s11042-014-1958-6
24. Alvarez, G., Li, S.: Some basic cryptographic requirements for chaos-based cryptosystems. Int. J. Bifurcation Chaos. **16**, 2129–2151 (2006). https://doi.org/10.1142/S0218127406015970
25. Song, C.-Y., Qiao, Y.-L., Zhang, X.-Z.: An image encryption scheme based on new spatiotemporal chaos. Optik – Int. J. Light Electron Optics. **124**, 3329–3334 (2013). https://doi.org/10.1016/j.ijleo.2012.11.002
26. http://vision.middlebury.edu/stereo/data/. Accessed 30 Aug 2019
27. Wang, Z., Bovik, A.C., Sheikh, H.R., Simoncelli, E.P.: Image quality assessment: from error visibility to structural similarity. IEEE Trans. Image Process. **13**, 600–612 (2004). https://doi.org/10.1109/TIP.2003.819861

Knowledge Information
and Engineering Systems

An Improvement of Applying Multi-objective Optimization Algorithm into Higher Order Mutation Testing

Quang-Vu Nguyen[1]([✉]) and Hai-Bang Truong[2]

[1] Korea-Vietnam Friendship Information Technology College, Quangnam, Vietnam
vunq@viethanit.edu.vn
[2] Faculty of Computer Science, University of Information Technology, Vietnam National
University Ho Chi Minh City (VNU-HCM), Ho Chi Minh City, Vietnam
bangth@uit.edu.vn

Abstract. In order to raise the quality of higher order mutation testing, in this paper, we propose an approach for effect improving of multi-objective optimization algorithms which can be used in the field of higher order mutation testing in order to reduce the number of generated mutant, generate the hard-to-kill mutant and construct the quality higher order mutants. We have performed an empirical evaluation with 20 real-word, open-source projects and 10 multi-objective optimization algorithms (including 5 original algorithms and 5 corresponding modification algorithms) to evaluate experimental results as well as bring out some opinions to effectiveness apply multi-objective optimization algorithms into higher order mutation testing. The study results indicate that our approach is an effectiveness one to get better the quality of higher order mutation testing.

Keywords: Mutation testing · Higher order mutation testing · Quality of higher order mutation testing · Multi-objective optimization algorithms · Mutant reduction · Quality mutants

1 Introduction

According to IEEE definitions (IEEE Standard Glossary of Software Engineering Terminology), *"software testing is the process of analyzing a software item to detect differences between existing and required conditions and to evaluate the features of the software items"*. In other words, to test software, we execute software using a designed set of test cases including a given set of test data to satisfy two distinct goals. The first is to demonstrate that designed and developed software includes all of customer requirements or not. Checking any situation in which behaviour of the software is incorrect, undesirable, or does not conform to its specification is the second goal.

Mutation Testing (MT), a technique that has been derived from two basic ideas: Competent Programmer Hypothesis (*"programmers write programs that are reasonably close to the desired program"*) and Coupling Effect Hypothesis (*"detecting simple faults will lead to the detection of more complex faults"*), was originally proposed in 1970s years

© Springer Nature Switzerland AG 2020
H. A. Le Thi et al. (Eds.): ICCSAMA 2019, AISC 1121, pp. 361–369, 2020.
https://doi.org/10.1007/978-3-030-38364-0_32

[1, 2]. The purpose of mutation testing is to assess the quality of test cases (TCs) and this is synonymous with supporting the testers in creating a good set of test cases (TCs). A set of TCs is called as good as it can be able to expose all of the potential defects of program under test (PUT). In MT, mutants are the modified version of original PUT by changing one operator by another. The mutant is called "killed" when the outputs of mutant and PUT are different with the same given set of test cases, otherwise the mutant is called "alive" [1, 2]. According to [3, 4], three big problems of MT are: (1) the generated mutants is too much (but not necessary); (2) the mutants are so easy to be killed and (3) do not represent actual faults [3, 4].

Mutant reduction, generating hard-to-kill mutant as well as constructing quality mutants are the issues that need to be considered when using higher order mutation testing (HOMT). In [5–8, 10–21], multi-objective optimization algorithms, which have been devised for solving optimization problems and making the decisions that satisfy multiple objectives, were used as an effectiveness approach to overcome above-mentioned three big problems.

Higher order mutation testing (also included second order mutation testing) [5–8, 10–22] have been considered as a promising solution for overcoming limitations of traditional (first order) mutation testing. However, higher order mutation testing (HOMT) applying is not yet widely used in software testing practice due to the quality problem which is necessary for further studying to improve. That is the aim which be considered of our research in this paper.

The next section introduces our proposed approach and summarizes the related works. Section 3 presents our empirical study in detail. Section 4 informs and analyzes the results of the experiment. The last is Sect. 5 that is the conclusions and future work.

2   Proposed Approach and Related Works

As we presented in our previous work [21], McConnell has concluded [29] that "*there are about 1–25 errors per 1000 lines of code for delivered software and about 10–20 defects per 1000 lines of code during in-house testing and 0.5 defects per 1000 lines of code in released product*". It means that, as our understanding, in the complete software projects versions written by programmers, who are experienced and good programmers, a line of code rarely has more than one defect. Derived from that statement, and with the aim of reducing the number of generated mutants which do not represent actual faults, we research and introduce an approach to modify the multi-objective optimization algorithm applied to construct the set of better mutants. We do not randomly combine the First Order Mutants (FOMs) to create HOMs. Instead of this, with the rule "*apply no more than one mutation operator to each line of code*", we create an initial list of HOMs by combining two-or-more FOMs that have mutation operators at different lines of code.

In [21], we modified the eNSGAII algorithm (the modified algorithm is named eNSGAII-DiffLOC) guiding by above-mentioned rule, because it is the best of all algorithms that we have used to construct the "*high quality and reasonable mutants*" (named H7). H7 [17, 18] is a higher order mutant (HOM) which is harder to be killed than its constituent first order mutants (FOMs). Moreover, a set of test cases that can kill H7, it also can kill all constituent FOMs of that H7. Our obtained results indicate that *"The eNSGAII-DiffLOC seems to be slightly better than original eNSGAII algorithm in terms of mutant reduction, generating harder-to-kill mutant and constructing H7"* [21].

The results in [21] were evaluated by means of comparison the following values:

- NoM: The ratio of the number of HOMs to the number of FOMs.
- NoT: The ratio of the number of TCs which kill HOMs to the number of TCs which kill constituent first order mutants of that HOMs.
- NoR: The ratio of generated *"reasonable higher order mutants"* [17, 18] to all HOMs.
- NoH7 is the ratio of H7 [17, 18] to all HOMs.
- NoH1 is the ratio of H1 (*"live (potentially equivalent) higher order mutants)"* [17, 18] to all HOMs.

In this paper, we focus on modification 5 multi-objective optimization algorithms in order to perform an empirical study to apply 10 algorithms (5 original multi-objective optimization algorithms and 5 corresponding modification algorithms) into HOMT with 20 real-word, open-source projects. Then, we evaluate the results based on 5 ratios: NoM, NoT, NoR, NoH7 and NoH1.

The multi-objective optimization algorithms are NSGA-II, eNSGA-II, NSGA-III [24, 25], eMOEA [26, 27] and SPEA-II [28] and the corresponding modification algorithms are named NSGAII-DiffLOC, eNSGAII-DiffLOC, NSGAIII-DiffLOC, eMOEA-DiffLOC and SPEAII-DiffLOC respectively.

We also use our objective and fitness functions which have been proposed by us and effectively applied in our previous works [17–22].

3   Supporting Tool and PUTs

Judy, a mutation testing tool, (http://www.mutationtesting.org/), is a tool which has been developed in Java and for Java by Madeyski et al. [23]. This tool supports a large set of Java mutation operators for execution mutation testing such as: mutants generation, mutants execution and mutation analysis. In addtion, there are many built-in multi-objective optimization algorithms in Judy [17–22].

That are the main reasons to explain why, in this paper, we continue to choose Judy for extending and using as the supporting tool in our experiment.

All of 20 PUTs were downloaded from the website https://github.com as well as http://sourceforge.net. They are 20 open source projects including the given set of TCs and also passed successfully their included (in each PUT) set of TCs.

Table 1 presents in short 20 selected PUTs. Name of all selected PUTs (PUTs, in column 1), lines of code (LOC, in column 2) and number of given test cases (#TCs, in column 3) are the information of that table.

4   Results Analysis

Experimental results are shown in Table 2 and Fig. 1. Table 2 reports the mean values of ratios of NoM, NoT, NoR, NoH7 and NoH1 (see in Sect. 2) for each selected PUTs:

Table 1. Projects under test (PUTs)

No	PUTs	LOC	#TCs
1	Antomology	1073	35
2	BeanBin	5925	68
3	Commons-csv	9972	292
4	Commons-cli-1.3.1-src	11375	247
5	Jettison	11690	116
6	CommonsFileUploads 1.3.1	12321	12
7	Commons-email	12495	175
8	CommonsChain 1.2	13410	17
9	JWBF	13572	305
10	Commons-dbutils	15022	228
11	Commons-chain-1.2-src	17702	105
12	Java-util	18690	379
13	Barbecue	23996	190
14	Commons-text	25298	455
15	CommonsValidator 1.4.1	25422	66
16	Commons-validator	32743	350
17	CommonsJxPath 1.3	41079	28
18	Commons-digester3-3.2-src	41986	175
19	CommonsLang3 3.0	122964	126
20	Commons-lang3-3.4-src	124151	2587

Antomology; BeanBin; Commons-csv; Commons-cli-1.3.1-src; Jettison; Commons-FileUploads 1.3.1; Commons-email; CommonsChain 1.2; JWBF; Commons-dbutils; Commons-chain-1.2-src; Java-util; Barbecue; Commons-text; CommonsValidator 1.4.1; Commons-validator; CommonsJxPath 1.3; CommonsJxPath 1.3; CommonsLang3 3.0; Commons-lang3-3.4-src named from 1 to 20 respectively in Column 1. These mean values are average number of original multi-objective optimization algorithms (in row named Original) and modification algorithms (in row named DiffLOC) for each PUT.

In column 2 of Table 2, the term "Original" represents the 5 original multi-objective optimization algorithms (NSGA-II, eNSGA-II, NSGA-III, eMOEA and SPEA-II), while the term "DiffLOC" represents 5 corresponding modification algorithms (NSGAII-DiffLOC, eNSGAII-DiffLOC, NSGAIII-DiffLOC, eMOEA-DiffLOC and SPEAII-DiffLOC).

Figure 1 illustrates the comparison between the values of the above-mentioned ratios (NoM, NoT, NoR, NoH7, NoH1) of HOMT execution applying original algorithms and modification algorithms.

Table 2. The mean ratios of NoM, NoT, NoR, NoH7, NoH1 (%)

PUT	Algorithm	NoM	NoT	NoR	NoH7	NoH1
1	Original	16.22	64.33	57.74	6.43	28.67
	DiffLOC	14.47	61.21	69.02	6.98	30.02
2	Original	17.16	63.48	56.78	6.12	30.34
	DiffLOC	15.34	60.12	58.04	6.45	33.67
3	Original	16.76	62.17	58.92	7.02	30.21
	DiffLOC	13.99	59.89	59.75	7.38	33.42
4	Original	15.79	65.74	57.56	6.24	31.95
	DiffLOC	14.01	61.05	59.01	6.78	34.50
5	Original	16.87	63.45	54..56	6.10	32.73
	DiffLOC	14.98	60.12	58.23	6.53	34.02
6	Original	15.95	64.94	60.12	6.22	32.11
	DiffLOC	15.75	62.36	62.04	7.01	34.35
7	Original	16.02	63.18	58.53	6.03	29.54
	DiffLOC	14.56	60.23	59.84	6.78	32.87
8	Original	16.78	64.58	53.78	6.17	31.98
	DiffLOC	15.43	61.29	55.92	6.28	33.45
9	Original	16.24	65.47	59.32	6.23	32.84
	DiffLOC	14.67	63.92	61.02	6.89	34.56
10	Original	15.94	65.49	58.37	5.98	32.17
	DiffLOC	14.22	62.36	59.98	6.24	35.23
11	Original	17.01	67.25	57.64	6.72	31.67
	DiffLOC	15.72	63.57	59.03	7.04	32.89
12	Original	16.04	64.58	58.12	6.38	30.09
	DiffLOC	14.05	61.87	59.45	6.84	34.20
13	Original	16.57	65.14	58.74	6.11	30.65
	DiffLOC	15.02	60.49	59.83	6.75	33.57
14	Original	15.92	64.52	54.57	6.43	31.29
	DiffLOC	14.01	61.04	57.39	6.95	34.02
15	Original	16.11	64.68	60.43	6.14	32.36
	DiffLOC	15.20	62.00	61.71	6.53	33.98
16	Original	16.90	63.79	55.89	6.04	30.73
	DiffLOC	14.12	61.22	58.22	6.58	33.70
17	Original	16.23	64.00	57.38	6.37	29.99
	DiffLOC	15.22	62.39	59.02	6.92	32.47
18	Original	16.35	66.24	56.87	5.99	32.11
	DiffLOC	14.32	61.98	59.01	6.74	34.05
19	Original	15.77	65.16	59.34	6.05	31.59
	DiffLOC	14.21	62.34	61.54	6.78	34.08
20	Original	15.93	66.03	58.60	6.75	30.12
	DiffLOC	13.98	62.54	62.15	6.94	33.45

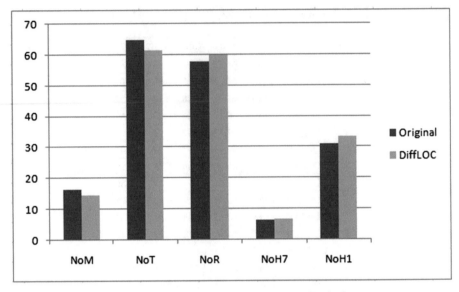

Fig. 1. Comparison the values of NoM, NoT, NoR, NoH7, NoH1

As we mentioned before, in this paper we focus on execution higher order mutation testing applying the original multi-objective algorithms (named EXEC1) and corresponding modified algorithms (named EXEC2) to evaluate and compare the effect
of mutant reduction, generating hard-to-kill mutant and constructing quality mutants
between them (EXEC1 and EXEC2).

The obtained results indicate that the modification algorithms are significantly better
than the original algorithms, more specifically as follows:

– The number of generated mutants of EXEC2 is smaller than the number of generated
  mutants of EXEC1 (evaluated basing on NoM).
– The number of generated hard-to-kill mutants is increased (evaluated basing on NoT,
  NoR and NoH1). By means of NoT, we realize that number of TCs which kill generated
  HOMs in EXEC2 is smaller than in EXEC1 (compared to the number of TCs which
  kill their constituent FOMs). The values of NoR indicate that number of of generated
  "reasonable HOMs" to al generated HOMs of EXEC2 is better than EXEC2. The
  problem of equivalent mutant is a main barrier of mutation testing [3] and so a larger
  number of H1 ("*live (potentially equivalent) mutants*") (NoH1) can be seen as a
  disadvantage. However, with the carefully analyzing, we realized that H1 mutants can
  be "hard-to-kill mutants" [17–21], which are valuable. Those HOMs cannot be killed
  by the given set of test cases which are obtained in the selected project under test, but
  they are fully capable of being killed by one or more new and better (in terms of able
  to expose the potential defects of the computer programs) test cases. Therefore, more
  specific studies are needed to confirm whether these H1 mutants are really. To do this,
  we can spend a lot of time and the error prone activity may is inevitable [9].

– The number of H7 (NoH7) in EXEC2 is slightly higher than in EXEC1. This leads to demonstration that constructing high quality and reasonable HOMs (H7) of modified algorithms is better than corresponding original multi-objective optimization algorithms.

5 Conclusions

The quality of mutation testing in general and higher order mutation testing in particular is the important problem that is needed to study for applying widely mutation testing in software testing practice.

In this paper, we introduced an empirical study with 20 projects under test and 5 original multi-objective optimization algorithms as well as 5 corresponding modification algorithms to try for bringing out the confirmation the effectiveness of our proposed method in the field of higher order mutation testing applying multi-objective optimization algorithms. The obtained results indicate that our method is an effectiveness one for improving the quality of mutation testing in general in terms of mutant reduction, generating hard-to-kill mutants and constructing high quality, reasonable mutants.

We know that, the 20 selected Java project in this paper may not be representative of Java programs in particular and other languages programs in general. In addition, the quality as well as coverage criteria of 20 test suites which are included in 20 PUTs are completely different (higher and lower) from actual projects. So, we also know that it hard to say for sure that our proposed solution is an absolutely good one.

However, with the above-presented initial satisfactory results in [21] and in this paper, we hope that our method is one of effect solutions for improving the quality of applying multi-oblective algorithms into higher order mutation testing. Based on the our belief in useful of our study results, we continue to perform other further researchs in the future to confirm the correctness and effectiveness of the presented method.

Acknowledgement. This paper is funded by Vietnam National University Ho Chi Minh City (VNU-HCM) under grant number C2018-26-09.

References

1. DeMillo, R.A., Lipton, R.J., Sayward, F.G.: Hints on test data selection: help for the practicing programmer. IEEE Comput. **11**(4), 34–41 (1978)
2. Hamlet, R.G.: Testing programs with the aid of a compiler. IEEE Trans. Softw. Eng. **SE-3**(4), 279–290 (1977)
3. Nguyen, Q.V., Madeyski, L.: Problems of mutation testing and higher order mutation testing. In Do, T., Le Thi, H.A., Nguyen, N.T. (eds.) ICCSAMA 2014, Advanced Computational Methods for Knowledge Engineering. Advances in Intelligent Systems and Computing, vol. 282, pp. 157–172. Springer (2014). https://doi.org/10.1007/978-3-319-06569-4_12
4. Jia, Y., Harman, M.: An analysis and survey of the development of mutation testing. IEEE Trans. Softw. Eng. **37**(5), 649–678 (2011)
5. Jia, Y., Harman, M.: Higher order mutation testing. Inf. Softw. Technol. **51**, 1379–1393 (2009)

6. Harman, M., Jia, Y., Langdon, W.B.: A manifesto for higher order mutation testing. In: Third International Conference on Software Testing, Verification, and Validation Workshops (2010)
7. Offutt, A.J.: Investigations of the software testing coupling effect. ACM Trans. Softw. Eng. Methodol. **1**, 5–20 (1992)
8. Polo, M., Piattini, M., Garcia-Rodriguez, I.: Decreasing the cost of mutation testing with second-order mutants. Softw. Test. Verif. Reliab. **19**(2), 111–131 (2008)
9. Madeyski, L., Orzeszyna, W., Torkar, R., Józala, M.: Overcoming the equivalent mutant problem: a systematic literature review and a comparative experiment of second order mutation. IEEE Trans. Softw. Eng. **40**(1), 23–42 (2014). https://doi.org/10.1109/TSE.2013.44
10. Jia, Y., Harman, M.: Constructing subtle faults using higher order mutation testing. In: Proceedings of Eighth International Working Conference on Source Code Analysis and Manipulation (2008)
11. Omar, E., Ghosh, S., Whitley, D.: Constructing subtle higher order mutants for Java and AspectJ programs. International Symposium on Software Reliability Engineering, pp. 340–349 (2013)
12. Omar, E., Ghosh, S., Whitley, D.: Comparing search techniques for fnding subtle higher order mutants. In: Proceedings of the 2014 Annual Conference on Genetic and Evolutionary Computation, pp. 1271–1278 (2014)
13. Fevzi Belli, F., Guler, N., Hollmann A., Suna, E., Yildiz, E.: Model based higher-order mutation analysis. In: Advances in Software Engineering. Communications in Computer and Information Science, vol. 117, pp. 164–173 (2010)
14. Akinde, A.O.: Using higher order mutation for reducing equivalent mutants in mutation testing. Asian J. Comput. Sci. Inf. Technol. **2**(3), 13–18 (2012)
15. Langdon, W.B., Harman, M., Jia, Y.: Multi-objective higher order mutation testing with genetic programming. In: Proceedings Fourth Testing: Academic and Industrial Conference Practice and Research (2009)
16. Langdon, W.B., Harman, M., Jia, Y.: Efficient multi-objective higher order mutation testing with genetic programming. J. Syst. Softw. **83**, 2416–2430 (2010)
17. Nguyen, Q.V., Madeyski, L.: Searching for strongly subsuming higher order mutants by applying multi-objective optimization algorithm. In: Le Thi, H.A., Nguyen, N.T., Do, T.V. (eds.) Advanced Computational Methods for Knowledge Engineering. Advances in Intelligent Systems and Computing, vol. 358, pp. 391–402. Springer, Cham (2015). https://doi.org/10.1007/978-3-319-17996-4_35
18. Nguyen, Q.V., Madeyski, L.: Empirical evaluation of multi-objective optimization algorithms searching for higher order mutants. Cybern. Syst. Int. J. **47**, 48–68 (2016)
19. Nguyen, Q.V., Madeyski, L.: Higher order mutation testing to drive development of new test cases: an empirical comparison of three strategies. Lecture Notes in Computer Science (2016)
20. Nguyen, Q.V., Madeyski, L.: On the relationship between the order of mutation testing and the properties of generated higher order mutants. Lecture Notes in Computer Science (2016)
21. Nguyen, Q.V., Madeyski, L.: Addressing mutation testing problems by applying multi-objective optimization algorithms and higher order mutation. J. Intell. Fuzzy Syst. **32**, 1173–1182 (2017). https://doi.org/10.3233/JIFS-169117
22. Nguyen Q.V., Pham D.T.H.: Is higher order mutant harder to kill than first order mutant? An experimental study. Lecture Notes in Computer Science, vol. 10751. Springer (2018). https://doi.org/10.1007/978-3-319-75417-8_62
23. Madeyski, L., Radyk, N.: Judy - a mutation testing tool for Java. IET Softw. **4**(1), 32–42 (2010). https://doi.org/10.1049/iet-sen.2008.0038
24. Deb, K., Pratap, A., Agarwal, S., Meyarivan, T.: A fast and elitist multi objective genetic algorithm: NSGA-II. IEEE Trans. Evol. Comput. **6**(2), 182–197 (2002)

25. Deb, K., Jain, H.: An evolutionary many-objective optimization algorithm using reference-point-based nondominated sorting approach, Part I: solving problems with box constraints. IEEE Trans. Evol. Comput. **18**(4), 577–601 (2014)
26. Kollat, J.B., Reed, P.M.: The value of online adaptive search: a performance comparison of NSGAII, ε-NSGAII and ε-MOEA. In: Coello, C.A.C., Aguirre, A.H., Zitzler, E. (eds.) Third International Conference on Evolutionary Multi-Criterion Optimization, EMO 2005, Guanajuato, Mexico, 9–11 March 2005 (2005)
27. Deb, K., Mohan, M., Mishra, S.: A Fast Multi-objective Evolutionary Algorithm for Finding Well-Spread Pareto-Optimal Solutions. KenGAL, Report No. 2003002. Indian Institute of Technology, Kanpur, India (2003)
28. Zitzler, E., Laumanns, M., Thiele, L.: SPEA2: improving the strength Pareto evolutionary algorithm for multi-objective optimization. In: Giannakoglou, K., Tsahalis, D., Periaux, J., Papailiou. K., Fogarty, T. (eds.) Evolutionary Methods for Design, Optimisations and Control, pp. 19–26 (2002)
29. McConnell, S.: Code Complete, 2nd edn. Microsoft Press, Redmond (2004)

Belief Merging for Possibilistic Belief Bases

Thi Thanh Luu Le[1,2] and Trong Hieu Tran[2(✉)]

[1] University of Finance and Accountancy, Quangngai, Vietnam
lt.thanhluu@gmail.com
[2] VNU - University of Engineering and Technology, Hanoi, Vietnam
hieutt@vnu.edu.vn

Abstract. Belief merging has received much attention from the research community with a large range of applications in Computer Science and Artificial Inteligence. In this paper, we represent a new belief merging approach for prioritized belief bases. The main idea of this method is to use two operators, namely connective strong operator and averagely increasing operator to merge possibilistic belief bases. By this way, the proposed method allows to keep more useful beliefs, which may be lost in other methods because of drowning effect. The logical properties of merging result are also analyzed and discussed.

Keywords: Belief merging · Possibilistic logic · Prioritized belief base

1 Introduction

The logic-based belief merging methods have attracted a lot of attention in many areas of computer science. In the logical model, each information source is often treated as a belief base and represented as a set of logical formulas. Belief merging by argumentation is often implemented in classical logic and possibilistic logic [1, 2]. In practice, we usually face with contradition, however in this case, classical logic is inapplicable. In order to deal with the contradition, possibilistic logic [3] is one of the most common tools. In possibilistic logic, each formula is attached to a weight indicating the necessity of the formula, which is understood as the priority of formulas. We can deduce some non-trivial results from a partially inconsistent belief base by using a possibilistic consequence relationship.

In literature, there are some typical approaches for merging possibilistic belief bases such as [4–9]. In these works, a merging operator is used to calculate a new possibility distribution from the possibilities from given belief bases. Then, the syntax representation of the new possibility distribution are generated [7]. Unlike propositional merging operators, possibilistic merging operators can lead to an inconsistent belief base even though the input belief bases are consistent. Further, these methods do not require that the input belief bases are consistent. However, they are affected by drowning effect. It causes to lose all information with the weight smaller than the inconsistence degree in the input belief bases.

© Springer Nature Switzerland AG 2020
H. A. Le Thi et al. (Eds.): ICCSAMA 2019, AISC 1121, pp. 370–380, 2020.
https://doi.org/10.1007/978-3-030-38364-0_33

We can face the problem of merging belief bases that an agent can have inconsistent belief and accept that. Many researchers proposed some methods to resolve the inconsistence problem, typically [10–14]. Some works improve semantic merging rules like [15–17]. In these works, the merging process includes two steps, weakening the conflict belief and them combining them. However, they are not recommended to integrate belief bases with strong conflicts, namely, the inconsistency level of their belief base is high. The approach based on argumentation in [1, 2, 4] is difficult to calculate to build arguments from an inconsistent belief base.

As we know, there is no clear method of merging inconsistent belief bases. They are always assumed that the input belief base is consistent when we face with merging problems [10, 11, 15, 17–20]. In some case, althrought the input belief base is inconsistent, it can still preserve some useful information about the real world. Thus, the result of the merging process may not be required to be a consistent belief base.

In this paper, we propose a new approach to merge priority belief bases in possibilistic logic by choosing an appropriate priority relationship between arguments. Our approach is defined semantically by the belief merging model in propositional logic. The integrated result is still a possibility distribution. In order to avoid useful information lost after merging due to the drowning effect we also propose a method based on combining many different operators. Moreover, this approach will also allow to merge potential information. We point out that our approach captures the minimal change and some good logical properties.

The structure of this paper is as follows: Sect. 2 gives the introduction, we introduce the basic knowledge of possibilistic logic and theory of argumentation. In Sect. 3, the belief merging model for inconsistent prioritized belief bases is proposed. The properties of the belief merging operators based on combining multiple operators are discussed in Sect. 4. Several conclusions and future work are in Sect. 5.

2 Preliminary

2.1 Possibilistic Logic

In this section, we recall about in possibilistic logic (more details can be found in [3]).

The semantics of possibilistic logic based on the concept of possibility distribution π, which is a map from the set of representations Ω to the interval $[0, 1]$. The possibility degree $\pi(\omega)$ represents the degree of the satisfaction of representation ω with the belief available about the real world. A possibility distribution is said to be standard if $\exists \omega_0 \in \Omega$, thus $\pi(\omega_0) = 1$. From a possibility distribution π, two measurements can be defined: the *possibility degree of the formula* φ, $\Pi_\pi(\varphi) = max\{\pi(\omega) : \omega \in \Omega, \omega \models \varphi\}$ and *the necessity degree of the formula* φ, $N_\pi(\varphi) = 1 - \Pi_\pi(\neg\varphi)$.

At the syntactic level, we present each possibilistic formula by a pair (φ, α), where φ is a propositional formula and the weight $\alpha \in [0, 1]$. It is said that the necessity degree of formula φ is greater than or equal to α, namely $N(\varphi) \geq \alpha$. A belief base is a finite set of formulas that have the form $T = a\{(\varphi_i, a_i) : i = 1, \ldots, n\}$. The classical belief base combined with T is denoted T^*, namely $T^* = \{\varphi_i | (\varphi_i, a_i) \in T\}$. The formulas of belief base T can be ordered according to their weights so that $\alpha_1 = 1 \geq \alpha_2 \geq \cdots \geq \alpha_n \geq 0$. It is easy to see that a possibilistic belief base T is consistent if and only if its classical belief base T^* is consistent. A possibilistic belief profile E is a multi-set of possibilistic belief bases.

With a feasible belief base T, by the principle of minimum specification entropy [3], we obtain only one possibility distribution, denoted by π_T as follows:

For $\forall \omega \in \Omega$,

$$\pi_T(\omega) = \begin{cases} 1 \ \text{nếu} \forall (\varphi_i, \alpha_i) \in T, \omega \vDash \varphi_i, \\ 1 - \max\{\alpha_i : (\varphi_i, \alpha_i) \in T \text{ và } \omega \nvDash \varphi_i\} otherwise. \end{cases} \quad (1)$$

Definition 1. Let T be a possibilistic belief base, and $\alpha \in [0, 1]$,

1. a-cut of T is $T_{\geq \alpha} = \{\varphi_i \in T^* | (\varphi_i, \beta_i) \in T \text{ và } \beta_i \geq \alpha\}$
2. strict a-cut of T is $T_{>\alpha} = \{\varphi_i \in T^* | (\varphi_i, \beta_i) \in T \text{ và } \beta_i > \alpha\}$
3. The inconsistency degree of T is $Inc(T) = \max\{\alpha_i : T_{\geq \alpha_i} \text{ is inconsistent}\}$

The inconsistency degree of T. is the largest weight α_i where α_i-cut of T is inconsistent.

Definition 2. Let T be a possibilistic belief base, the formula φ is said to be the result of T with degree α, denoted by $T \vdash_\pi (\varphi, \alpha)$, if and only if (i) $T_{\geq \alpha}$ is consistent; (ii) $T_{\geq \alpha} \vdash \varphi$; (iii) $\forall \beta > \alpha, T_{\geq \beta} \nvdash \varphi$.

According to Definition 2, an inconsistent belief base may still deduce non-trivial results. However, it is affected by drowning effect. That is, with an inconsistent belief base T, all formulas, whose certainty degrees are not greater than $Inc(T)$, are completely useless for non-trivial inference. For example, for $T = \{(p, 0.9), (\neg p, 0.8), (r, 0.3), (q, 0.5)\}$, obviously T is equivalent to $T = \{(p, 0.9), (\neg p, 0, 8)\}$ because of $Inc(T) = 0.8$. Therefore, $(q, 0.5)$ and $(r, 0.3)$ are not used in possibilitic reasoning.

Two possibilistic belief bases T and T' are equivalent, denoted by $T \equiv T'$, if and only if $\forall \alpha \in (0, 1], T_{\geq \alpha} \equiv T'_{\geq \alpha}$. Two possibilistic belief profiles $E = \{T_1, T_2, \ldots, T_n\}$ and $E' = \{T'_1, T'_2, \ldots, T'_n\}$ are equivalent, denoted by $E \equiv E'$ if and only if there exists a permutation σ on $[1,.., n]$ so that $T_i \equiv T'\sigma(i)$ for all $i \in [1, .., n]$.

Many operators have been proposed for merging possibilistic belief bases. Give a possibilistic belief profiles $\{T_1, T_2, \ldots, T_n\}$ with possible distributions $\{\pi_{T_1}, \ldots, \pi_{T_n}\}$, the semantic combination of possibility distributions by a aggregation function $\oplus$ results in a new possibilistic distribution $\pi_\oplus(\omega) = (\ldots ((\pi_1(\omega) \oplus \pi_2(\omega)) \oplus \pi_3(\omega)) \oplus \ldots) \oplus \pi_n(\omega)$.

The possibilitic belief base, which is the result of merging process is calculated by the following equation [7]:

$$T_\oplus = \{(K_j, 1 - \oplus(x_1, \ldots, x_n)) : j = 1, \ldots, n\}, \quad (2)$$

where K_j is the conjunction with size j of the formulas taken from different T_i $(i = 1, \ldots, n)$ and x_i equal to $1 - \alpha_i$ or 1 depending on whether φ_i belongs to K_j or not.

Operator $\oplus$ should satisfy the following attributes:

1. $\oplus(0, \ldots, 0) = 0$
2. If $\forall i = 1, \ldots, n, \alpha_i \geq \beta_i$ then $\oplus(\alpha_1, \ldots, \alpha_n) \geq \oplus(\beta_1, \ldots, \beta_n)$

2.2 Belief Merging

Belief merging is a process to integrate the beliefs from multiple belief sources into a common belief base. Many approaches for belief mergings are proposed, analysed and discussed in past thirty years. Therefore, because of the lack of space, in this paper we do not discuss more about them. We just focus on the set of axioms to characterise the result of merging process. These axioms are stated as follows:

Let ε, ε_1, ε_2 be belief profiles and μ, μ_1, μ_2 be formulas from $\mathcal{L}_{PS}$. A belief merging operator Δ is a map from set of belief profiles to a propositional belief base. The following axioms should be satisfied:

(IC0) $\Delta_\mu(\varepsilon) \vdash \mu$.
(IC1) If μ is consistent then $\Delta_\mu(\varepsilon)$ is also consistent.
(IC2) If $\wedge \varepsilon \wedge \mu$ is consistent then $\Delta_\mu(\varepsilon) \equiv \wedge \varepsilon \wedge_\mu$.
(IC3) If $\varepsilon_1 \equiv_s \varepsilon_2$ and $\mu_1 \equiv \mu_2$ then $\Delta_{\mu_1}(\varepsilon_1) \equiv \Delta_{\mu_2}(\varepsilon_2)$.
(IC4) If $K_1 \vdash \mu$ and $K_2 \vdash \mu$, then $\Delta_{IC}(\{K_1, K_2\}) \wedge K_1$ is consistent if and only if $\Delta_{IC}(\{K_1, K_2\}) \wedge K_2$ is consistent.
(IC5) $\Delta_\mu(\varepsilon_1) \wedge \Delta_\mu(\varepsilon_2) \vdash \Delta_\mu(\varepsilon_1 \sqcup \varepsilon_2)$.
(IC6) If $\Delta_\mu(\varepsilon_1) \wedge \Delta_\mu(\varepsilon_2)$ is consistent then $\Delta_\mu(\varepsilon_1 \sqcup \varepsilon_2) \vdash \Delta_\mu(\varepsilon_1) \wedge \Delta_\mu(\varepsilon_2)$
(IC7) $\Delta_{\mu_1}(\varepsilon) \wedge \mu_2 \vdash \Delta_{\mu_1 \wedge \mu_2}(\varepsilon)$.
(IC8) If $\Delta_{\mu_1}(\varepsilon) \wedge \mu_2$ is consistent then $\Delta_{\mu_1 \wedge \mu_2}(\varepsilon) \vdash \Delta_{\mu_1}(\varepsilon) \wedge \mu_2$.

3 Belief Merging for Possibilistic Logic

In [9], authors proposed a syntax-based approach for merging a set of n consistent possibilistic belief bases $T_1, \ldots, T_n$. A possibilistic belief merging operator, denoted by $\oplus$., is a map from $[0, 1]^n$ to $[0, 1]$ to integrate the certainty of belief from multiple belief sources. The result of merging process $\mathcal{T}_\oplus$ is as follows:

$$\mathcal{T}_\oplus = \{(\varphi, \oplus(\alpha_1, \ldots, \alpha_n)) : T_i \vdash_\pi (\varphi, \alpha_i)\} \tag{3}$$

This method considers all the formulas in T_i even if T_i is inconsistent. Let $T_1 = \{(\varphi_i, \alpha_i) : i = 1, \ldots, n\}$ and $T_2 = \{(\partial_j, \beta_j) : j = 1, \ldots, m\}$ be two possibilistic belief bases. The merging result $\mathcal{T}_\oplus$ from T_1 and T_2 according to (3) is equivalent as follows:

$$\mathcal{T}_\oplus = \{(\varphi_i, \oplus(\alpha_i, 0)) : (\varphi_i, \alpha_i) \in T_1 \text{ and } \varphi_i \notin T_2^*\} \cup \{(\partial_j, \oplus(0, \beta_j)) : (\partial_j, \beta_j) \in$$
$$T_2 \text{ and } \partial_j \notin T_1^* \cup \{(\varphi_i \vee \partial_j, \oplus(\alpha_i, \beta_j)) : (\varphi_i, \alpha_i) \in T_1 \text{ and } (\partial_j, \beta_j) \in T_2\}$$
$$\tag{4}$$

Let T_1 and T_2 be two possibilistic belief bases, and $\pi_\oplus$ be combined from π_{T_1} and π_{T_2} based on belief merging operator $\oplus$. The belief merging result is as follows:

$$T_1 \oplus T_2 = \{(\varphi_i, 1 - ((1 - \alpha_i) \oplus 1)) : (\varphi_i, \alpha_i) \in T_1\} \cup \{(\partial_j, 1 - (1 \oplus (1 - \beta_j))) : (\partial_j, \beta_j) \in T_2\}$$
$$\cup \{(\varphi_i \vee \partial_j, \oplus(\alpha_i, \beta_j)) | (\varphi_i, \alpha_i) \in T_1 \text{ and } (\partial_j, \beta_j) \in T_2\} \tag{5}$$

When $\oplus = min$, we can easy to show that $T_1 \oplus T_2 = T_1 \cup T_2$

Let T_1 and T_2 be two possibilistic belief bases. If operator $\oplus$ in (3) is *maximum* and operator $\oplus$ in (5) is *minimum*, we have:

$$\mathcal{T}_\oplus \equiv T_1 \otimes T_2 \tag{6}$$

There are two important group of possibilistic belief merging operators: group of conjunctive operators *minimum* including *product* $(\alpha \times \beta)$ and *linear product* $(max(0, \alpha + \beta - 1))$ and group of disjunctive operators *maximum* including *probabilistic sum* $(\alpha + \beta - \alpha \times \beta)$ and *bounded sum* $(min(1, \alpha + \beta))$ [21].

Definition 6. Let $\oplus_1$ and $\oplus_2$ be two belief merging operators satisfied above properties. $\oplus_1$ and $\oplus_2$ are dual if and only if $\oplus_1(\alpha, \beta) = 1 - (1 - \alpha) \oplus_2 (1 - \beta)$.

The typical dual operators are conjunctive and disjunctive operators in [22].

For each formula φ, if $(\varphi, \alpha) \in T_1$ and $(\varphi, b) \in T_2$, such that $\alpha, \beta > 0$ then $(\varphi, \oplus(\alpha, \beta)) \in \mathcal{T}_\oplus$. On the other hand, φ will be in $T_1 \otimes T_2$ in several forms, namely $(\varphi, \otimes(\alpha, 0))$, $(\varphi, \otimes(0, \beta))$ và $(\varphi, \otimes(\alpha, \beta))$. Obviously, $(\varphi, \otimes(\alpha, 0))$ and $(\varphi, \otimes(0, \beta))$ are redundancy, we can remove them to simplify the merging result.

Example 1. Let $T_1 = \{(\varphi, 0.3), (\partial, 0.6), (\neg\varphi, 0.5), (\lambda, 0.5)\}$ and $T_2 = \{(\varphi, 0.1), (\partial \vee \lambda, 0.7)\} \oplus (\alpha, \beta) = min(1, \alpha + \beta)$ According to (4), the merging result of T_1 and T_2 is

$$\mathcal{T}_\oplus = \left\{ \begin{array}{c} (\varphi, 0.4), (\neg\varphi, 0.5), (\partial, 0.6), (\lambda, 0.5), (\varphi \vee \lambda, 0.6), (\varphi \vee \partial, 0.7), (\partial \vee \lambda, 1), (\varphi \vee \partial \vee \lambda, 1), \\ (\neg\varphi \vee \partial \vee \lambda, 1) \end{array} \right\}$$

If we merge T_1 and T_2 by using (5) with conjunctive operator $(\otimes (\alpha, \beta) = max0, \alpha + \beta - 1)$, the merging result is

$$T_1 \otimes T_2 = \{(\varphi, 0.3), (\neg\varphi, 0.5), (\partial, 0.6), (\lambda, 0.5), (\partial \vee \lambda, 0.7), (\neg\varphi \vee \partial \vee \lambda, 0.2)\}$$

In Example 1, φ occurs in $\mathcal{T}_\oplus$ with the weight 0.4, however it occurs in $T_1 \otimes T_2$ with three weights 0.3, 0.1 and 0. Formulas $(\varphi, 0.1)$ and $(\varphi, 0)$ are redundent. Similarly $\partial \vee \lambda$ occurs in $\mathcal{T}_\oplus$ with the weight 1 and 0.7, along with it also occurs in $T_1 \otimes T_2$ with three weights 0.7, 0.3 and 0.2. So we only hold the largest weighted formula in the merging result.

In [23], several belief merging operators are proposed to merge two possibilistic belief bases by using (4). It is pointed out that *maximum* is suitable for conflict belief bases, and *minimum* is meaningful when the belief bases are consistent. When *maximum* is chosen, the merging result is

$$T_1 \oplus_{max} T_2 = \{(\varphi_i \vee \partial_j, min(\alpha_i, \beta_j)) : (\varphi_i, \alpha_i) \in T_1 \text{and} (\partial_j, \beta_j) \in T_2\}$$

Clearly, $T_1 \oplus_{max} T_2$ is the weak merging result, i.e. too much belief is lost.

The above belief merging approaches utilizes only one operator even if possibilistic belief bases are conflict. The following example is about such operator.

Example 2. Let $T_1 = \{(\varphi, 0.5), (\partial, 0.6), (\neg\partial, 0.4), (\xi, 0.7)\}$ and $T_2 = \{(\neg\varphi, 0.7), (\varphi, 0.3), (\partial, 0.7), (\xi, 0.5), (\lambda, 0.4)\}$ Two belief ∂ and λ are supported by T_1 and T_2 with high degrees and they are not related to the inconsistency of $T_1 \cup T_2$, thus it is necessary to increase the possibility for them. Suppose that belief merging operator *maximum* is *probabilistic sum* defined by $\oplus(\alpha, \beta) = \alpha + \beta - \alpha \times \beta$ According to (4), the merging result of T_1 and T_2 is as follows:

$$T_\oplus =$$

$\{(\neg\varphi, 0.7), (\neg\partial, 0.4), (\lambda, 0.4), (\varphi, 0.65), (\varphi \vee \partial, 0.85), (\varphi \vee \xi, 0.75), (\varphi \vee \lambda, 0.7),$
$(\neg\varphi \vee \partial, 0.88), (\varphi \vee \partial, 0.72), (\partial, 0.88), (\partial \vee \xi, 0.8), (\partial \vee \lambda, 0.76), (\neg\varphi \vee \neg\partial, 0.82),$
$(\varphi \vee \neg\partial, 0.58), (\neg\partial \vee \xi, 0.7), (\neg\partial \vee \lambda, 0.64),$
$\quad (\neg\varphi \vee \xi, 0.91), (\varphi \vee \xi, 0.79), (\partial \vee \xi, 0.91), (\xi, 0.85), (\xi \vee \lambda, 0.82)\}$

In this example, the necessities of ∂ and ξ increase because *probabilistic sum* make the weight increased. Because formulas φ and $\neg\varphi$ are strong conflict, the necessity degrees of both φ and $\neg\varphi$ should be descreased. However, in the merging result, the necessity degree of φ increases up to 0.65 and the necessity degree of $\neg\varphi$ is still 0.7, it irrational. This issue is caused by using a unique operator to merge two inconsistent formulas.

Given n possibilistic belief bases $\{T_1, T_2, \ldots, T_n\}$ from n different sources. For the formulas, related to the conflicts in $T_1 \cup T_2 \cup \ldots \cup T_n$, the necessity degrees of them will be decreased. In contrast, the necessity degrees of them will be increased if they are supported from these sources.

Now, we introduce a belief merging operator proposed in [9].

Definition 3. A belief merging operator $\oplus$ is *strongly connective* on [0, 1] if $(\alpha_1, \ldots, \alpha_n) \in [0, 1]$

$$\oplus(\alpha_1, \ldots, \alpha_n) \geq max(\alpha_1, \ldots, \alpha_n)$$

A strongly connective operator is rational because it is satisfied almost axioms introduced in [9]. If a strongly connective operator satisfies $\oplus(\alpha_1, \ldots, \alpha_n) \geq max(\alpha_1, \ldots, \alpha_n)$ with $\forall \alpha_i \neq 1$ and $\oplus(\alpha_1, \ldots, \alpha_n) = 1$ when $\exists i, \alpha_i = 1$ is called *monotonic operator*. A strongly connective operator is suitable to merge conflict-free formulas.

We proposed a new merging operator as follows:

Definition 4. An operator $\oplus$ is an *averagely increasing operator* on on [0, 1] if $(\alpha_1, \ldots, \alpha_n) \in [0, 1]$

$$\oplus(\alpha_1, \ldots, \alpha_n) \leq max(\alpha_1, \ldots, \alpha_n)$$

This operator represents that the necessity of a formula in merging result do not excess the maximal necessity of this formula in the input possibilistic belief bases. An example about averagely increasing operator is average of weight, defined by $\oplus(\alpha, \beta) = x\alpha + y\beta$, such that $x, y \in [0, 1]$ and $x + y = 1$ If $x = y = 1/2$, this operator is the standard average operator and if $x > y$ (or $x < y$) it means that the source associated with x is more (less resp.) reliable than the other. An averagely increasing operator is suitable for merging conflict-related formulas.

Definition 5. [8] Let $\mathcal{E}$ be a part of a classical belief base T, $\mathcal{E}$ is minimal inconsistent set if and only if it meets the following conditions:

- $\mathcal{E} \vdash false$ and
- $\forall \varphi \in \mathcal{E}, \mathcal{E} - \{\varphi\} \nvdash false$.

Definition 6. A formula φ is in a conflict in a classical belief base T if and only if it belongs to a minimal inconsistent set of T. The set of formulas in the conflicts in T is denoted by $Conflict(T)$.

Here, we propose a new method for belief merging based on multiple operators. We suppose that with a possibilistic belief bases T, if a formula φ is not in T^*, then $(\varphi, 0)$ is added in T.

Definition 7. Let $T_1 = \{(\varphi_i, \alpha_i) : i = 1, \ldots, n\}$ and $T_2 = \{(\partial_j, \beta_j) : j = 1, \ldots, m\}$ be two possibilistic belief bases. Given strongly connective operator $\oplus_{st}$ and averagely increasing operator $\oplus_{ua}$. The merging result of T_1 and T_2 is defined as $\Delta_{\oplus_{st}, \oplus_{ua}}(T_1, T_2) = A \cup B$, such that

$$A = \{\varphi, \oplus_{ua}(\alpha, \beta) | \varphi \in (Conflict(T_1 \cup T_2))^*, (\varphi, \alpha) \in T_1 \text{ and } (\varphi, \beta) \in T_2\}$$
$$B = \{\varphi, \oplus_{st}(\alpha, \beta) | \varphi \notin (Conflict(T_1 \cup T_2))^*, (\varphi, \alpha) \in T_1 \text{ and } (\varphi, \beta) \in T_2\}$$

In Definition 7, we use two operators, a strongly connective operator and an averagely increasing operator, to merge possibilistic belief bases. With the conflict-free formulas in $T_1 \cup T_2$ we choose the strongly connective operator, and with the conflict-related formulas in $T_1 \cup T_2$, we use averagely increasing operator to merge them.

An other important point in Definition 7 is that $\Delta_{\oplus_{st}, \oplus_{ua}}(T_1, T_2)$ not only merges the formulas in T_1 and T_2 but also considers the formulas deduced from T_1 and T_2. In this paper we study implicit belief, the belief is inferred from the input belief bases.

Example 3 (Continue Example 2). Suppose that we use merging operators $\oplus_{st}(\alpha, \beta) = \alpha + \beta - \alpha\beta$ and $\oplus_{ua}(\alpha, \beta) = (\alpha + \beta)/2$. By Definition 7, the merging result of T_1 and T_2 is

$$\Delta_{\oplus_{st}, \oplus_{ua}}(T_1, T_2) = \{(\varphi, 0.4), (\neg\varphi, 0.35), (\partial, 0.65), (\neg\partial, 0.2), (\xi, 0.6), (\lambda, 0.4)\}$$

In Example 3, after merging process, the necessity degrees of both φ and $\neg\varphi$ are decreased, and the necessity degree of φ is larger than the necessity degree of $\neg\varphi$. The necessity degrees of other formulas are similar to in Example 2. However, the disjunctive formulas in Example 2 does not exist in Example 3. Althrough we can infer these formulas from $\Delta_{\oplus_{st},\oplus_{ua}}(T_1, T_2)$ with smaller necessity degrees than in Example 2, $\Delta_{\oplus_{st},\oplus_{ua}}(T_1, T_2)$ is simpler than $\mathcal{T}_\oplus$ in Example 2.

In Definition 7, all conflict-related formulas is weaken to obtain the smaller necessity degrees in the merging result. However, in some particular cases, it is more rational if the necessity degrees of some of these formulas are increased. Example, suppose that we have two possibilistic belief bases $T_1 = \{(\varphi, 0.7), (\partial, 0.7)\}$ and $T_2 = \{(\neg\varphi, 0.4), (\varphi, 0.7), (\partial, 0.4), (\xi, 0.5), (\lambda, 0.4)\}$ from two sources. Clearly, φ is supported by T_1. Although φ is conflict-related formula in T_2, the necessity degrees of φ is larger than $\neg\varphi$, thus φ can be considered totally by T_2. Therefore, both of sources support φ, and the necessity degrees of φ should be increased.

Definition 8. Let T be an inconsistent belief base. Formula φ is a conflict-free formulas in T and it is *weakly supported* by T if and only if $\exists(\varphi, \alpha) \in T$ such that $\alpha > \beta$ for all $(\neg\varphi, \beta) \in T$.

Definition 9. Let T_1 and T_2 be two possibilistic belief bases. Formula φ is a *weakly conflict-related formula* of T_1 and T_2 if and only if φ is weakly supported by T_1 and T_2, respectively. The set of weakly conflict-related formulas in $T_1 \cup T_2$ is denoted by $Weak(T_1 \cup T_2)$.

Here, we define the merging method based on multiple operators as follows:

Definition 10. Let $T_1 = \{(\varphi_i, \alpha_i) : i = 1, \dots, n\}$ and $T_2 = \{(\partial_j, \beta_j) : j = 1, \dots, m\}$ be two possibilistic belief bases. Given strongly connective operator $\oplus_{st}$ and averagely increasing operator $\oplus_{ua}$. The merging result of T_1 and T_2 is defined as $\Delta_{\oplus_{st},\oplus_{ua}}(T_1, T_2) = A \cup B$, such that

$$A = \{(\varphi, \oplus_{ua}(\alpha, \beta))|\varphi \in (Conflict(T_1 \cup T_2)\backslash Weak(T_1 \cup T_2))^*, (\varphi, \alpha)$$
$$\in T_1 \text{ and } (\varphi, \beta) \in T_2\}B = \{(\varphi, \oplus_{st}(\alpha, \beta))|\varphi \notin (Conflict(T_1 \cup T_2))^* or \varphi$$
$$\in (Weak(T_1 \cup T_2))^*, (\varphi, \alpha) \in T_1 \text{ and } (\varphi, \beta) \in T_2\}$$

In $\Delta_{\oplus_{st},\oplus_{ua}}(T_1, T_2)$, the necessity degrees of formulas conflict-related and not weakly supported by both sources will be decreased. In contrast, the necessity degrees of formulas conflict-free or weakly supported by both sources will be increased. Obviously, if $\Delta_{\oplus_{st},\oplus_{ua}}$ is conjunctive, the belief merging method can easily generalize to n sources.

Example 4. Given $T_1 = \{(\varphi, 0.5), (\partial, 0.6)\}$ and $T_2 = \{(\neg\varphi, 0.4), (\varphi, 0.3), (\partial, 0.5), (\xi, 0.5), (\lambda, 0.4)\}$. Suppose that belief merging operators are $\oplus_{st}(\alpha, \beta) = \alpha + \beta - \alpha\beta$ and $\oplus_{ua}(\alpha, \beta) = (\alpha + \beta)/2$. Because φ is weakly conflict-related in $T_1 \cup T_2$ by Definition 10, the merging result is $\Delta_{\oplus_{st},\oplus_{ua}}(T_1, T_2) = \{(\varphi, 0.65), (\neg\varphi, 0.2), (\partial, 0.8), (\xi, 0.5), (\lambda, 0.4)\}$

Given a possibilistic belief profile $\mathcal{E}$ and two operator, a strongly connective operator and an averagely increasing operator, the merging result of our method is a possibilistic belief base $\mathcal{T}_{\mathcal{E}, \oplus_{st},\oplus_{ua}}$. With the conflict-free formulas in $T_1 \cup T_2$ we choose the strongly connective operator, and with the conflict-related formulas in $T_1 \cup T_2$, we use averagely increasing operator to merge them.

Belief merging algorithm for possibilistic belief bases based on combining multiple operators is as follows:

Input: Given a possibilistic belief profile $\mathcal{E} = \{T_1, \dots, T_n\}$; a strongly connective operator $\oplus_{st}$ and an averagely increasing operator $\oplus_{ua}$; belief merging operator $\Delta_{\oplus_{st},\oplus_{ua}} (T_1, \dots, T_n)$;

Output: A possibilistic belief base $\mathcal{T}_{\mathcal{E},\oplus_{st},\oplus_{ua}}$

Begin

1. $i = 1$;

2. $\mathcal{T}_{\mathcal{E},\oplus_{st},\oplus_{ua}} \leftarrow \{(\varphi, \alpha): \varphi \in T_1, (\psi, \beta): \partial \in T_2\}; \alpha, \beta \in [0,1]$

3.   **while** $i \leq n$ **do**

4. $A_{\oplus_{st}} = \{(\varphi, \oplus_{st}(\alpha, \beta)) | \varphi \notin (Conflict(T_i \cup T_{i+1}))^* \text{ or } \varphi$
$\in (Weak(T_i \cup T_{i+1}))^*, (\varphi, \alpha) \in T_i \text{ and } (\varphi, \beta)$
$\in T_{i+1}\}$
$B_{\oplus_{ua}} = \{(\varphi, \oplus_{ua}(\alpha, \beta)) | \varphi$
$\in (Conflict(T_i \cup T_{i+1}) \backslash Weak(T_i \cup T_{i+1}))^*, (\varphi, \alpha)$
$\in T_i \text{ and } (\varphi, \beta) \in T_{i+1}\}$
$\Delta_{\oplus_{st},\oplus_{ua}} (T_i, T_{i+1}) = A_{\oplus_{st}} \cup B_{\oplus_{ua}}$;

5. $\mathcal{T}_{\mathcal{E},\oplus_{st},\oplus_{ua}} \leftarrow \Delta_{\oplus_{st},\oplus_{ua}} (T_i, T_{i+1})$;

6.   If $Cn(T_i) = \{\varphi_i \in \mathcal{L}: T_i \vdash \varphi_i\}$ or $Cn(T_{i+1}) = \{\varphi_{i+1} \in \mathcal{L}: T_{i+1} \vdash \varphi_{i+1}\}$ then

$\mathcal{T}_{\mathcal{E},\oplus_{st},\oplus_{ua}} \leftarrow \mathcal{T}_{\mathcal{E},\oplus_{st},\oplus_{ua}} \cup \{(\varphi, \alpha): \varphi$
$\in \mathcal{E} \backslash (T_i \cup T_{i+1})\} \cup Cn(T_i) \cup Cn(T_{i+1})$

7.   Else $\mathcal{T}_{\mathcal{E},\oplus_{st},\oplus_{ua}} \leftarrow \mathcal{T}_{\mathcal{E},\oplus_{st},\oplus_{ua}} \cup \{(\varphi, \alpha): \varphi \in \mathcal{E} \backslash (T_i \cup T_{i+1})\}$;

8. $i = i + 1$;

9.   **end-while**

10.   **return** $\mathcal{T}_{\mathcal{E},\oplus_{st},\oplus_{ua}}$

End

4   Logical Properties

In this section, we examine several logical properties of merging result obtained from our method.

Proposition 1. Let $\mathcal{T} = \{T_1, \dots, T_n\}$ be a set of n possibilistic belief bases which are jointly consistent, the merging result of $\mathcal{T}$ satisfies Eq. (3).

Proof
If $\bigcup_{i=1}^{n} T_i$ is consistent, for any formula (φ, α_i), T $_i \vdash_\pi (\varphi, \alpha_i)$ if and only if there exists an inference for (φ, α_i) in $\mathcal{T}_\oplus$ Therefore, the merging result of $\mathcal{T}$ satifies Eq. (3).

Proposition 2. Let $\mathcal{T} = \{T_1, \ldots, T_n\}$ be a set of n possibilistic belief bases, if $\mathcal{T}_\oplus$ is the merging result of $\mathcal{T}$ which satisfies Eq. (4) and $(T_1 \otimes T_2 \ldots \otimes T_n)$ is the merging result of $\mathcal{T}$ which satisfies Eq. (5), then $\mathcal{T}_\oplus \equiv (T_1 \otimes T_2 \ldots \otimes T_n)$ and $(T_1 \otimes T_2 \ldots \otimes T_n)^* \subseteq (\mathcal{T}_\oplus)^*$

Proof

It is easy to see that $\mathcal{T}_\oplus \equiv (T_1 \otimes T_2 \ldots \otimes T_n)$ and $(T_1 \otimes T_2 \ldots \otimes T_n)^* \subseteq (\mathcal{T}_\oplus)^*$. In order to show the reverse direction is not true, we consider the following counter example: Given $T_1 = \{(\varphi, 0.7), (\neg\varphi, 0.5), (\lambda, 0.5)\}$ and $T_2 = \{(\partial, 0.6)\}$ Because of $Inc(T_1) = 0.5$, λ is omited in $(T_1 \otimes T_2)$. However, we have $(\lambda, \oplus(0.5, 0)) \in \mathcal{T}_\oplus$

Proposition 2 shows that the merging results may omit some beliefs. In the counter example, formula λ is conflict-free with other formulas, it should not be ommited in merging result. In practice, λ may be some important belief and it may recovered from $\mathcal{T}_\oplus$.

The following proposition compares $\mathcal{T}_\oplus$ and $(T_1 \otimes T_2 \ldots \otimes T_n)$.

Proposition 3. Let $\mathcal{T} = \{T_1, T_2, \ldots, T_n\}$ be a set of possibilistic belief bases. If $\mathcal{T}_\oplus$ is a merging result which satisfies Eq. (4) and $(T_1 \otimes T_2 \ldots \otimes T_n)$ is a merging result of $\mathcal{T}$ which satisfies Eq. (5), where $\otimes$ is a dual operator $\oplus$, then $\mathcal{T}_\oplus \equiv (T_1 \otimes T_2 \ldots \otimes T_n)$ and $\mathcal{T}_\oplus \subseteq (T_1 \otimes T_2 \ldots \otimes T_n)$.

The proof of Proposition 3 is obvious, thus, we skip to give it.

Proposition 4. Let $\mathcal{T} = \{T_1, T_2, \ldots, T_n\}$ be a set of possibilistic belief bases. If $\mathcal{T}_\oplus$ is a merging result which satisfies Eq. (4), it satisfies (IC1), (IC2), (IC4), (IC5), (IC6), (IC7), (IC8) and it do not satisfy (IC0) and (IC3).

5   Conclusion

In this paper, we propose a model to merge prioritized belief bases in possibilistic logic. The main contributions are as follows: Firstly, we introduced an algorithm to obtain a standard possibilistic distribution from a belief profile by using different operators to achieve the best merging results. We use two family of operators, connective strong operators and averagely increasing operators to merge possibilistic belief bases. Syntax copies of semantic merging methods are also considered. Secondly, we pointed out that our method can avoid drowning effect and keep implicit beliefs. It is the preeminent point of this method compared to other methods. Lastly, we investigated and analysed the good logical properties of merging results obtained by our method. For future work, we will propose a set of axioms to characterise merging result and discuss about the complexity of our algorithm.

References

1. Nguyen, T.H.K., et al.: Merging possibilistic belief bases by argumentation. In: Nguyen, N., Tojo, S., Nguyen, L., Trawiński, B. (eds.) Intelligent Information and Database Systems, p. 24–34. Springer, Cham (2017)

2. Tran, T.H., et al.: Argumentation framework for merging stratified belief bases. In: Nguyen, N.T., Trawiński, B., Fujita, H., Hong, T.P. (eds.) Intelligent Information and Database Systems, p. 43–53. Springer, Heidelberg (2016)
3. Dubois, D., Prade, H.: Possibilistic logic - an overview. In: Computational Logic, pp. 283–342 (2014)
4. Amgoud, L., Kaci, S.: An argumentation framework for merging conflicting knowledge bases. Int. J. Approx. Reason. **45**(2), 321–340 (2007)
5. Amor, N.B., et al.: Possibilistic games with incomplete information, pp. 1544–1550 (2019)
6. Benferhat, S., et al.: Inconsistency management and prioritized syntax-based entailment. In: Proceedings of the 13th International Joint Conference on Artificial Intelligence, vol. 1, pp. 640–645. Morgan Kaufmann Publishers Inc., Chambery (1993)
7. Benferhat, S., et al.: Possibilistic merging and distance-based fusion of propositional information. Ann. Math. Artif. Intell. **34**(1), 217–252 (2002)
8. Benferhat, S., Dubois, D., Prade, H.: Some syntactic approaches to the handling of inconsistent knowledge bases: a comparative study part 1: the flat case. Stud. Logica. **58**(1), 17–45 (1997)
9. Benferhat, S., Kaci, S.: Fusion of possibilistic knowledge bases from a postulate point of view. Int. J. Approx. Reason. **33**(3), 255–285 (2003)
10. Everaere, P., et al.: On egalitarian belief merging. In: Proceedings of the Fourteenth International Conference on Principles of Knowledge Representation and Reasoning, pp. 121–130. AAAI Press, Vienna (2014)
11. Konieczny, S., Pino Pérez, R.: Logic based merging. J. Philos. Logic **40**(2), 239–270 (2011)
12. Tran, T.H., Nguyen, N.T., Vo, Q.B.: Axiomatic characterization of belief merging by negotiation. Multimedia Tools Appl. **65**(1), 133–159 (2013)
13. Tran, T.H., Vo, Q.B.: An axiomatic model for merging stratified belief bases by negotiation. In: Nguyen, NT., Hoang, K., Jędrzejowicz, P. (eds.) Computational Collective Intelligence. Technologies and Applications, vol. 7653, pp 174–184. Springer, Heidelberg (2012)
14. Konieczny, S., Lang, J., Marquis, P.: DA^2 merging operators. Artif. Intell. **157**(1–2), 49–79 (2004)
15. Konieczny, S.: Belief base merging as a game. J. Appl. Non-Class. Logics **14**(3), 275–294 (2004)
16. Qi, G., et al.: A split-combination approach to merging knowledge bases in possibilistic logic. Ann. Math. Artif. Intell. **48**(1), 45–84 (2006)
17. Tran, T.H., Vo, Q.B., Kowalczyk, R.: Merging belief bases by negotiation. In: König, A., Dengel, A., Hinkelmann, K., Kise, K., Howlett, R.J., Jain, L.C. (eds.) Knowledge-Based and Intelligent Information and Engineering Systems, pp. 200–209. Springer, Heidelberg (2011)
18. Liberatore, P., Schaerf, M.: Arbitration (or how to merge knowledge bases). IEEE Trans. Knowl. Data Eng. **10**(1), 76–90 (1998)
19. Baral, C., et al.: Combining knowledge bases consisting of first order theories. In: Ras, Z.W., Zemankova, M. (eds.) Methodologies for Intelligent Systems, vol. 542, pp. 92–101. Springer, Heidelberg (1991)
20. Lin, J., Mendelzon, A.O.: Merging databases under constraints. Int. J. Coop. Inf. Syst. **7**, 55–76 (2000)
21. Liu, W., Qi, G., Bell, D.: Adaptive merging of prioritized knowledge bases. Fundam. Inform. **73**, 389–407 (2006)
22. Klement, E.P., Mesiar, R.: Triangular norms. Tatra Mt. Math. Publ. **13** (1997)
23. Benferhat, S., Dubois, D., Prade, H.: A computational model for belief change and fusing ordered belief bases. In: Williams, M.-A., Rott, H. (eds.) Frontiers in Belief Revision, pp. 109–134. Springer, Dordrecht (2001)

Discrete Time Sliding Mode
Control of Milling Chatter

Satyam Paul$^{(\boxtimes)}$(iD) and Magnus Löfstrand(iD)

School of Science and Technology, Örebro University, Örebro, Sweden
{satyam.paul,magnus.lofstrand}@oru.se

Abstract. The technique of mitigating chatter phenomenon in an effective manner is an important issue from the viewpoint of superior quality machining process with quality production. In this paper, an innovative solution to control chatter vibration actively in the milling process is presented. The mathematical modelling associated with the milling technique is presented in the primary phase of the paper. In this paper, an innovative technique of discrete time sliding mode control (DSMC) is blended with Type 2 fuzzy logic system. Superior mitigation of chatter is the outcome of developed active controller. The Lyapunov scheme is implemented to validate the stability criteria of the proposed controller. The embedded nonlinearity in the cutting forces and damper friction are compensated in an effective manner by the utilization of Type-2 fuzzy technique. The vibration attenuation ability of DSMC-Type-2 fuzzy (DSMC-T2) is compared with the Discrete time PID (D-PID) and DSMC-Type-1 fuzzy (DSMC-T1) for validating the effectiveness of the controller. Finally, the numerical analysis is carried out to validate that DSMC-T2 is superior to D-PID and DSMC-T1 in the minimization of the milling chatter.

Keywords: Milling chatter · Type-2 fuzzy · Sliding mode control

1 Introduction

Superior quality production is hampered by self-generated vibration in machining process. This type of vibration effects the quality of the final product and should be looked into with greater importance due its connection with manufacturing industries [1]. The self-generated vibration is termed as chatter which is an important phenomenon that degrades the machining process thus resulting in dimensional inaccuracy and minimization in the removal rate of the material (MRR). Chatter phenomenon also results in low quality finish with significant tool wear [2]. The active control system is comprised of effective controllers, efficient sensors and dampers which when implemented improves the performance of machining with less vibration in machine tool [3]. The innovative mechanism of active damper for mitigation of chatter which is justified experimentally has been investigated by Harms et al. [4]. Chen et al. [5] developed an adaptive algorithm on the basis of

© Springer Nature Switzerland AG 2020
H. A. Le Thi et al. (Eds.): ICCSAMA 2019, AISC 1121, pp. 381–390, 2020.
https://doi.org/10.1007/978-3-030-38364-0_34

Fourier series for active control of chatter. Weremczuk et al. [6] proposed an algorithm to control milling chatter using active approach and harmonic excitation methodology. Alharbi et al. [7] implemented the concept of PID controller for chatter mitigation in milling process. Cutting forces associated with machining process involves nonlinearities [8]. Moradi et al. [9] demonstrated that the cutting forces are combination of square as well as cubic polynomial terms which are nonlinear in nature. In situation where the model of the milling procedure is unknown, fuzzy logic comes very handy and effective. Fuzzy logic has earned great research reputations due to its capacity to do nonlinear mapping thus maintaining robustness and simplicity. Liang et al. [10] proposed an innovative solution to control chatter in end milling process by utilizing the concept of fuzzy logic system. Type-2 fuzzy logic system performs significantly better than Type-1 fuzzy logic system due to its possession of additional DOF which is known as footprint of uncertainty [11]. The Type-2 fuzzy technique is applied to handle the unknown uncertainties, and is combined with the PD/PID control for vibration mitigation in the work of Paul et al. [12]. Sliding Mode Control (SMC) is a superior control mechanism for vibration attenuation in milling tool since the SMC exhibit the same movement pattern as that of the tool vibration pattern. SMC is most suited for nonlinear systems due to its specific design criteria [13]. The innovative combined mechanism of Discrete Time Sliding Mode Control (DSMC) with Type 1 fuzzy system for the control of structural vibration is presented by Paul et al. [14]. Moradi et al. [15] investigated the chatter phenomenon in turning process and proposed SMC for chatter attenuation. Ma et al. [16] developed an active sliding mode control strategy to mitigate the chatter in turning process utilizing the concept of dynamic output feedback sliding surface combined with an adaptive law for noise approximation. SMC has various stability condition criteria depending on the continuous systems. An important condition is [17]:

$$| s(k+1) | < | s(k) | \tag{1}$$

in this case, $s(k) = 0$ which is termed as sliding surface. Another important condition given by Bartoszewicz et al. [18]:

$$| s(k) | < g \tag{2}$$

where g is stated as quasi-sliding mode band width. Discrete Time Sliding Mode Control (DSMC) is an efficient controller for vibration attenuation due to its criteria of sampling period which is an important aspect in vibration control. This work is carried out by implementing the third option "active control of chatter". In the first instance along x and y component, the mathematical modeling of milling process is done. Then the nonlinearities are identified for efficient compensation. The actual outcomes of Active Vibration Damper (AVD) was simulated using Matlab/Simulink for chatter suppression in milling process. The modeling is accomplished by taking into consideration the dynamics of AVD. Discrete Time Sliding Mode (DSMC) generates the control signals which is used for the suppression of chatter. DSMC is combined with Type 2 fuzzy logic (DSMC-T2 fuzzy) for efficient strategy. The implementation of Type 2 fuzzy system ensures that the nonlinearities are compensated in effective manner. The approach of Lyapunov stability analysis is implemented to prove that DSMC-T2 fuzzy controller is a stable one. The chatter attenuation in milling process is achieved by combined action of DSMC-T2 fuzzy with AVD. The wide significance of the concept and methodology is validated using numerical analysis. The results of DSMC-T2 fuzzy is compared with Discrete Time Sliding Mode Control with Type 1 fuzzy (DSMC-T1 fuzzy) and Discrete time PID (D-PID) to prove the effectiveness of most suited controller.

2   Modeling of Milling Process with Active Control

In case of milling tool with n evenly spaced teeth which is almost flexible to the rigid workpiece, a generalized 2-DoF mathematical model is [19]:

$$M_m \ddot{\mathbf{x}}_c(t) + C_m \dot{\mathbf{x}}_c(t) + K_m \mathbf{x}_c(t) = \mathbf{F}_m(\mathbf{t}) \tag{3}$$

the equivalent mass, damping and stiffness matrices are illustrated by the term M_m, C_m and K_m respectively.

$$
\begin{aligned}
M_m &= \begin{bmatrix} m_{mx} & 0 \\ 0 & m_{my} \end{bmatrix} \in \Re^{2 \times 2} \\
C_m &= \begin{bmatrix} c_{mx} & 0 \\ 0 & c_{my} \end{bmatrix} \in \Re^{2 \times 2} \\
K_m &= \begin{bmatrix} k_{mx} & 0 \\ 0 & k_{my} \end{bmatrix} \in \Re^{2 \times 2}
\end{aligned} \tag{4}
$$

Again, $\mathbf{x}_c(t) = [x \ y]^T$ illustrates the displacement of the tool along x and y components. $\mathbf{F}_m(\mathbf{t}) = [F_{fx} \ F_{fy}]^T$ illustrates x and y components cutting forces. The Fig. 1 illustrates the dynamics of milling process [20]. The closed form equations representing the nonlinear cutting forces along x and y components is illustrated as [9]:

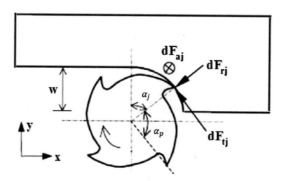

Fig. 1. Illustration of milling process dynamics.

$$F_x = +\frac{N}{2\pi} \left\{ \begin{array}{l} \zeta_1 \Delta x^3 + \eta_1 \Delta y^3 + \zeta_2 \Delta x^2 + \eta_2 \Delta y^2 + \zeta_3 \Delta x + \eta_3 \Delta y \\ +3\gamma_1 \Delta x^2 \Delta y + 3\gamma_2 \Delta x \Delta y^2 + 2\gamma_3 \Delta x \Delta y + \gamma_4 \end{array} \right\}$$

$$F_y = -\frac{N}{2\pi} \left\{ \begin{array}{l} \zeta_1^* \Delta x^3 + \eta_1^* \Delta y^3 + \zeta_2^* \Delta x^2 + \eta_2^* \Delta y^2 + \zeta_3^* \Delta x + \eta_3^* \Delta y \\ +3\gamma_1^* \Delta x^2 \Delta y + 3\gamma_2^* \Delta x \Delta y^2 + 2\gamma_3^* \Delta x \Delta y + \gamma_4^* \end{array} \right\} \tag{5}$$

where $\Delta x + x(t-\tau) = x(t)$ and $\Delta y + y(t-\tau) = y(t)$. The time delay is illustrated as $\tau = \frac{2\pi}{n\Omega}$, $\Omega = $ spindle speed (rad/s).

2.1   Active Vibration Damper (AVD) for Active Control Mechanism

As seen from the Fig. 2, AVD is placed on the top of the spindle and is utilized to minimize the tool chattering generated by the external force. The position of the AVD is at the centre of mass (CM). Also, it makes an inclination φ with CM. The efficient placement of damper is a cost effective thus mitigating the requirement of two dampers. The combination of modeling equation and control force $\mathbf{u_c}$ yields:

$$M_m \ddot{\mathbf{x}}_c(t) + C_m \dot{\mathbf{x}}_c(t) + K_m \mathbf{x}_c(t) = \mathbf{F}_m(t) + \mathbf{u_c} - \mathbf{d_c} \tag{6}$$

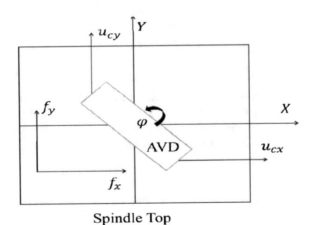

Fig. 2. AVD on Spindle Top.

the control signals impinged to the damper along both axis is illustrated by $\mathbf{u_c} = [u_{cx}, u_{cy}]^T \in \Re^{2\times 1}$, $\mathbf{d_c} = [d_{cx}, d_{cy}] \in \Re^{2\times 1}$ is the combined damping-fricton effect resolved along two axis.It is very important to consider the damper friction which can be resolved as:

$$\begin{array}{l} d_{cx} = \Lambda \dot{x}_{i,x} + \Gamma m_{dg} \tanh\left[\Upsilon \dot{x}_{i,x}\right] \\ d_{cy} = \Lambda \dot{x}_{i,y} + \Gamma m_{dg} \tanh\left[\Upsilon \dot{x}_{i,y}\right] \end{array} \tag{7}$$

where Λ, Υ and Γ are termed as damping coefficients associated with the Coulomb friction [21], m_d = mass of the damper, $\dot{x}_{i,x}$ and $\dot{x}_{i,y}$ represents the relative velocity of the damper along x component and the relative velocity of the damper along y component respectively. Also, $\mathbf{d_c} = [d_{cx}\ d_{cy}]^T$. The combination of the control methodology with closed loop system is given by:

$$m_{mx}\ddot{x} + c_{mx}\dot{x} + k_{mx}x = F_{fx} + u_{cx} - \Lambda\dot{x}_{i,x} - \Gamma m_d g\tanh[\Upsilon\dot{x}_{i,x}]$$
$$m_{my}\ddot{y} + c_{my}\dot{y} + k_{my}y = F_{fy} + u_{cy} - \Lambda\dot{x}_{i,y} - \Gamma m_d g\tanh[\Upsilon\dot{x}_{i,y}] \quad (8)$$

In Eq. (8), the nonlinear terms are $\Lambda\dot{x}_{i,x} + \Gamma m_d g\tanh[\Upsilon\dot{x}_{i,x}]$, $\Lambda\dot{x}_{i,y} + \Gamma m_d g\tanh[\Upsilon\dot{x}_{i,y}]$, F_{fx} and F_{fy}. The intelligent technique is incorporated to deal with the involved nonlinearities.

3 Discrete Time Sliding Mode Control with Type-2 Fuzzy Compensation

The continuous time model of the milling process which is a closed loop system from Eq. (8) is illustrated as:

$$M_m\ddot{\mathbf{x}}_c(t) + C_m\dot{\mathbf{x}}_c(t) + \mathbf{f}_{kn}(\mathbf{x}) = \mathbf{F}_m(t) + \mathbf{u_c}(t) - \mathbf{d_c} \quad (9)$$

The stiffness $\mathbf{f}_{kn}$ will be considered as nonlinear. It is very important to discretize the milling process model for digitalization and making it appropriate for the design of computer based control. For this, following steps are implemented: $V_1(t) = \mathbf{x}_c$ and $V_2(t) = \dot{\mathbf{x}}_c$. The model represented by (9) is illustrated as state space model by:

$$\dot{Z}(t) = AZ + Bu + F_{kn} + f_n \quad (10)$$

Again, $V(t) = \begin{bmatrix} V_1(t) \\ V_2(t) \end{bmatrix}$, $A_D = \begin{bmatrix} 0 & 0 \\ 0 & -M_m^{-1}C_m \end{bmatrix}$, $B_D = \begin{bmatrix} 0 \\ M_m^{-1} \end{bmatrix}$, $F_{kn} = M_m^{-1}\mathbf{f}_{kn}$, $f_n = M_m^{-1}[\mathbf{F}_m + \mathbf{d_c}]$. Considering $V(k)$ to be a state vector with A_{dis} as a state matrix, also, $A_{dis} = e^{A_D T}$, and B_{dis}=input vector, $B_{dis} = \left(\int e^{A_D \tau}d\tau\right)B_D$, $\mathbf{u}_c(k)$ =scalar input, $F_{kn}(Z(k))$=model uncertainty matrix and $f_{dn}(k)$ =nonlinearity involved in cutting forces and frictional forces, using (10), the discretized model is [22]:

$$V(k+1) = A_{dis}Z(k) + B_{dis}\mathbf{u}_c(k) + F_{kn}(Z(k)) + f_{dn}(k) \quad (11)$$

From (11), the discrete time model is:

$$V(k+1) = A_{dis}V(k) + B_{dis}\mathbf{u}_c(k) + F_{kn}(Z(k)) + f_{dn}(k) \quad (12)$$

From the viewpoint of preciseness and for the introduction fuzzy system to compensate nonlinearities, the following step is considered:

$$V(k+1) = j[z(k)] + h[z(k)]\mathbf{u}_c(k) \quad (13)$$

where $j[z(k)] = A_{dis}V(k)+F_{kn}(Z(k))+f_{dn}(k)$ and $h[z(k)] = B_{dis}$, A_{dis} and B_{dis} are unknown and $F_{kn}(Z(k))+f_{dn}(k)$ is nonlinear. So the term $j[z(k)]$ and $h[z(k)]$ will be modeled using Type-2 fuzzy logic technique. The Type-2 fuzzy logic system with jth output can be expressed as:

$$fuz_j = \frac{y_{rj} + y_{lj}}{2} = \frac{1}{2}\left[(\phi_{rj}^T(z)w_{rj}(z(k)) + \phi_{lj}^T(z)w_{lj}(z(k)))\right] \qquad (14)$$

where $y_{lj} = \frac{\sum_{i=1}^{p} f_{lj}^i y_{lj} + \sum_{i=1}^{p} f_{rj}^i y_{lk}}{\sum_{i=1}^{q} f_r^i + \sum_{i=1}^{q} f_l^i}$, $y_{rj} = \frac{\sum_{i=1}^{p} f_{lj}^i y_{rj} + \sum_{i=1}^{p} f_{rj}^i y_{rk}}{\sum_{i=1}^{q} f_r^i + \sum_{i=1}^{q} f_l^i}$,

$q_{lj}^i = \frac{f_l^i}{\sum_{i=1}^{q} f_r^i + \sum_{i=1}^{q} f_l^i}$, and $q_{rj}^i = \frac{f_r^i}{\sum_{i=1}^{q} f_r^i + \sum_{i=1}^{q} f_l^i}$. Again, f_l^i and f_r^i are the firing strengths associated with y_{lj}^i and y_{rj}^i of i-th rule. So the compensation technique for $j[z(k)]$ and $h[z(k)]$ are applied as follows:

$$\begin{aligned}\hat{j} &= \tfrac{1}{2}w_{rf}(k)\phi_{rf}^T[z(k)] + \tfrac{1}{2}w_{lf}(k)\phi_{lf}^T[z(k)] \\ \hat{h} &= \tfrac{1}{2}w_{rg}(k)\phi_{rg}^T[z(k)] + \tfrac{1}{2}w_{lg}(k)\phi_{lg}^T[z(k)]\end{aligned} \qquad (15)$$

$\Omega_1(k)$ and $\Omega_2(k)$ satisfies the following:

$$\Omega_1(k) = \begin{cases} \dfrac{\Omega_1(k)}{1+\Phi_1(k)} & \text{if } \| e_m(k+1) \| > \tfrac{1}{\beta_1} \| e_m(k) \| \\ 0 & \text{if } \| e_m(k+1) \| < \tfrac{1}{\beta_1} \| e_m(k) \| \end{cases}$$
$$\Omega_2(k) = \begin{cases} \dfrac{\Omega_2(k)}{1+\Phi_2(k)} & \text{if } \| e_m(k+1) \| > \tfrac{1}{\beta_2} \| e_m(k) \| \\ 0 & \text{if } \| e_m(k+1) \| < \tfrac{1}{\beta_2} \| e_m(k) \| \end{cases} \qquad (16)$$

where $0 < \Omega_1(k) \le 1$ and $0 < \Omega_2(k) \le 1$. Also the dead-zone parameter are illustrated by β_1 and β_2. Again,

$$\begin{aligned}\Phi_1(k) &= \| \phi_{rf}^T[z(k)] \|^2 + \| \phi_{rg}^T[z(k)]\mathbf{u}_c(k) \|^2 \\ \Phi_2(k) &= \| \phi_{lf}^T[z(k)] \|^2 + \| \phi_{lg}^T[z(k)]\mathbf{u}_c(k) \|^2\end{aligned} \qquad (17)$$

Now the modeling error $e_m(k)$ is represented as:

$$e_m(k) = \hat{v}(k) - v(k) \qquad (18)$$

where the state of the fuzzy model is represented by $\hat{v}(k)$, therefore:

$$(\beta_1 + \beta_2)\hat{v}(k+1) = \hat{j}[z(k)] + \hat{h}[z(k)]\mathbf{u}_c(k) \qquad (19)$$

where β_1 and β_2 are positive constant and $\beta_1, \beta_2 > 1$ which is a design parameter. In case of active vibration control it considered that $v^d(k) = 0$. The equation validating control error is:

$$\mathbf{e}_c(k) = v^d(k) - v(k) = -v(k) \qquad (20)$$

The SMC can be illustrated as:

$$\mathbf{u}_c(k) = \frac{2[K^T\mathbf{e}_c(k) - \tfrac{1}{2}(\omega_{rf}(k)\phi_{rf}^T[z(k)] + \omega_{lf}(k)\phi_{lf}^T[z(k)]) - \sigma sign[s(k)]]}{[\omega_{rg}(k)\phi_{rg}^T[z(k)] + \omega_{lg}(k)\phi_{lg}^T[z(k)]]}$$
$$(21)$$

where $\mathbf{e}_c(k) = [e_c(k+1-n)\cdots e_c(k)]^T$, The vector representing feedback gain is $K_G = [k_n \cdots k_1]^T \in R^n$. The sliding mode gain and switching function are represented by σ and $s(k)$ respectively, where switching function is:

$$s(k) = K_G^T \left[\mathbf{e}_c(k-1) + \frac{1}{K_G^T} \mathbf{e}_c(k) \right] \tag{22}$$

Theorem 1. *If the fuzzy model (19) is implemented for the compensation of the the nonlinear system mentioned by (13) with the updated laws*

$$\begin{aligned}
w_{rf}(k+1) - w_{rf}(k) &= -\Omega_1(k)e_m(k)\phi_{rf}^T[z(k)] \\
w_{lf}(k+1) - w_{lf}(k) &= -\Omega_2(k)e_m(k)\phi_{lf}^T[z(k)] \\
w_{rg}(k+1) - w_{rg}(k) &= -\Omega_1(k)\mathbf{u}_c(k)e_m(k)\phi_{rg}^T[z(k)] \\
w_{lg}(k+1) - w_{lg}(k) &= -\Omega_2(k)\mathbf{u}_c(k)e_m(k)\phi_{lg}^T[z(k)]
\end{aligned} \tag{23}$$

then the uniform stability of the closed loop system is assured and bounded provided that identification error $e_m(k)$ is within the range

$$\| e_m(k) \|^2 \geq \frac{2\psi(k)\bar{\xi}}{\Phi_1(k) + \Phi_2(k)} \tag{24}$$

and control error satisfies

$$\|\mathbf{e}_c(k)\|_U^2 \leq \sigma^2 \|Z\| \left(1 + \frac{(\beta_1 + \beta_2)G}{\sigma} \right) \tag{25}$$

with the gain σ of the discrete-time sliding mode controller (21) establishing

$$\sigma \geq \frac{G}{\|K_G\|}(\beta_1 + \beta_2) \tag{26}$$

The above theorem validates that both the identification error and the control error are bounded. It justifies that the control system is a stable one. The proof for the Theorem 1 will be presented in the expanded version of this paper. The Lyapunov candidate of the form

$$\begin{aligned}
L(k) = \tfrac{1}{2} \| \tilde{\omega}_{rf}(k) \|^2 &+ \tfrac{1}{2} \| \tilde{\omega}_{lf}(k) \|^2 + \tfrac{1}{2} \| \tilde{\omega}_{rg}(k) \|^2 \\
&+ \tfrac{1}{2} \| \tilde{\omega}_{lg}(k) \|^2 + \tfrac{1}{\sigma^2}\mathbf{e}_c^T(k)Z\mathbf{e}_c(k)
\end{aligned} \tag{27}$$

is utilized to prove the stability of the controller.

4 Numerical Analysis

In the first instance, the cutting conditions of milling process illustrated in [23] are extracted for tool vibration simulation so as to validate the effectiveness of the developed control mechanism. For the validation of the significant performances of DSMC-T2 fuzzy, the results of DSMC-T2 fuzzy is compared with DSMC-T1 fuzzy and Discrete time PID (D-PID). *Matlab/Simulink* is utilized as the software environment. Simulation results are presented to validate that the tool chatter can be mitigated significantly by implementing AVD in combination with effective control system involving DSMC-T2 fuzzy. The proposed control strategy result is then compared with the strategies like DSMC, D-PID, DSMC-T1 fuzzy to prove the effectiveness of DSMC-T2 fuzzy in superior vibration mitigation. In this paper, for the control of chatter considering Type-2 fuzzy logic concept, six fuzzy rules for $\hat{j}$ as well as four fuzzy rules for $\hat{h}$ are sufficient for effective control. Considering Type-1 fuzzy logic concept, for the control of chatter, nine fuzzy rules for $\hat{j}$ and six fuzzy rules for $\hat{h}$ are sufficient. The membership functions are designed using Gaussian function. IF-THEN rules are applied for both the types of fuzzy system. The chosen learning rates are $\Omega_1 = \Omega_2 = 0.9$. Theorem 1 is utilized for selecting σ which is 0.17. The tool vibration attenuation along x-direction are presented using the plots Figs. 3, 4 and 5. The results validate that DSMC-T2 fuzzy controller performed better than all the controller used in this research.

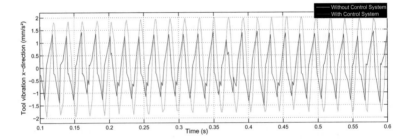

Fig. 3. Tool vibration along x-direction using D-PID.

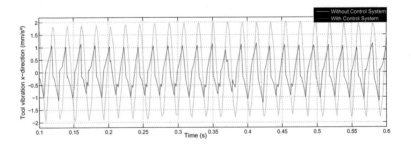

Fig. 4. Tool vibration along x-direction using DSMC-T1 fuzzy.

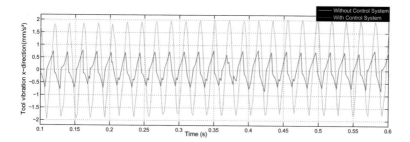

Fig. 5. Tool vibration along x-direction using DSMC-T2 fuzzy.

5 Conclusion

In this paper, a novel technique for milling chatter mitigation is demonstrated using an active control strategy. Using Lyapunov analysis technique, theorem is laid down to prove that the systems states of the DSMC-T2 fuzzy controller is bounded. The efficient approach of Type-2 fuzzy system is implemented to handle the nonlinearties in suitable manner. The result from numerical analysis establish that the most superior controller is DSMC-T2 fuzzy. In this research, the setup is made cost effective by installing a single AVD in an inclined position to control the forces along x and y axis.

Acknowledgments. This work is carried out within the projects: - Production Centred Maintenance (PCM) for real time predictive maintenance decision support to maximise production efficiency, funded by the Swedish Knowledge Foundation (Stiftelsen för kunskaps- och kompetensutveckling), and - A digital twin to support sustainable and available production as a service (DT-SAPS), funded by Produktion2030, the Strategic innovation programme for sustainable production in Sweden. We gratefully acknowledge the support and funding.

References

1. Rusinek, R., Wiercigroch, M., Wahi, P.: Modelling of frictional chatter in metal cutting. Int. J. Mech. Sci. **89**, 167–176 (2014)
2. Parus, A., Powalka, B., Marchelek, K., Domek, S., Hoffmann, M.: Active vibration control in milling flexible workpiece. J. Vib. Control **19**, 1103–1120 (2013)
3. Quintana, G., Ciurana, J.: Chatter in machining processes: a review. Int. J. Mach. Tools Manuf. **51**, 363–376 (2011)
4. Harms, A., Denkena, B., Lhermet, N.: Tool Adaptor for Active Vibration Control in Turning Operation. In: 9th International Conference on New Actuators, Bremen, Germany, pp. 694–697 (2004)
5. Chen, Z., Zhang, H.-T., Zhang, X., Ding, H.: Adaptive active chatter control in milling processes. J. Dyn. Syst. Meas. Control **136**(2), 0210007 (2014)
6. Weremczuk, A., Rusinek, R., Warminski, J.: The concept of active elimination of vibrations in milling process. Procedia CIRP **31**, 82–87 (2015)

7. AlharbiI, W.N., Batako, A., Gomm, B.: PID controller design for viibratory milling. In: Proceedings of ISER International Conference, Marrkech, Morocco (2017)
8. Melkote, S.N., Endres, W.J.: The importance of including size effect when modeling slot milling. J. Manuf. Sci. Eng. **120**(1), 68–75 (1998)
9. Moradi, H., Bakhtiari-Nejad, F., Movahhedy, M.R., Vossoughi, G.R.: Stability improvement and regenerative chatter suppression in nonlinear milling process via tunable vibration absorber. J. Sound Vib. **331**, 4668–4690 (2012)
10. Liang, M., Yeap, T., Hermansya, A.: A fuzzy system for chatter suppression in end milling. Proc. Inst. Mech. Eng. Part B J. Eng. Manuf. **218**(4), 403–417 (2004)
11. Mendel, J.M.: Uncertain Rule-Based Fuzzy Logic Systems: Introduction and New Directions. Prentice Hall PTR, Upper Saddle River (2001)
12. Paul, S., Yu, W., Li, X.: Bidirectional active control of structures with type-2 fuzzy PD and PID. Int. J. Syst. Sci. **49**(4), 766–782 (2018)
13. Utkin, V.I.: Sliding Modes in Control and Optimization. Springer, Berlin (1992)
14. Paul, S., Yu, W., Li, X.: Discrete-time sliding mode for building structure bidirectional active vibration control. Trans. Inst. Meas. Control **41**(2), 433–446 (2019)
15. Moradi, H., Movahhedy, M.R., Vossoughi, G.: Sliding mode control of machining chatter in the presence of tool wear and parametric uncertainties. J. Vib. Control **16**(2), 231–251 (2009)
16. Ma, H., Wu, J., Yang, L., Xiong, Z.: Active chatter suppression with displacement-only measurement in turning process. J. Sound Vib. **401**, 255–267 (2017)
17. Sarpturk, S.Z., Istefanopolos, Y., Kaynak, O.R.: On the stability of discrete-time sliding mode control systems. IEEE Trans. Autom. Control **32**(10), 930–932 (1987)
18. Bartoszewicz, A.: Discrete-time quasi-sliding mode control strategies. IEEE Trans. Industr. Electron. **45**(4), 633–637 (1998)
19. Zhang, H.-T., Wu, Y., He, D., Zhao, H.: Model predictive control to mitigate chatters in milling processes with input constraints. Int. J. Mach. Tools Manuf. **91**, 54–61 (2015)
20. Altintas, Y., Stepan, G., Merdol, D., Dombovari, Z.: Chatter stability of milling in frequency and discrete time domain. CIRP J. Manuf. Sci. Technol. **1**(1), 35–44 (2008)
21. Roldán, C., Campa, F.J., Altuzarra, O., Amezua, E.: Automatic identification of the inertia and friction of an electromechanical actuator. In: New Advances in Mechanisms, Transmissions and Applications. Mechanisms and Machine Science, vol. 17, pp. 409–416 (2014)
22. Kim, J., Oh, S., Cho, D., Hedrick, J.: Robust discrete-time variable structure control methods. ASME. J. Dyn. Sys. Meas. Control **122**(4), 766–775 (2000)
23. Moradi, H., Movahhedy, M.R., Vossoughi, G.: Dynamics of regenerative chatter and internal resonance in milling process with structural and cutting force nonlinearities. J. Sound Vib. **331**(16), 3844–3865 (2012)

Efficient Processing of Recursive Joins on Large-Scale Datasets in Spark

Thuong-Cang Phan[1(✉)], Anh-Cang Phan[2], Thi-To-Quyen Tran[3], and Ngoan-Thanh Trieu[1]

[1] Can Tho University, Can Tho, Vietnam
{ptcang,ttngoan}@cit.ctu.edu.vn
[2] Vinh Long University of Technology Education, Vinh Long, Vietnam
cangpa@vlute.edu.vn
[3] University of Rennes 1, Lannion, France
thi-to-quyen.tran@irisa.fr

Abstract. MapReduce has become the dominant programming model for analyzing and processing large-scale data. However, the model has its own limitations. It does not completely support iterative computation, caching mechanism, and operations with multiple inputs. Besides, I/O and communication costs of the model are so expensive. One of the most notably complex operations extensively and expensively used in MapReduce is recursive joins. It requires processing characteristics that are the limitations of a MapReduce environment. Therefore, this research proposes efficient solutions for processing recursive joins in Spark, a next-generation data processing engine of MapReduce. Our proposal eliminates a large amount of redundant data generated in repeated join steps and takes advantages of in-memory computing means and cache mechanism. Through experiments, the present research shows that our solutions significantly improve the execution performance of recursive joins on large-scale datasets.

Keywords: Big data processing · Recursive join · MapReduce · Spark

1 Introduction

In the growing up of information technology, the amount of information on the Internet rapidly increases. Therefore, the term "Big Data" has widely been used to describe a large amount of complex data on a dataset. That has also posed many challenges for researchers in variety fields of study, such as search-engines, social network analysis, web-data analysis, etc. There is a need of new distributed programming models running on computer clusters to process huge amount of data for the applications mentioned above. The idea of distributed computing is to divide a problem into small problems and execute each of them on a node of a cluster.

MapReduce [1] was developed by Google since 2004. The motivation is to process huge amount of input data in a reasonable time. Nowadays, it has become a

© Springer Nature Switzerland AG 2020
H. A. Le Thi et al. (Eds.): ICCSAMA 2019, AISC 1121, pp. 391–402, 2020.
https://doi.org/10.1007/978-3-030-38364-0_35

popular standard model for processing large datasets on parallel and distributed systems. A computer cluster operate MapReduce tasks can include thousands of computing nodes with high fault-tolerant capacity. It is appropriate for processing huge amount of data on a parallel and distributed fashion. MapReduce is compatible with many algorithms, e.g., web-scale document analysis [1], relational query evaluation [2], and large-scale image processing [3]. However, it is not designed for operations with multiple inputs. It does not directly support iterative computations for applications like PageRank [4], HITS (Hypertext Induced Topic Search) [5], recursive relational queries [6], data clustering [7], neural network analysis [8], social network analysis [9], and Internet traffic analysis [10]. These applications are involved in repeated computations on large-scale datasets until a fix-point is reached. Programmers must set up an iterative algorithm in a MapReduce environment and control iterative tasks themselves. As a result, the operation of reading/writing data needs to be processed many times so that the costs of I/O, CPU, and communication obviously increase. Those problems altogether propose big challenges for large-scale data processing in a MapReduce environment.

Join [11,12] is a combination operation of two or more tuples on a database. Join is an operation that is usually used in a typical data query with expensive costs and complexity. There are several kinds of join: two-way join, multi-way join [13], chain join [14], and recursive join [15–17]. Join query on tuples become more complicated on Big Data world. In this research, we focus on recursive joins, a complex computation that has expensive performing cost but still need to be used in a majority of fields, e.g., PageRank, graph mining, network monitoring, social network, and bioinformatics.

A typical example of recursive join is a query to discover the relationships of a person in social network. It is defined as follows:

$$\text{Friend(x, y)} \leftarrow \text{Know(x, y);}$$
$$\text{Friend(x, y)} \leftarrow \text{Friend(x, z)} \bowtie \text{Know(z, y);}$$

A person x is a friend of a person y if x knows y. Person x is also a friend of y if x is a friend of z and z knows y. This is a query to find all friends of friends of a user.

Obviously, processing of iterative join operations on large-scale datasets is too heavyweight because MapReduce is completely unaware of the nature of iterative calculations which waste much bandwidth, I/O, and CPU cycles. Consequently, we have conducted to improve it by several solutions. We first leverage the advantages of Spark RDD for efficient iterative computations. Then, we remove most non-joining data in repeated join steps to minimize the amount of data transmitted over the network. In addition, caching technique for a persistent dataset in each iteration step is also considered.

2 Related Works

2.1 Recursive Join in MapReduce

A recursive join computes the transitive closure of a relation that involves a number of repeating join operations until a fix-point is reached [17]. In fact, there are many algorithms that have been introduced to solve the problem of recursively defined relation in traditional database such as Naïve [23], Semi-Naïve [1–3], Smart [24,25], Minimal evaluations [25], Warshall [26], and Warren [27]. However, the algorithms mentioned above are not always appropriate to work in MapReduce.

Currently, there are studies that focus on recursions in clustering environment. Afrati et al. [27,28] have proposed a recursive evaluation on computer clusters to compute transitive closure for a recursive query. The research used two kinds of tasks: Join tasks and Dup-elim tasks. Nevertheless, the problem of this algorithm is that tasks recursively run for a long time will increase the failure rate. Moreover, it leaves an open question of minimization of data-volume cost since it is impressively higher than that of the linear transitive closure.

HaLoop [29] is an extended version of Hadoop that was designed to support applications using MapReduce. It enhances the efficiency of these programs by an inner-iteration cache mechanism and a loop-aware task scheduler to improve data locality. Ideally, HaLoop can efficiently support processing recursions on huge datasets. However, HaLoop stills stay in research level and cannot be practical in use. Currently, HaLoop is no longer developed and supported.

Recently, Shaw et al. [31] present an optimization solution for recursive joins using the Semi-Naïve algorithm in a Hadoop MapReduce environment. It repeats two types of MapReduce tasks: a join job and a computing incremental dataset job. Reducer Input Cache of HaLoop is used to reduce costs of related tasks. Nevertheless, this solution is inevitable the limitations of HaLoop. The cost of read/write cache is become significant because all incremental datasets of previous iterations need to be re-written, re-indexed, and re-read to detect duplication.

Therefore, in order to cover the above drawbacks, we have conducted to consider another framework that can give better support for caching mechanism and iterative computation. On that framework, we have optimized performance of recursive joins by eliminating non-joining data.

2.2 Apache Spark

Apache Spark [18] is a framework that was written in Scala language that allows processing huge amount of distributed data in a fast and efficient manner. It is compatible with many file system such as HDFS (Hadoop Distributed File System), Cassandra [19], HBase [20], and Amazon S3 [21]. The striking feature of Spark is that in comparison with Hadoop, it has the processing speed 100 times faster in memory and 10 times faster on disk.

Spark's Iterative Processing Capacity. Spark is a powerful tool to support iterative processing on large-scale datasets using RDD (Resilient Distributed Datasets). RDD is a fault-tolerant parallel data structure that can be cached in memory and used again for future transformations [22]. Basically, evaluation of RDD is lazy in nature. It means a series of transformations on an RDD is be delayed until it is really needed. This saves a lot of time and improves efficiency.

The iterative processing of Hadoop MapReduce is carried out as a sequence of operations in which the intermediate results must be written to HDFS then being read as input to the following task. Meanwhile, Spark will read data from HDFS, perform repeated operations with RDDs, and finally write to HDFS.

Spark's Cache Mechanism. Due to the fact that RDD will be recalculated after an Action, it is costly and waste time if a RDD is being reused many times. Therefore, Spark provides a mechanism called persist, allowing RDDs to be reused efficiently. Indeed, the caching mechanism is an optimization technology for (iterative and interactive) Spark computations. It helps saving interim partial results so it can be reused in subsequent stages. These interim results as RDDs are thus kept in memory (default) or more solid storages like disk and/or replicated.

2.3   Intersection Bloom Filter

Bloom Filter. In 1970, Burton Bloom introduced Bloom Filter - BF [39]. It is a space efficient probabilistic data structure that is used to test membership of an element in a set. Here BF(S) is the abbreviation for a Bloom Filter with m bits, k independent hash functions, and a set S containing n elements.

Intersection Bloom Filter. Intersection Bloom filter (IBF) was introduced by our research group in [17,32]. IBF is a probabilistic data structure that was designed for performing the intersection of two sets and used to denote common elements of sets with a probability of false positive.

2.4   Intersection Bloom Join Algorithm

IntersectionBloomJoin [32] is a join algorithm that has improved BloomJoin by using IBF instead of standard BF. This algorithm uses IBF to filter most of the tuples that does not participate in join. It dramatically reduces cost of related tasks and is proved to be more efficient than other join algorithms [17]. Therefore, in this research, IntersectionBloomJoin will be used for evaluating recursive joins.

3   Optimizing Recursive Joins

One of the most popular recursive join algorithms, suitable for implementation in a MapReduce environment, is Semi-Naïve [17,30]. Besides, it is also used to process on datasets that are presented in rows and columns, the problem that we are interested in. Hence, in this section, we present recursive join evaluation based on the Semi-Naïve algorithm in MapReduce, then propose solutions to improve the evaluation in Spark.

For the convenience of presentation, we introduce additional notations for this work in Table 1.

Table 1. Notations

Notation	Description
F	The result dataset Friend(x,y)
ΔF	The incremental dataset
K	The grounding dataset Know(x,y)
O	The output of the join job of ΔF and K
BF	Bloom filter
IBF	Intersection Bloom filter

To start with, we recall the traditional Semi-Naïve algorithm, whose basic idea is quite simple and avoids recomputing already generated tuples in addition to returning the same result as the Naïve one. As illustrated in Table 2, the algorithm works as follows:

- begin by assuming the result dataset F set to empty and the incremental dataset ΔF initialized to be the grounding dataset K;
- repeatedly perform two activities: add the incremental dataset ΔF to the result F and recompute ΔF by join and difference operations using the previous ΔF;
- stop when the incremental dataset ΔF in the i-th iteration is empty.

Table 2. The Semi-Naïve algorithm for evaluating recursive joins

Algorithm 1 - Semi-Naïve evaluation for recursive joins	
$F_0 = \emptyset, \Delta F_0 = K(x,y), i = 0;$	(1)
Repeat	(2)
$\quad$ i++;	(3)
$\quad F_{i-1} = (\Delta F_0 \cup ... \cup \Delta F_{i-1})$	(4)
$\quad \Delta F_i = \Pi_{xy}(\Delta F_{i-1} \bowtie_z K) - F_{i-1};$	(5)
Until $\Delta F_i \neq \emptyset$	(6)

Recursive Join Evaluation in MapReduce. As shown in Table 2, the Semi-Naïve algorithm is used to evaluate a recursive join by iterations of two MapReduce jobs: a join job and a difference job to compute an incremental dataset. At step i, the join job will perform a join operation of ΔF_{i-1} and K to generate set O_i. The second one will read O_i and calculate $\Delta F_i = \Delta F_{i-1} - O_i$ by eliminating duplicated tuples in O_i with previous results. However, these jobs take a lot of costs for disk I/O operations and communication. Besides, the join job emits much non-joining intermediate data, thus substantially increasing the related costs.

Proposed Solutions to Improve the Recursive Join Evaluation. Considering line 5 of the Semi-Naïve algorithm, $\Delta F_i = \Pi_{xy}(\Delta F_{i-1} \bowtie_Z K) - F_{i-1}$, computation of the incremental dataset based on two join and difference operations, we propose several solutions to optimize the operations in a Spark environment. (1) Use in-memory processing capability of Spark RDD for the repeated operations to reduce slow disk I/O. (2) Utilize Spark's data caching mechanism for a consistent dataset K to speed up the operations that access the same K multiple times. When data does not fit in memory, Spark will spill data to hard drive as this order: MEMORY_AND_DISK. (3) Eliminate non-joining data to reduce the read/write and communication costs related to those useless data. We use IBF to remove the redundant data in the join job of ΔF and K. The join algorithm used in this work is IntersectionBloomJoin [41] as mentioned above. Figure 1 presents a flowchart for processing the recursive join using the proposed solutions in Spark.

Preprocessing. Before join two datasets K and ΔF, both will need to be transferred to Spark data type - PairRDD and eliminate null key/value pairs. At the same time, we construct an intersection bloom filter IBF(K, ΔF).

Iterative Evaluation of the Recursive Join. Dataset K and ΔF are filtered with the IBF(K, ΔF) to remove tuples without participation in join operation. Then, we join the filtered datasets: after each join task we process the result to create a new ΔF; the following join task will be done between K and the new ΔF. This join task will be repeated until a fix-point is reached.

In the process of generating new datasets, we perform partitioning datasets for the purpose of splitting up data. We aim to avoid overflow memory, reducing data transmits via network, and increasing processing speed.

Stop conditions for recursive join (fix-point):

- Number of processing cycles reaches the maximum limitation.
- OR ΔF_i is empty, there is no new data generated.
- OR intersection of two filters BF(K) and BF(ΔF_i) is empty, new generated result has no data participated in join in the next iteration.

4 Experiment and Evaluation

4.1 Describe Clusters and Datasets

We conduct an experiment on a computer cluster with 14 nodes (1 master and 13 slaves) at the Mobile Network and BigData Laboratory of Information and Communication Technology Faculty, Can Tho University. The configuration of each computer is 4 CPUs Intel Core i5 3.2 Ghz, 4 GB of RAM, 500 GB of HDD, and Ubuntu operating system 14.04 LTS 64 bits. Versions of applications used: java 1.8, Hadoop 2.7.1, and Spark 1.6.

Standard datasets generated by Purdue MapReduce Benchmarks Suite have size of 5 GB, 10 GB, 20 GB, and 30 GB. Data is saved as text file with each row has maximum 39 fields separated by a comma and each field contains 19 characters.

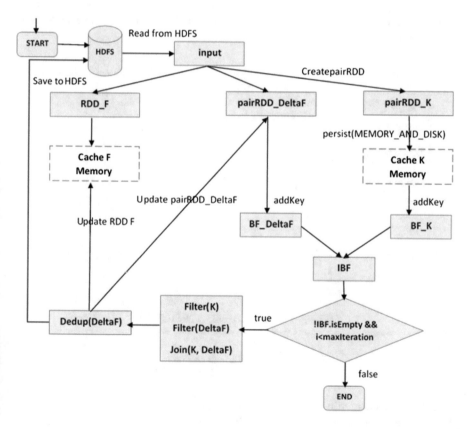

Fig. 1. Flowchart for optimized recursive join algorithm in Spark.

4.2   Evaluation Method

Our work applies three approaches: the Semi-Naïve algorithm in Hadoop MapReduce, the Semi-Naïve algorithm in Spark, and the optimized Semi-Naïve algorithm in Spark for implementing a recursive join. On each experimental dataset,
we evaluate the approaches based on intermediate data needed to transmit over
network and execution time.

4.3   Evaluate the Approaches

Here we compare the Semi-Naïve evaluation in Hadoop MapReduce, the evaluation in Spark, and the optimized evaluation using cache and filters in Spark
to get the amount of intermediate data and execution time. Thereby, we can
evaluate the level of improvement bringing by the proposed approach.

Figure 2 shows the amount of intermediate data that is needed to transmit
over the network of the three approaches. Figure 3 presents the execution time
of the approaches.

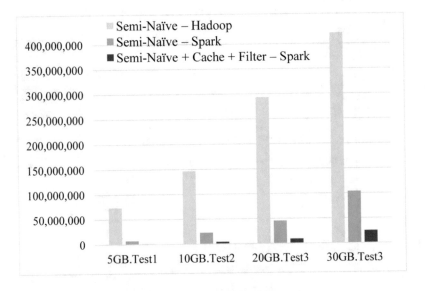

Fig. 2. Intermediate data (records)

Figure 2 clearly illustrates the improvement of Spark compare to Hadoop.
Iterative processing and partitioning on memory greatly reduce the amount of
intermediate data transmitting over the network. Bloom filter and cache mechanism continue to reduce redundantly non-joining data, which optimizes the
recursive join in Spark.

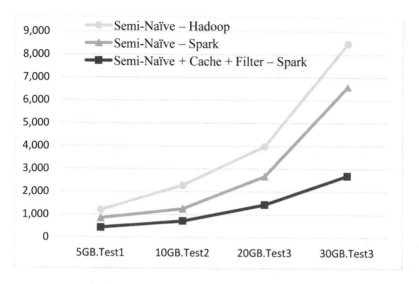

Fig. 3. Execution time (seconds)

Figure 3 shows that Spark improves the speed of data processing through more effective use of memory and cache when compared with Hadoop. The proposed optimization helps to improve execution time for the Semi-Naïve algorithm. For the optimized Semi-Naïve algorithm, the larger dataset the better performance and small dataset is costly for processing filters.

5   Conclusion and Future Work

5.1   Conclusion

A recursive join is an operation that is costly in time and resources. The research has provided efficient and simple solutions for evaluating recursive joins on large-scale datasets in Spark. The noticeable results of this work include:

1. Investigation about available solutions for recursive joins on large-scale datasets in a MapReduce environment. It points out the limitations and necessity of the previous studies.
2. Optimization for recursive joins in Spark. The solutions are proposed to efficiently compute recursive joins as follows: (a) Iterative processing mechanism on memory with Spark RDD to enhance execution performance; (b) Spark's caching mechanism to cache the constant dataset over iterations and reduces costs of repeatedly read/write data; (c) Intersection filters and the Intersection Bloom join algorithm to eliminate non-joining data, significantly reducing read/write and communication costs related to those useless data.
3. Experiments and evaluations for the recursive join in Hadoop MapReduce and Spark.

Through experiments, we demonstrate that the proposed solutions bring significant efficiency in comparison with current available solutions for recursive joins on large-scale datasets. Performing a recursive join in Spark, we maximum exploit the capacities of Spark such as distributed and parallel processing, iterative processing, caching mechanism, and fast computing on memory. Moreover, using Bloom filters to remove redundant elements of input datasets has alleviated the burden of read/write and process too much data over network.

Optimizing recursive joins in Spark provides many benefits for a variety of fields, e.g., large-scale database, social network, bioinformatics, sensor network, network monitoring, machine learning, etc. Finally, this research is an important step contributes to the context of optimization management large-scale databases on cloud infrastructure.

5.2   Future Work

The work remarkably reduces the costs for read/write data and speed up the execution process. However, optimizing recursive joins on large-scale datasets in Spark still exist many limitations. To have efficiency as expectation, we are required to have computer clusters that are stably and strong enough to processing data after each iteration. Furthermore, skewed data is a big challenge for this research in particular and for problem of processing large-scale datasets in general. It is also a future work that we want to research in the near future.

References

1. Dean, J., Ghemawat, S.: MapReduce: simplified data processing on large clusters. Commun. ACM **51**(1), 107 (2008)
2. Apache Hive TM. https://hive.apache.org/. Accessed 14 Jun 2016
3. Wiley, K., Connolly, A., Krughoff, S., Gardner, J., Balazinska, M., Howe, B., Kwon, Y., Bu, Y.: Astronomical Image Processing with Hadoop. ResearchGate, July 2011
4. Page, L., Brin, S., Motwani, R., Winograd, T.: The PageRank citation ranking: bringing order to the web. ResearchGate, January 1998
5. Kleinberg, J.M.: Authoritative sources in a hyperlinked environment. J. ACM **46**(5), 604–632 (1999)
6. Bancilhon, F., Ramakrishnan, R.: An amateurs introduction to recursive query processing strategies (1986)
7. Jain, A.K., Murty, M.N., Flynn, P.J.: Data clustering: a review (1999)
8. Hagan, M.T., Demuth, H.B., Beale, M.H., De Jess, O.: Neural Network Design, 2nd edn. Martin Hagan, Atlanta (2014)
9. Wasserman, S., Faust, K.: Social Network Analysis: Methods and Applications. Cambridge University Press, Cambridge (1994)
10. Moore, A.W., Zuev, D.: Internet traffic classification using bayesian analysis techniques. In: Proceedings of the 2005 ACM SIGMETRICS International Conference on Measurement and Modeling of Computer Systems, New York, NY, USA, pp. 50–60 (2005)
11. Codd, E.F.: A relational model of data for large shared data banks. Commun. ACM **13**(6), 377–387 (1970)

12. Codd, E.F.: Relational completeness of data base sublanguages. In: Rustin, R. (ed) Database System, pp. 65–98. Prentice Hall (1972). IBM Research report RJ 987 San Jose, California
13. Tan, K.-L., Lu, H.: A note on the strategy space of multiway join query optimization problem in parallel systems. SIGMOD Rec. **20**(4), 81–82 (1991)
14. Lin, X., Orlowska, M.E.: An efficient processing of a chain join with the minimum communication cost in distributed database systems. Distrib. Parallel Databases **3**(1), 69–83 (1995)
15. Ordonez, C.: Optimizing recursive queries in SQL. In: Proceedings of the 2005 ACM SIGMOD International Conference on Management of Data, New York, NY, USA, pp. 834–839 (2005)
16. Idreos, S., Liarou, E., Koubarakis, M.: Continuous multi-way joins over distributed hash tables. In: Proceedings of the 11th International Conference on Extending Database Technology: Advances in Database Technology, New York, NY, USA pp. 594–605 (2008)
17. Phan, T.-C., d'Orazio, L., Rigaux, P.: A theoretical and experimental comparison of filter-based equijoins in mapreduce. In: Hameurlain, A., Kung, J., Wagner, R. (eds.) Transactions on Large-Scale Data- and Knowledge-Centered Systems XXV, pp. 33–70. Springer, Berlin Heidelberg (2016)
18. Apache SparkTM - Lightning-Fast Cluster Computing. http://spark.apache.org/. Accessed 14 Jun 2016
19. The Apache Cassandra Project. http://cassandra.apache.org/. Accessed 14 Jun 2016
20. Apache HBase - Apache HBaseTM Home. https://hbase.apache.org/. Accessed 14 Jun 2016
21. Amazon Simple Storage Service (S3) - Cloud Storage. https://aws.amazon.com/s3/. Accessed 14 Jun 2016
22. Zaharia, M., Chowdhury, M., Das, T., Dave, A., Ma, J., McCauley, M., Franklin, M.J., Shenker, S., Stoica, I.: Resilient distributed datasets: a fault-tolerant abstraction for in-memory cluster computing. In: Proceedings of the 9th USENIX Conference on Networked Systems Design and Implementation, Berkeley, CA, USA, p. 2 (2012)
23. Bancilhon, F.: Naive evaluation of recursively defined relations. In: Brodie, M.L., Mylopoulos, J. (eds.) On Knowledge Base Management Systems, pp. 165–178. Springer, New York (1986)
24. Ullman, J.D.: Principles of Database and Knowledge-Base Systems, vol. I. Computer Science Press Inc., New York (1988)
25. Ioannidis, Y.E.: On the computation of the transitive closure of relational operators. In: Proceedings of the 12th International Conference on Very Large Data Bases, San Francisco, CA, USA, pp. 403–411 (1986)
26. Warshall, S.: A theorem on boolean matrices. J. ACM **9**(1), 11–12 (1962)
27. Warren Jr., H.S.: A modification of warshall's algorithm for the transitive closure of binary relations. Commun. ACM **18**(4), 218–220 (1975)
28. Afrati, F.N., Borkar, V., Carey, M., Polyzotis, N., Ullman, J.D.: Map-reduce extensions and recursive queries. In: Proceedings of the 14th International Conference on Extending Database Technology, New York, NY, USA, pp. 1–8 (2011)
29. Bu, Y., Howe, B., Balazinska, M., Ernst, M.D.: The HaLoop approach to large-scale iterative data analysis. VLDB J. **21**(2), 169–190 (2012)
30. Tran, T.T.Q.: Traitement de la jointure recursive en MapReduce. Universite Blaise Pascal-Clermont-Ferrand II, Clermont-Ferrand (2014)

31. Shaw, M., Koutris, P., Howe, B., Suciu, D.: Optimizing large-scale semi-naïve datalog evaluation in hadoop. In: Proceedings of the Second International Conference on Datalog in Academia and Industry, Berlin, Heidelberg, pp. 165–176 (2012)
32. Phan, T.-C., d'Orazio, L., Rigaux, P.: Toward intersection filter-based optimization for joins in mapreduce. In: Proceedings of the 2nd International Workshop on Cloud Intelligence, New York, NY, USA, p. 2:1–2:2 (2013)

New Feed Rate Optimization Formulation in a Parametric Domain for 5-Axis Milling Robots

Chu Anh My[1]($\boxtimes$), Duong Xuan Bien[1], Bui Hoang Tung[1], Nguyen Van Cong[1], and Le Chi Hieu[2]

[1] Le Quy Don Technical University, Hanoi, Vietnam
mychuanh@yahoo.com
[2] Faculty of Science and Engineering, University of Greenwich, London, UK

Abstract. When producing a numerical control (NC) program for a 5-axis CNC machine (the so-called milling robot) to mill a sculptural surface, a constant feed rate value is usually assigned based on programmer's experiences. For this reason, the feed rate in most of NC programs is often not optimized, it is much lower than maximum reachable value. To increase the productivity of the machining process, the feed rate in NC programs for the milling robots need to be maximized. This paper proposes a new feed rate optimization model, of which the objective function and all the kinematic constraints are transformed and expressed explicitly in a parametric domain which is commonly used in the tool path generation process performed by current CAM systems. Thus, the optimal feed rate values along a parametric tool path can be computed in an effective and simplified manner. Numerical examples demonstrate the effectiveness of the proposed method.

Keywords: High speed milling · 5-axis milling robot · Kinematic modelling · Feed rate interpolation

1 Introduction

In recent years, the high speed 5-axis CNC machines (the milling robots) are widely used in manufacturing industries to machine complex parts comprised of sculptured surfaces such as molds, dies and impellers. A 5-axis CNC machine is similar to two cooperative robots, one robot carrying the tool and one robot carrying the workpiece. The structure diagram of a typical 5-axis CNC machine is shown in Fig. 1, and the two kinematic chains of the machine can be illustrated in Fig. 2.

During the time milling a part, the two robots always cooperate and contact along the machining tool path. Therefore, the machines are also called the milling robots. In practice of programming for the machines, in order to machine a part, CAD/CAM systems are usually used to generate tool paths as parameterized curves (CL files). After that the CL files are postprocessed to produce G-code files to control the machines. In the postprocessing step, as usual, a conservative constant feed rate is selected for sections of the generated tool path or sometimes a single feed rate for the whole tool path. For this

H. A. Le Thi et al. (Eds.): ICCSAMA 2019, AISC 1121, pp. 403–411, 2020.
https://doi.org/10.1007/978-3-030-38364-0_36

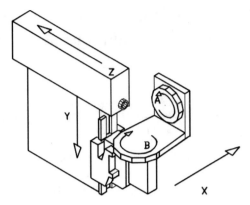

Fig. 1. The structure diagram of a typical 5-axis CNC machine

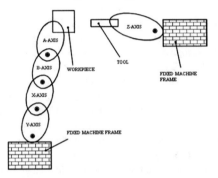

Fig. 2. The kinematic chain diagram of a typical 5-axis CNC machine

reason, the axis drive limits are never reached, which may lead to lengthy machining times, since the programmed feed rate is never reached by the machine tool. Therefore, optimizing the feed rate profile along the tool path to minimize the machining time without violating the limits of the drives of the machine is an important need.

Years ago, the feed rate optimization for high speed 5-axis CNC machining is a known issue. However, in most of the previous investigations, the formulations of the feed rate optimization models were mainly expressed either in terms of the arc length parameter of the tool path [2–6] or in an implicit and complex form [7–13]. Note that the use of the arc length re-parameterization of the tool path increases the computational complexity when solving the feed rate optimization problem. In the researches [7–9], the optimization model was formulated and expressed in a parametric domain. However, it is difficult to determine analytically the high order derivatives of the parameter along with the arc length in such optimization models. Though the objective function and the derive constraints were derived with respect to the parametric domain [10–12], the actual feed rate is not presented in the formulations, it is implicitly calculated via the derivative of the parameter. It is clear that little attention has been paid to formulate the feed rate optimization problem by using explicit inverse kinematics equations and their derivatives expressed in the parametric domain. For this reason, in this study, a new formulation of the feed rate optimization model is developed. The main advantage of the proposed formulation is that all constraints are transformed and explicitly expressed in terms of a common parameter, with the use of the Jacobian matrix and its time derivatives. Thus, optimal values of the feed rate along a parametric tool path which is usually generated by CAMs can be calculated in an effective and simplified manner, as compared with other previous methods.

2 Differential Inverse Kinematics of 5-Axis CNC Machines in a Parametric Domain

In this section, based on the forward kinematic equation of a general 5-axis CNC machine that has been derived in our previous investigations [1, 14, 15], the differential inverse kinematic equations are transformed in a parametric domain. These equations are of importance when formulating and solving the feed rate optimization problem for the machines.

Let $X = \begin{bmatrix} x\ y\ z\ \emptyset\ \varphi \end{bmatrix}^T$ denotes the position of the tool, in which (x, y, z) are the tool tip coordinates and $(\emptyset, \varphi)$ are the orientation angles of the tool axis in the workpiece coordinate system $O_w x_w y_w z_w$. Let $q = \begin{bmatrix} q_1\ q_2\ q_3\ q_4\ q_5 \end{bmatrix}^T$ be a vector of the five joint variables of a machine. The forward kinematic equation is as follows

$$X = f_{(q)} \tag{1}$$

For example, the forward kinematic equations for the 5-axis CNC machine Spinner U5-620 (Fig. 3) are expressed as follows

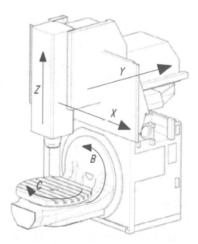

Fig. 3. The machine spinner U5-620

$$
\begin{aligned}
x &= X \cos B \cos C - Y \sin C + Z \sin B \cos C - d \sin B \cos C; \\
y &= X \cos B \sin C + Y \cos C + Z \sin B \cos C - d \sin B \cos C; \\
z &= -X \sin B + Z \cos B - d \cos B + d;
\end{aligned}
\tag{2}
$$

$$\emptyset = B;$$
$$\varphi = C;$$

Note that the joint variables of the machine are denoted as $q = [X \ Y \ Z \ B \ C]^T$; d is the distance from the table center point to the center line of the rotary axis B, when $B = C = 0^0$. Given a parametric tool path $X_{(u)} = [x_{(u)} \ y_{(u)} \ z_{(u)} \ \emptyset_{(u)} \ \varphi_{(u)}]^T \in \Re^5$, $u \in [0, 1]$ and a feed rate profile $f(t) = \dot{s}(t)$ where s is the arc length of the curve $X(s) = [x(s) \ y(s) \ z(s)]^T$. The inverse kinematic equation at position level can be written as follows

$$q_{(u)} = f^{-1}_{X_{(u)}} \tag{3}$$

To determine the inverse kinematic equations at velocity, acceleration and jerk level, the following mathematical transformations are introduced.

Taking a time derivative both sides of Eq. (1) yields

$$\dot{X} = J\dot{q} \tag{4}$$

where $J = \frac{\partial f}{\partial q}$ is the Jacobian matrix. Thus

$$\dot{q} = J^{-1}\dot{X} = J^{-1}\frac{dX}{ds}\frac{ds}{dt} = J^{-1}X'_s \dot{s} \tag{5}$$

Since,

$$ds \cong \left|\frac{dX}{du}\right| du \tag{6}$$

$$X'_s = \frac{dX}{ds} = \frac{dX/du}{ds/du} = \frac{dX/du}{|ds/du|} = \frac{X'_u}{|X'_u|} \tag{7}$$

Substituting Eq. (6) into Eq. (5) yields

$$\dot{q} = J^{-1}\frac{X'_u}{X'_n}f \tag{8}$$

Equation (8) is the inverse kinematic equation at velocity level. In other words, at any value of the parameter u, the joint (axis) velocities of a 5-axis machine can be calculated according to the given feed rate f, the geometrical characteristics of the desired tool path X'_u, and the kinematic structure of a machine J. Continuously taking a time derivative of Eq. (4) yields

$$\ddot{X} = J\ddot{q} + \dot{q}\ddot{q} \tag{9}$$

Thus

$$\ddot{q} = J^{-1}\left(\ddot{X} - J\dot{q}\right) \tag{10}$$

Since $\dot{X} = X'_s \dot{s}$,

$$\ddot{X} = \frac{dX'_s}{ds}\dot{s}^2 + X'_s\ddot{s} = X''_s\dot{s}^2 + X'_s\ddot{s} \tag{11}$$

Rewriting Eq. (6) in the following form Eq. (12)

$$s'_u = X'_{u'} \tag{12}$$

and substituting it into Eq. (7) yields

$$X'_s = \frac{X'_u}{s'_u} \tag{13}$$

Taking derivative of Eq. (13) yields

$$X''_s = \frac{(dX'_u/ds)s'_u + X'_u(ds'_u/ds)}{\left|X'_u\right|^2} \tag{14}$$

Note that

$$\frac{dX'_u}{ds} = \frac{X''_u}{s'_u} = \frac{X''_u}{\left|X'_{u'}\right|}; \quad \frac{ds'_u}{ds} = \frac{s''_u}{s'_u} = \frac{\left(X'_u\right)^T X''_u}{\left|X'_u\right|^2} \tag{15}$$

Thus

$$X''_s = \frac{X''_u}{\left|X'_u\right|^2} + \frac{X'_u\left(\left(X'_u\right)^T X''_u\right)}{\left|X'_u\right|^4} \tag{16}$$

Substituting Eqs. (13 and 15) into Eq. (11) and recalculating Eq. (10) yield

$$\ddot{q} = J^{-1}\left(\left(\frac{X'_u}{\left|X'_u\right|^2} + \frac{X'_u\left(\left(X'_u\right)^T X''_u\right)}{\left|X'_u\right|^4}\right)f^2 + \frac{X'_u}{\left|X'_u\right|}\dot{f} - \dot{J}\dot{q}\right) \tag{17}$$

It is interesting that, in a parametric domain, the axis accelerations of a machine can be totally determined with Eq. (15) which depends on the feed rate square f^2 and the feed acceleration $\dot{f}$, the geometrical characteristics of the tool path X'_u and X''_u, and the kinematic structure of a machine J and $\dot{J}$. For the jerk calculation, the third order time derivative of Eq. (1) is derived as follows

$$\dddot{X} = J\dddot{q} + 2\dot{J}\ddot{q} + \ddot{J}\dot{q} \tag{18}$$

Thus

$$\dddot{q} = J^{-1}\left(\dddot{X} - 2\dot{J}\ddot{q} - \ddot{J}\dot{q}\right) \tag{19}$$

Note that

$$\ddot{X} = \frac{d\dot{X}}{ds}\dot{s} = \frac{d}{ds}\left(X_s''\dot{s}^2 + X_s'\ddot{s}\right) = \left(X_s'''\dot{s}^2 + X_s'\ddot{s}\right)\dot{s} \tag{20}$$

Substituting Eq. (19) into Eq. (18), and then substituting Eqs. (8, 15 and 18) into Eq. (17) yield

$$\ddot{\vec{q}} = J^{-1}\left(\begin{array}{c} \dfrac{X_u''}{\left|X_u'\right|^3}f^3 + \dfrac{X_u''\left(\left(X_u'\right)^T X_u''\right) + X_u'\left(\left(X_u''\right)^T X_u'' + \left(X_u'\right)^T X_u'''\right)}{\left|X_u'\right|^5}f^3 + \dfrac{X_u''}{\left|X_u'\right|^2}\dot{} \\[4mm] + \dfrac{X_u'\left(\left(X_u'\right)^T X_u''\right)}{\left|X_u'\right|^4}\dot{f}f - 2\dot{J}\ddot{\vec{q}} - \ddot{J}\dot{\vec{q}} \end{array}\right) \tag{21}$$

Equation (21) implies that the jerk of the machine axes is calculated in a parametric domain as well. Note that in Eqs. (17 and 21), the terms $\dot{J}$ and $\ddot{J}$ can be determined as follows

$$\dot{J} = \begin{bmatrix} \sum_{i=0}^{5}\frac{\partial J_{11}}{\partial q_i}\dot{q}_i & \cdots & \sum_{i=0}^{5}\frac{\partial J_{15}}{\partial q_i}\dot{q}_i \\ \vdots & \ddots & \vdots \\ \sum_{i=0}^{5}\frac{\partial J_{51}}{\partial q_i}\dot{q}_i & \cdots & \sum_{i=0}^{5}\frac{\partial J_{55}}{\partial q_i}\dot{q}_i \end{bmatrix} = \left[\frac{\partial J_{ij}}{\partial q}\right] \otimes \dot{q} \tag{22}$$

And,

$$\ddot{J} = \begin{bmatrix} \sum_{i=0}^{5}\left(\sum_{k=0}^{5}\frac{\partial J_{11}}{\partial q_i \partial q_k}\dot{q}_i\dot{q}_k + \frac{\partial J_{11}}{\partial q_i}\ddot{q}_i\right) & \cdots & \sum_{i=0}^{5}\left(\sum_{k=0}^{5}\frac{\partial J_{15}}{\partial q_i \partial q_k}\dot{q}_i\dot{q}_k + \frac{\partial J_{15}}{\partial q_i}\ddot{q}_i\right) \\ \vdots & \ddots & \vdots \\ \sum_{i=0}^{5}\left(\sum_{k=0}^{5}\frac{\partial J_{51}}{\partial q_i \partial q_k}\dot{q}_i\dot{q}_k + \frac{\partial J_{51}}{\partial q_i}\ddot{q}_i\right) & \cdots & \sum_{i=0}^{5}\left(\sum_{k=0}^{5}\frac{\partial J_{55}}{\partial q_i \partial q_k}\dot{q}_i\dot{q}_k + \frac{\partial J_{55}}{\partial q_i}\ddot{q}_i\right) \end{bmatrix}$$

$$= \left[\frac{\partial}{\partial q}\left(\left[\frac{\partial J_{ij}}{\partial q}\right] \otimes \dot{q}\right)\right] \otimes \begin{bmatrix} \dot{q} \\ \cdots \\ \dot{q} \end{bmatrix}_{1x5} + \left[\frac{\partial J_{ij}}{\partial q}\right] \otimes \ddot{q} \tag{23}$$

3 Optimization of the Feed Rate

In this section, the feed rate optimization model for 5-axis machining is formulated with an objective function of the feed rate maximization while respecting machine tool constraints. The machine tool constraints are the velocity, acceleration and jerk limits of all active linear and rotary feed drives. In other worlds, the feed rate optimization problem can be defined as the maximization of the feed rate values computed along the entire curved toolpath, while respecting a set of physical limits of the machine. The formulation of the problem is expressed as follows

$$\max_{u \in [0,1]} f(u) \tag{24}$$

Subject to

$$|\dot{q}(u)| \leq V_{max} \tag{25}$$

$$|\ddot{q}(u)| \leq A_{max} \tag{26}$$

$$|\dddot{q}(u)| \leq J_{max} \tag{27}$$

where V_{max}, A_{max}, and J_{max} are vectors of the velocity, acceleration and jerk limits of the axes of a machine, respectively.

$$\begin{aligned}
V_{max} &= \begin{bmatrix} V_{xmax} & V_{ymax} & V_{zmax} & V_{bmax} & V_{cmax} \end{bmatrix}^T \\
A_{max} &= \begin{bmatrix} A_{xmax} & A_{ymax} & A_{zmax} & A_{bmax} & A_{cmax} \end{bmatrix}^T \\
J_{max} &= \begin{bmatrix} J_{xmax} & J_{ymax} & J_{zmax} & J_{bmax} & J_{cmax} \end{bmatrix}^T
\end{aligned} \tag{28}$$

The constraints Eqs. (25–27) are calculated with Eqs. (8, 17, 21) respectively. Suppose that the parameter domain $u \in [0, 1]$ is subdivided into n intervals of equal length, $u_1, u_2, \ldots, u_n$, Thus, at a value u_i, $(i = 1 \div n)$, the optimal value of the feed rate f_i can be obtained by solving Eqs. (24–27). For n steps, all optimal values f_i, $(i = 1 \div n)$ are yielded in the same manner. For example, we consider again the machine Spinner U5-620 as shown in Fig. 3. A desired tool path needs to be cut with the machine is represented in the workpiece coordinate system $O_w x_w y_w z_w$ as a Bezier curve as follows

$$X_{(u)} = P_0(1-u)^3 + 3P_1(1-u)^2 u + 3P_2(1-u)u^2 + P_3 u^3 \tag{29}$$

Where $P_0 = \begin{bmatrix} 0 & 0.05 & 0.05 \end{bmatrix}^T$, $P_1 = \begin{bmatrix} 0.10 & 0.05 & 0.15 \end{bmatrix}^T$.

$P_2 = \begin{bmatrix} 0, & 20 & 0.05 & 0.05 \end{bmatrix}^T$, and $P_3 = \begin{bmatrix} 0.30 & 0.05 & 0.15 \end{bmatrix}^T$. To machine the given tool path, by solving the feed rate optimization model, the optimal feed rate values along the path are shown in Fig. 4. Note that, for an individual machine, a maximum allowable feed rate f_{max} is usually given. The optimal value of the feed rate calculated with Eqs. (25–27) must satisfy $f_i < f_{max}$. For the case of the machine Spinner U5-620, $f_{max} = 0.1$ m/s.

For the machine, the limits of velocities, accelerations and jerks are as follows

$$\begin{aligned}
V_{max} &= \begin{bmatrix} 0.25 & 0.25 & 0.25 & 2.5 & 4.0 \end{bmatrix}^T \\
A_{max} &= \begin{bmatrix} 2.0 & 2.0 & 2.0 & 10.0 & 10.0 \end{bmatrix}^T \\
J_{max} &= \begin{bmatrix} 3.0 & 3.0 & 3.0 & 4.36 & 3.49 \end{bmatrix}^T
\end{aligned} \tag{30}$$

It is clearly seen that, as compared with the often use constant feed rate $f = 0.02$ m/s, the feed rate values calculated in this study are significantly increased at every point along the tool path on the machining surface. Thus, by using this optimal feed rate profile for postprocessing the NC program, the machining time will be significantly reduced, and thereby the productivity will be increased when machining complex and big parts by 5-axis CNC machines.

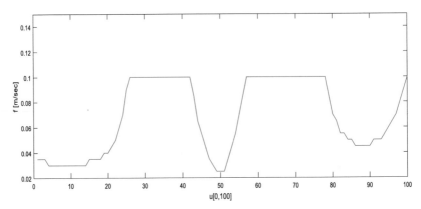

Fig. 4. The optimal feed rates

4 Conclusions

A new feed rate optimization model was formulated in a parametric domain. The main idea of the proposed method is based on a mapping of the differential kinematic equations of the 5-axis CNC machines into a parametric domain. Since, in practice, the curved tool path for 5-axis CNC machining is usually calculated and represented in the parametric domain, the use of the inverse parametric kinematic equations yielded in this study to investigate the behavior of the machine's drives is advantageous and applicable. All the constraints formulated in this study can be evaluated effectively at every point on the tool path without any complex arc length re-parameterization of the cutter trajectory. Finally, the optimal feed rate profile for the required tool path can be obtained that plays an important role in machining optimization for the 5-axis CNC machines.

Experimental investigation and validation will be the future works of the proposed research.

Funding Statement. Research under Vingroup Innovation Foundation (VINIF) annual research support program in project code VINIF. 2019. DA08.

References

1. My, C.A., Bohez, E.: A novel differential kinematics model to compare the kinematic performances of 5-axis CNC machines. Int. J. Mech. Sci. **163**, 105–117 (2019)
2. Vulliez, M., Lavernhe, S., Bruneau, O.: Dynamic approach of the feedrate interpolation for trajectory planning process in multi-axis machining. Int. J. Adv. Manuf. Technol. **88**(5–8), 2085–2096 (2001)
3. Beudaert, X., Lavernhe, S., Tournier, C.: Feedrate interpolation with axis jerk constraints on 5-axis NURBS and G1 tool path. Int. J. Mach. Tools Manuf **57**, 73–82 (2012)
4. Beudaert, X., Pechard, P., Tournier, C.: 5-axis tool path smoothing based on drive constraints. Int. J. Mach. Tools Manuf **51**(12), 958–965 (2011)
5. Sencer, B., Altintas, Y., Croft, E.: Feed optimization for five-axis CNC machine tools with drive constraints. Int. J. Mach. Tools Manuf **48**(7–8), 733–745 (2008)

6. Liang, F., Zhao, J., Ji, S.: An iterative feed rate scheduling method with confined high-order constraints in parametric interpolation. Int. J. Adv. Manuf. Technol. **92**(5–8), 2001–2015 (2017)
7. Sun, Y., Zhao, Y., Bao, Y., Guo, D.: A novel adaptive-feedrate interpolation method for NURBS tool path with drive constraints. Int. J. Mach. Tools Manuf **77**, 74–81 (2014)
8. Sun, Y., Zhao, Y., Xu, J., Guo, D.: The feedrate scheduling of parametric interpolator with geometry, process and drive constraints for multi-axis CNC machine tools. Int. J. Mach. Tools Manuf **85**, 49–57 (2014)
9. Sun, Y., Bao, Y., Kang, K., Guo, D.: An adaptive feedrate scheduling method of dual NURBS curve interpolator for precision five-axis CNC machining. Int. J. Adv. Manuf. Technol. **68**(9–12), 1977–1987 (2013)
10. Dong, J., Ferreira, P., Stori, J.: Feed-rate optimization with jerk constraints for generating minimum-time trajectories. Int. J. Mach. Tools Manuf **47**(12–13), 1941–1955 (2007)
11. Fan, W., et al.: Time-optimal interpolation for five-axis CNC machining along parametric tool path based on linear programming. Int. J. Adv. Manuf. Technol. **69**(5–8), 1373–1388 (2013)
12. Zhou, J., Sun, Y., Guo, D.: Adaptive feedrate interpolation with multi-constraints for five-axis parametric toolpath. Int. J. Adv. Manuf. Technol. **79**(9–12), 1873–1882 (2014)
13. Dong, J., Wang, T., Li, B., Ding, Y.: Smooth feedrate planning for continuous short line tool path with contour error constraint. Int. J. Mach. Tools Manuf **76**, 1–12 (2014)
14. My, C.A., Bohez, E.: New algorithm to minimize kinematic tool path errors around 5-axis machining singular points. Int. J. Prod. Res. **54**(20), 5965–5975 (2016)
15. My, C.A.: Integration of CAM systems into multi-axes computerized numerical control machines. In: Proceedings of the Second International Conference on Knowledge and Systems Engineering, pp. 119–124 (2010)

Opensource Based IoT Platform and LoRa Communications with Edge Device Calibration for Real-Time Monitoring Systems

Ha Duyen Trung$^{(\boxtimes)}$, Nguyen Tai Hung, and Nguyen Huu Trung

School of Electronics and Telecommunications, Hanoi University of Science and Technology, 405/C9, No.1, Dai Co Viet Rd., Hai Ba Trung, Hanoi, Vietnam
{trung.haduyen,hung.nguyentai,trung.nguyenhuu}@hust.edu.vn

Abstract. Internet of Things (IoT) is an attractive part of our daily life. However, the development of IoT applications still faces many challenges, such as handling unstructured data, intelligent analytics, connectivity, compatibility and interoperability, integration, and data security and privacy. In this paper, first, a real-time monitoring system is proposed based on opensource platform (called BKThings) and long-range (LoRa) communication technology. Specifically, a LoRa gateway prototype is implemented to collect, store and forward monitored data from the IoT devices. In addition, LoRa end devices are developed, where data calibrations on the low-cost sensors to accurately measure environmental and gas air parameters are integrated. Second, an opensource based IoT platform functional architecture is developed. The implementation includes main functionalists of a typical IoT platform such as user management, device registration, monitoring, connectivity, data acquisition, processing, storage, and visualization on the dashboards. The achieved results demonstrate that the tested implementation of the opensource IoT platform and the LoRa IoT gateway were suitable and scalable for sensing environmental parameters in certain data visualization.

Keywords: Opensource · IoT platform · LoRa · Edge devices calibration · Real-time monitoring

1 Introduction

Internet of Things (IoT) technology has been added significant intelligence to things and communication networks that brings ordinary computers into

---

This research is funded by the Ministry of Science and Technology of Vietnam under the framework of "research and development of information technology products for e-Government" (Ref: KC.01/16-20) with the research project titled "Research and development of Internet of Things platform (IoT), application to management of high technology, industrial zones", the mission code is KC.01.17/16-20.

© Springer Nature Switzerland AG 2020
H. A. Le Thi et al. (Eds.): ICCSAMA 2019, AISC 1121, pp. 412–423, 2020.
https://doi.org/10.1007/978-3-030-38364-0_37

daily life. With the rapid development of embedded hardware platform and software packages, multiple sensors transmit monitored data through gateway and connectivity networks to a cloud platform for processing, storage and analysis. Based no the analytical results, the human needs can be fulfilled by changing and managing their environment [1,2].

A proposal of general requirements were released by ITU-T in 2014 [3]. Shen, et al. in [4] outlined the common requirements of IoT technologies and the individual functions in each requirement. These requirements consist of non-functional, applications, services, connectivity, device management, and security and privacy. In [5], a concept for future IoT architecture, including definition, review of developments, various main requirements and technical design for enable implementation are presented. Palattella et al. [6] described the protocol stacks for the power-efficient PHY layer. In this regard, challenges in the design of a home machine-to-machine (M2M) network using IEEE 802.15.4 protocol for the ZigBee communication are proposed in [7].

Opensource cloud platform plays a crucial component in IoT. It brings valuable services in various application areas. There are a number of opensource platform vendors for IoT to take advantage of specific and appropriate IoT-based services. Several works [8]–[13] have been investigated the application development of IoT solutions based on the existing opensource platforms. Various requirements for both cloud and IoT integration were described in [8]. In this study, an agent-oriented and cloud-assisted model is considered based on a reference architecture. In [9], a general architecture is presented after depicting analyzing various studies. In this study, a smart device supported by an IoT cloud is evaluated for data collecting, processing and monitoring. A survey of sensing services in cloud-centric IoT and its challenges are mentioned in A brief survey in sensing services over cloud-centric IoT, and recent challenges are analyzed in [10]. In [11], a cloud IoT platform is proposed by an integration of cloud and IoT. In [12], an M2M remote telemetry station in cooperation with a big data processing platform and various sensors is implemented. It demonstrated the use of IoT cloud and data processing in disease prediction and alerting application. Wang et al. in [13] described different notions (i.e., data center, cloud computing, data management across data centers, benchmark, application kernels, standards, etc.) to visualize how distributed IoT data could be processed in the clouds. Although various possible participation of these IoT open clouds, however to the best of knowledge, no comparative and analytical study on standardization has been formed through the literature.

The contributions of this paper are as follows. First, we develop the LoRa gateway prototype based on open hardware platform in order to collect, store and forward data monitored from the end devices. We also develop the LoRa end devices, where data calibration is applied on the low-cost sensors to accurately measure environmental and gas air parameters. Second, the open source based IoT platform functional architecture is developed. The prototype implementation consists of main functionalists of a typical IoT platform, including user management, device registration and management, data storage, processing and visualization.

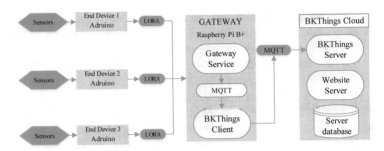

Fig. 1. BKThings cloud and LoRa communications-based the monitoring systems.

The remainder of paper is organized as follows. Section 2 provides the overall system description. The open source based IoT platform architecture for in according with the characteristics of IoT applications is presented and discussed in Sect. 3. The case-studied data processing and visualization results, which is evaluated in an environmental monitoring application scenario is presented in Sect. 4. The paper is concluded in Sect. 5.

2   System Description

2.1   System Model

Figure 1 presents the open platform-BKThings cloud and LoRaWAN for the monitoring systems. A gateway is implemented based on Raspberry Pi B+, which is responsible for bridging between server and cloud. In this paper, the gateway will accept the responsibility of connectivity between the end devices and the open platform (BKThings) Server. The gateway message received from the end devices follows an existing structure that will be standardized, converted, corrected or coded and then sent to the BKThings Client. The Client is a pre-designed firmware. The Client implements its algorithms at the gateway to match the platform before uploading data to the Server.

BKThings Client is a software that can identify end devices and send/receive their data to BKThings Server. We can use the Client interface device for end devices to communicate with the platform. Figure 1 also shows the relationship between end devices, Client and Server. We can treat the client as a proxy before the end devices for the agent to work with the Server. Client can be any software application that supports IoT protocols (such as MQTT, CoAP, etc.) to exchange data with the Server. It is noted that we do not use the concept of device or things here. Because like the end device concept, a physical device can correspond to multiple end devices, we can also open multiple connections simultaneously from one device, corresponding to multiple Clients. In this paper, the MQTT protocol is used. It is a lightweight publish/subscribe messaging protocol using for M2M IoT connectivity. It is used for connection between the end devices and the MQTT broker and the gateway. Its advantages are small code size, minimized

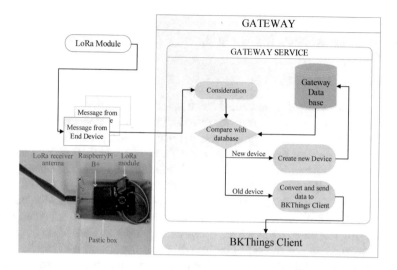

Fig. 2. The diagram of LoRa gateway functionalizes with the prototype.

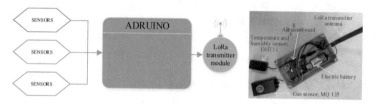

Fig. 3. The functionalized diagram and a prototype of the LoRa end device.

data packet size, open source implementation, and can distribute packets to multiple applications [14].

As shown in the diagram of LoRa gateway (Fig. 2), the gateway receives packets from the end devices via the LoRa gateway. These packets will be sent to the gateway service through the following two steps. (1) Consideration by checking if the packets are properly formatted. When the end device receives packets that do not fit in exact structure will be rejected. Then, depending on the user application, the server may request end device to re-send the packets or skip the packets and wait for the packets on the next time. (2) Comparison by checking the device ID defined in the database. Parameters defined in the database include device ID and device type. If the device ID already exists, data will be packed then send it to the server. If the device is not available, it must be defined at the server to create a new end device. In order to end devices transmitting data to the platform Server, the end device works as the Client. In this paper, an integration of Arduino board and LoRa module is also implemented for the end device. It is responsible for data collection and calibration from the low-cost sensors. The DHT11 sensor is used to measure temperature and humidity. Whereas, the MQ-135 sensor is employed to estimate the environmental air quality. The functionalized diagram and a prototype of the LoRa end device is shown in the Fig. 3.

2.2   LoRa and LoRaWAN

The LoRa physical layer is described in 2014 [15]. Its modulation produces chirp signals with the same time duration, practically [16]. In LoRa communications, two different types of chirps have been defined including the base chirp whose frequency time profile, from $f_{min} = -BW/2$ to $f_{max} = +BW/2$, where BW being the spreading bandwidth of the signal. Modulated chirps that are cyclically time shifted base chirps [15]. The chirp starts with frequency $f_1 = +BW/2$ and ends with $f_2 = -BW/2$, is referred to as a down-chirp, practically being the complex conjugate of the base chirp. In each chirp, the time shift is calculated by multiplying the chirp itself with a down-chirp. Suppose that N is the length of symbol's chirp, then there are N possible different cyclic shifts of the base chirp. The value of the cyclic shift is coded using $\log_2 N$ bits to create the spreading factor (SF) of the LoRa communication. LoRa also uses diagonal interleaving and forward error correction (FEC) codes to improve the robustness against noise and burst interference. The code rates (CRs) usually applied from 4/5 to 4/8. The symbol rate (R_s) and data rate (R_b) depend on the uses the SF and the bandwidth, are respectively given as $R_s = SF\left(\frac{BW}{2^{SF}}\right)$ and $R_b = SF\left(\frac{\frac{4}{4+CR}}{\frac{2^{SF}}{BW}}\right)$, where SF, BW, and CR are the spreading factor, the bandwidth in KHz, and the code rate, respectively. It can be noticed that the symbol time is increased by increasing the SF and/or decreasing the symbol rate. In addition, increasing the SF less bits per symbol encoded, decreasing the data rate. The packet structure consists of a preamble, an optional header and the data payload. The preamble is used to synchronize the receiver with the transmitter. LoRa transceiver should know the SF to detect the preamble [15].

The LoRaWAN medium access control (MAC) protocol is an open source protocol standardized by the LoRa Alliance [17]. It provides the medium access control mechanism for enabling connectivity between different end devices and gateway(s). The network architecture of the star topology allows the end devices only connect to LoRaWAN gateway(s), not directly with each other. Multiple gateways connect to the central server. The gateways are only responsible for forwarding data packets from end devices towards the server encapsulating them in UDP/IP packets. Further, the connectivity can be terminated at the application servers by third parties. In this paper, the gateway is designed specifically for LoRa. However, the gateway can be activated with different types of connections such as Bluetooth, ZigBee, Z-Wave, Cellular, etc., by typically needs to install the suitable environment and the library.

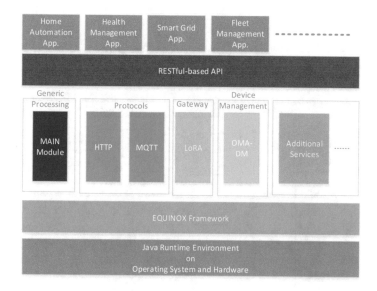

Fig. 4. BKThings IoT platform functional architecture.

3   The Opensource Platform

3.1   Opensource Based IoT Platform - BKThings

The proposed BKThings opensource platform is basically built on the M2M service platform in compliance with the ETSI standard. It provides RESTful API for exchanging XML data through unreliable connections in a distributed environment. It provides a modular architecture that operates on top of OSGi and Equinox [18] (Fig. 4). This platform provides a flexible (SCL) that can be deployed in M2M software modules, gateway or end devices [19]. A SCL consists of tightly linked small plugins, each plugin, providing specific functions. The plugin can be remotely installed, started, stopped, updated and uninstalled without reboot requirements. SCL can also detect additions or removals of services through the service registry and make appropriate adjustments that facilitate its expansion.

The CORE plugin module provides an independent protocol service to handle REST requests. Mapping plugins with specific interfaces can be added to support protocol constraints such as HTTP and MQTT. This platform can be extended through specific device management mapping plugins to update device management (DM) programs using existing protocols such as Open Mobile Alliance (OMA) - DM [20]. It can also be extended for different internetworking proxy plugins to enable communications with end devices using Bluetooth, Wi-Fi, Zig-Bee, Z-Wave, LoRa, and even 3G/4G/5G technologies.

The software modules of the BKThings platform are described in Fig. 5. The CORE plugin executes the *SCL_Service* interface to handle common RESTful requests. It receives request and response of protocol-independent indication

418   H. D. Trung et al.

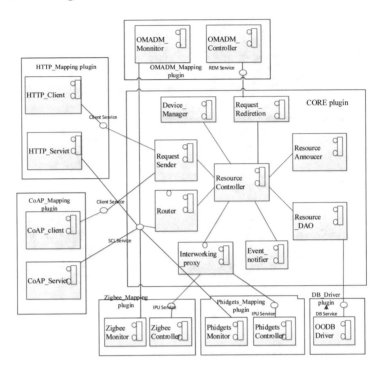

Fig. 5. The software modules of BKThings platform.

and confirmation, respectively. *Resource_Controller* executes resource methods
of Create, Get, Update and Delete (CRUD). It performs necessary checks such
as access authorization and syntax verification of resources. When a resource
is created, updated or deleted, *Event_Notifier* sends notifications to all concerned subscribers. It performs the filtering process to remove unwanted events
to a subscriber. *Resource_Announcer* informs the resource to the remote SCL
explicitly so that other machines can easily access. It also handles notification of resource cancel process. *Resource_DAO* provides a discrete interface
to encapsulate all access rights to persistent resource storage without revealing
the database details. The Router module identifies a unique route to handle
every request in the *Resource_Controller* using only the requested URI and the
method. *Request_Sender* is set to hold clients dedicated to the detected protocol executing the *Client_Service* interface. It plays as a proxy to send a general
request through the exact protocol. *Interworkₚroxy* holds the detected internetworking proxy unit (IPU) that executes the *IPUₛervice* interface. It performs
as a proxy to correctly call the IPU device controller. *Device_Manager* holds
Remote Entity Managers (REMs) that execute the *REM_Service* interface and
perform as a proxy to call the exact device manager controller.

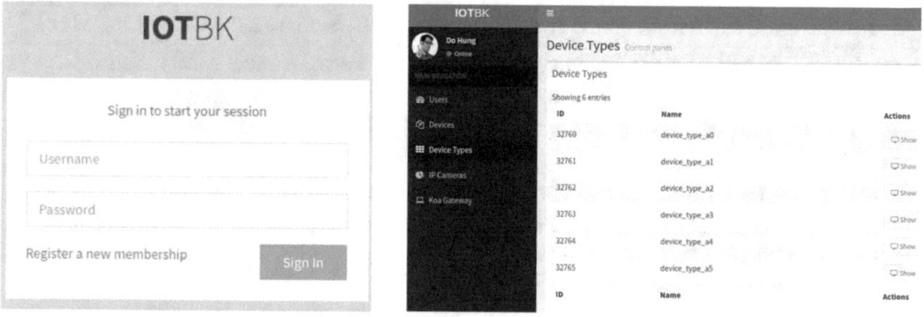

(a) User management

(b) Device registration and management

Fig. 6. User management and device registration & management of BKThings.

The *OMADM_Mapping* plugin software module provides bidirectional mapping to manage devices that support OMA-DM. The *OMADM_Monitor* listens for OMA-DM permitted devices, and calls the CORE plugin creating required resources based on the SCL. The *OMADM_Controller* executes the *REM_Service* interface to transform common requests into OMA-DM management. The *HTTP_Mapping* plugin software module supplies bidirectional links to the HTTP protocol. *HTTP_Servlet* receives and converts an HTTP request into a common request and invokes the CORE plugin through the *SCL_Service* interface. The *HTTP_Client* executes the *Client_Service* interface to send a common request using HTTP. The *Phidgets_Mapping* plugin software module supplies bidirectional mapping for interacting with older Phidgets devices. *Phidgets_Monitor* detects Phidgets devices, and calls the CORE plugin creating required resources based on the SCL. The *Phidgets_Controller* executes the *IPU_Service* interface to seamlessly execute a common request through the Phidgets API. The *DB_Driver* plugin software module supplies an Object Oriented Data Base (OODB) accessible through the *DB_Service* interface. By doing the same approach, other plugin software modules can be deployed to interact with other additional protocols or integrate new capabilities.

3.2   Implementation

We had implemented, in the scope of this paper, the prototype of this architecture to complete our IoT Testbed. The implementation prototype consists of major functionalities of the typical IoT platform, which includes user management, device registration and management, and data storage and visualization in the dashboard (see Fig. 6). The developed platform will be improved in the future to include new capabilities of utilizing the AI (Artificial Intelligence) and ML (Machine Learning) mechanisms in order to provide the IoT data analytical features that needed to provide more smart applications.

Algorithm 1. Calibration for the Temperature & Humidity sensor, DHT11

Input: C_v: Calibrated data of the gas sensor, t: Temperature, H_r: Humidity value,
A_v: Current analog value read from the gas sensor, R_l: the external load resistance,
A_m: the maximum analog read value, R_s/R_0: Resistance ratio

1: Read data from sensor pins (A_v, t, and H_r)
2: Convert the measured values to dependency values (H_r, θ, and σ)
3: Calculate the calibrated value of temperature and humidity, D_{th}, using Eq. (1)
4: Calculate the R_s/R_0
5: Calculate the calibrated value C_v

Algorithm 2. Calibration for the Gas sensor, MQ-135

1: Calculate the R_0 value
2: Calculate the R_s value
3: Read analog pins from the Gas sensor
4: Collect multiple samples and calculate the average (S)
5: Calculate $R_0 = (S)$/clean air factor
6: Extrapolate coefficients a and b
7: Calculate the ppm value, ppm $= a(R_s/R_0)b$

---

4   Data Processing and Visualization Results

4.1   Data Processing

In the implementation, characteristics of the low-cost sensor differ from each sensor production. Therefore, each sensor used in the monitoring system needs pre-calibration to accurately measure environmental and gas air parameters. In the designed system, the air temperature and humidity sensor of DHT11 and the gas sensor of MQ-135 are used. Temperature and humidity affect low-cost environmental sensors. Therefore, calibrated values require to be adjusted with respect to the these parameters to validate sensing accuracy. Algorithm 1 presents how to apply the auto-calibration of the DHT11 sensor. Assume that C_v is the calibrated data value of the gas sensors. It is calculated as $C_v = (R_s/R_0)D_{th}$, where R_s and R_0 are the sensor resistances in the certain gas presence and in the clean air, respectively; D_{th} is the calibrated value that depends on the temperature and humidity. The sensor resistances ratio can be estimated by $R_s/R_0 = R_l(A_m/A_v) - R_l$, where R_l is the external resistance. Finally, the calibrated value can be calculated by

$$D_{th} = \theta t^2 - t + \sigma, \tag{1}$$

where t is the current temperature, θ and σ are the temperature and humidity dependency values, respectively. Algorithm 2 introduces the steps for calibration of the Gas sensor, MQ-135. Coefficients of a and b are extrapolated from the curves provided in the sensor's data sheet. It is noted that R_0 and R_s are the sensor resistances in the clean air and in the presence of certain gas, respectively.

Fig. 7. Measured data collected at the LoRa gateway from end devices.

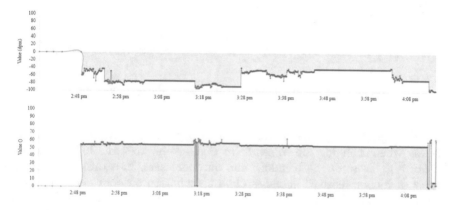

Fig. 8. RSSI (above) and SNR (below) at the LoRa IoT gateway.

4.2   Data Visualization

In order to visualize effectiveness of the monitoring system for case studies, edge-calibrated end devices, LoRa gateway and BKThings platform were integrated. As metrics to indicate the air quality index (AQI), it is calculated by measuring six main air pollutants directly from the sensor MQ-135. Whereas, the temperature and humidity parameters are measured from the sensor DHT11. As presented in the Sect. 2.2, it is a fact that when changing the SF will change the receiver's sensitivity. Therefore, increasing the ability to demodulate weak received signals due to long-range transmission. This means that data can be transmitted further by increasing the SF factor. Figure 7 shows measured packets received at the LoRa gateway for two IoT devices. In the experimental tests for different data visualization, various measurement scenarios were formulated and shown in Figs. 8, 9 and 10. In the Fig. 8, RSSI (Received Signal Strength

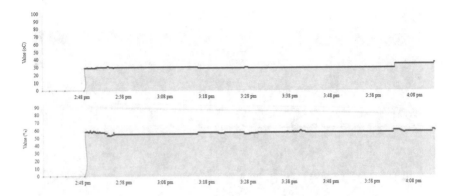

Fig. 9. Air temperature and humidity visualization in the open platform.

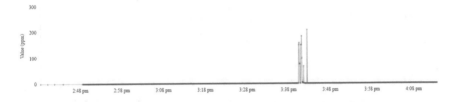

Fig. 10. Air quality index visualization in the open platform.

Indicator) and SNR (Signal-to-Noise Ratio) parameters were measured at the gateway when changing the transmission distance between the IoT devices and the LoRa gateway for both indoor and outdoor cases. It is observed that the quality of indoor received signal is almost similar to outdoor conditions with the same distance. Then, the air temperature and humidity ware measured and visualized the Fig. 9. Finally, measurement and visualization of air quality index was performed in the Fig. 10.

5   Conclusions

In this paper, we developed an IoT architecture based on opensource platform and LoRa technology with edge calibration for IoT devices, applications for the air monitoring system in real-time. The designed system integrates open, license-free access via LoRa technology. The system's prototype was performed with a small size, low-cost and easy to implement and develop in the open platform. The system prototype is able to monitor multiples gases, especially six main air pollutants, temperature and humidity. Algorithms for calibration at the IoT devices were employed to avoid temporary errors and unnecessary data before transmitting to the gateway and cloud. The future work includes capabilities of utilizing the machine learning and big IoT data analytics in the platform that can provide more smart services.

References

1. Ray, P.P.: A survey on Internet of Things architectures. J. King Saud Univer. Comput. Inf. Sci. **30**, 291–319 (2018)
2. Syafrudin, M., et al.: Performance analysis of IoT-based sensor, big data processing, and machine learning model for real-time monitoring system in automotive manufacturing. Sensors **18**, 1–24 (2018)
3. ITU-T Recommendation Y, 2066. Common requirements of the Internet of Things. Accessed 28 Aug 2019
4. Shen, S., Yang, Z.: Architecture of Internet of Things and its standardization. J. Nanjing Univ. Posts Telecommun. **35**, 1–18 (2015)
5. Uckelmann, D., Harrison, M., Michahelles, F.: An architectural approach towards the future Internet of Things. In: Uckelmann, D., Harrison, M., Michahelles, F. (eds.) Architecting the IoT, pp. 1–24. Springer, Heidelberg (2011)
6. Palattella, M.R., et al.: Standardized protocol stack for the Internet of (Important) Things. IEEE Commun. Surv. Tutorials **15**(3), 1389–1406 (2013)
7. Starsinic, M.: System architecture challenges in the home M2M network. In: Proceedings Application and Technology Conference, Long Island, USA (2010)
8. Babu, S.M., Lakshmi, A.J., Rao, B.T: A study on cloud based internet of things: cloud IoT. In: Proceedings of GlobalCOM, pp. 60–65 (2015)
9. Emeakaroha, V.C., et. al.: A cloud-based IoT data gathering and processing platform. In: Proceedings of 3rd International Conference on Future IoT and Cloud, pp. 50–57 (2015)
10. Kantarci, B., Mouftah, H.T.: Sensing services in cloud-centric internet of things: a survey, taxonomy and challenges. In: Proceedings of IEEE ICC, pp. 1865–1870 (2015)
11. Alessio, B., Walter de, D., et. al.: On the integration of cloud computing and internet of things. In: Proceedings of International Conference on Future IoT and Cloud, pp. 23–30 (2014)
12. Suciu, G., Vulpe, A., Fratu, O., Suciu, V.: M2M remote telemetry and cloud IoT big data processing in viticulture. In: Proceedings of IEEE IWCMC, pp. 1171–1121 (2015)
13. Wang, L., Ranjan, R.: Processing distributed internet of things data in clouds. IEEE Cloud Comput. **2**(1), 76–80 (2015)
14. Eclipse Paho. https://www.eclipse.org/paho/
15. Seller, O.B., Sornin, N.: Low power long range transmitter. US Patent 9,252,834, 2 Feb 2016
16. Sforza, F.: Communications system. US Patent 720139, 2 July 2009
17. LoRa Alliance. LoRaWAN R1.0. Open Standard Released for the IoT; LoRa Alliance: Fremont, CA, USA (2015)
18. The Equinox OSGi framework. https://www.eclipse.org/equinox/
19. The OM2M project. http://www.onem2m.org/
20. ETSI: Machine-to-Machine communications (M2M); Functional architecture. Technical Specification (2013)

Author Index

© Springer Nature Switzerland AG 2020
H. A. Le Thi et al. (Eds.): ICCSAMA 2019, AISC 1121, pp. 425–426, 2020.
https://doi.org/10.1007/978-3-030-38364-0

Printed in the United States
By Bookmasters